Radar High-Speed Target Detection via Coherent Integration Transform

This book offers a systematic presentation of high-speed radar target detection methods using coherent integration transforms including the signal model, derivations of coherent integration transforms, and definitions of related key concepts.

The authors present the mathematical models and design principles necessary to analyze the behavior of each type of coherent integration transform, and based on this, they introduce and convey new approaches and techniques for designing such transforms, which will help to achieve efficient signal integrators and detectors, especially in the challenging low signal-to-noise ratio (SNR) environments.

The book will be of interest to graduate students and engineering professionals in statistical signal processing, signal detection and estimation, and radar signal processing.

Xiaolong Li is an associate professor at the School of Information and Communication Engineering, University of Electronic Science and Technology of China. His research interests include radar moving target detection and multi input multi output (MIMO) radar signal processing, with an emphasis on high-speed maneuvering target detection.

Guolong Cui is a professor at the School of Information and Communication Engineering, University of Electronic Science and Technology of China. His research interests include cognitive radar, array signal processing, MIMO radar, and through-the-wall radar.

Lingjiang Kong is a professor at the University of Electronic Science and Technology of China. His research interests focus on signal processing for radar. He has authored and co-authored more than 150 publications in international journals, conferences, and books.

Zhi Sun is an associate researcher at the School of Information and Communication Engineering, University of Electronic Science and Technology of China. His research interests include weak target detection, time-frequency analysis, and clutter suppression.

Radar High-Speed Target Detection via Coherent Integration Transform

Xiaolong Li, Guolong Cui, Lingjiang Kong and Zhi Sun

CRC Press
Taylor & Francis Group
Boca Raton London New York

CRC Press is an imprint of the
Taylor & Francis Group, an **informa** business

This book is published with financial supported from the National Natural Science Foundation of China under Grant 62371113, the Young Elite Scientists Sponsorship Program by CAST under Grant YESS20200082, the Natural Science Foundation of Sichuan Province under Grant 2023NSFSC1386 and the Aeronautical Science Foundation under Grant 2023Z017080001.

Designed cover image: © Yaorusheng (Shutterstock Photo ID: 274395989)

MATLAB® and Simulink® are trademarks of The MathWorks, Inc. and are used with permission. The MathWorks does not warrant the accuracy of the text or exercises in this book. This book's use or discussion of MATLAB® or Simulink® software or related products does not constitute endorsement or sponsorship by The MathWorks of a particular pedagogical approach or particular use of the MATLAB® and Simulink® software.

First edition published 2025
by CRC Press
2385 NW Executive Center Drive, Suite 320, Boca Raton FL 33431

and by CRC Press
4 Park Square, Milton Park, Abingdon, Oxon, OX14 4RN

CRC Press is an imprint of Taylor & Francis Group, LLC

© 2025 Xiaolong Li, Guolong Cui, Lingjiang Kong and Zhi Sun

ISBN: 978-1-032-67176-5 (hbk)
ISBN: 978-1-032-86770-0 (pbk)
ISBN: 978-1-003-52910-1 (ebk)

DOI: 10.1201/9781003529101

Typeset in Latin Modern font
by KnowledgeWorks Global Ltd.

To my wife and daughter.
– Xiaolong Li
To my lovely family.
– Guolong Cui
To my beloved parents.
– Lingjiang Kong
To my lovely family and girlfriend.
–Zhi Sun

Contents

List of Abbreviations

ACCF	Adjacent cross correlation function
AR	Axis rotation
BSSL	Blind speed sidelobe
CFAR	Constant false alarm ratio
CFCR	Centroid frequency chirp rate
CI	Coherent integration
CPF	Cubic phase function
CZT	Chirp-transform
DFM	Doppler frequency migration
DPT	Discrete polynomial phase transform
DS	Doppler spread
EGRFT	Extended generalized radon Fourier transform
FBRFT	Frequency bin radon Fourier transform
FDDKT	Frequency-domain deramp-keystone transform
FFT	Fast Fourier transform
FRFT	Fractional Fourier transform
FRM	First-order range migration
FT	Fourier transform
GDP	Generalized dechirp process
GKT	Generalized keystone transform
GRFT	Generalized radon Fourier transform
HI	Hybrid integration
HT	Hough transform

IAR	Improved axis rotation
IFFT	Inverse fast Fourier transform
IFT	Inverse Fourier transform
KT	Keystone transform
LFM	Linear frequency modulated
LTCI	Long-time coherent integration
LVD	Lv's distribution
MART	Modified axis rotation transform
MFP	Match filtering process
MLRT	Modified location rotation transform
MSEs	Mean square errors
MTD	Moving target detection
PC	Pulse compression
PD	Pulse Doppler
PFT	Polynomial Fourier transform
PRF	Pulse repetition frequency
RCS	Radar cross section
RFRFT	Radon-fractional Fourier transform
RFT	Radon Fourier transform
RLVD	Radon-Lv's distribution
RM	Range migration
RMSEs	Root-mean-square errors
RW	Range walk
SAF	Symmetric autocorrelation function
SFT	Scale Fourier transform
SGRFT	Special generalized radon Fourier transform
SHI	Subspace hybrid integration
SIFT	Scaled inverse Fourier transform

SKT	Second-order keystone transform
SNR	Signal-to-noise ratio
SRM	Second-order range migration
STGRFT	Short time generalized radon Fourier transform
TRM	Third-order range migration
TRT	Time-reversing transform
WFRFT	Window fractional Fourier transform
WRFRFT	Window radon fractional Fourier transform

CHAPTER 1

Basic Concept

1.1   PULSE DOPPLER RADAR

A pulse-Doppler radar is a radar system that determines the range of a target using pulse-timing techniques and uses the Doppler effect of the returned signal to determine the target object's velocity. It combines the features of pulse radars and continuous-wave radars, which were formerly separate due to the complexity of the electronics.

The earliest radar systems failed to operate as expected. The reason was traced to Doppler effects that degrade the performance of systems not designed to account for moving objects. Fast-moving objects cause a phase shift on the transmit pulse that can produce signal cancellation. Doppler has maximum detrimental effect on moving target indicator systems, which must use reverse phase shift for Doppler compensation in the detector.

Pulse-Doppler radar was developed during World War II to overcome limitations by increasing pulse repetition frequency [1]. This required the development of the klystron, the traveling wave tube, and solid state devices. Early pulse-Dopplers were incompatible with other high-power microwave amplification devices that were not coherent, but more sophisticated techniques were developed that recorded the phase of each transmitted pulse for comparison to returned echoes.

Early examples of military systems include the AN/SPG-51B developed during the 1950s specifically for the purpose of operating in hurricane conditions with no performance degradation. It became possible to use pulse-Doppler radar on aircraft after digital computers were incorporated in the design. Pulse-Doppler radar provided look-down/shoot-down capability to support air-to-air missile systems in most modern military aircraft by the mid-1970s.

The signal processing enhancement of pulse-Doppler radar allows small high-speed objects to be detected. To achieve this, the transmitter must be coherent and should produce low phase noise during the detection interval, and the receiver must have large instantaneous dynamic range.

DOI: 10.1201/9781003529101-1

1.2 PULSE COMPRESSION

According to the radar equation, the detection power of the radar system is positively correlated with the transmitted power [2]. By increasing the pulse duration time, the average transmitted power can be increased and the maximum detection range of the radar can be increased. However, this decreases the radar range resolution. If the radar system transmits a wide pulse signal, the pulse compression processing output signal will be a narrow pulse signal with peak power. Then a large detection range and a high range resolution can be obtained in the same time.

Pulse compression is realized by pulse compression filtering of the transmitted signal with large time-bandwidth product. In this case, the radar transmitted signal is a wide pulse with a nonlinear phase spectrum. The frequency characteristic of the pulse compression filter is opposite to the changing regular of the transmitted signal, that is, the phase frequency characteristic of the pulse compression filter and the transmitted signal are conjugate and deconvolution in phase. Therefore, the ideal pulse compression filter is a matched filter. And pulse compression can be achieved not only in time domain but also in frequency domain.

First, consider pulse compression processing in time domain [3]. Assume that the input signal of the pulse compression is the target echo signal $s_r(t)$ and the system function of the matched filter is $h(t)$, then the pulse compression in time domain can be obtained as the convolution of the echo signal $s_r(t)$ and the matched filter $h(t)$, i.e.,

$$y(\tau) = s_r(t) * h(t) = \int_{-\infty}^{\infty} s_r(t) h(\tau - t) dt \tag{1.1}$$

where the notation $*$ represents the convolution operation.

The system function of the matched filter in time domain corresponding to the transmitted signal $s(t)$ is

$$h(t) = s^*(-t) \tag{1.2}$$

In addition, pulse compression can also be completed in frequency domain. Fig. 1.1 shows the flow chart of pulse compression processing in frequency domain. First, the input pulse signal in time domain is transformed by the fast Fourier transform (FFT) to frequency domain. Then the complex multiplier is used to multiply the echo signal in frequency domain with the matched filter, and finally the signal is converted into

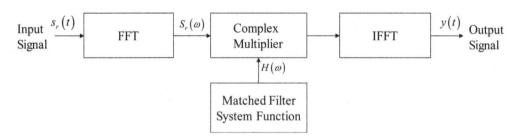

Figure 1.1   Flowchart of the pulse compression processing in frequency domain.

time domain by the inverse fast Fourier transform (IFFT) for the subsequent signal processing. The corresponding formula is described as

$$y(t) = \text{IFFT}\left[S_r(\omega) H(\omega)\right] \tag{1.3}$$

where $S_r(\omega)$ and $H(\omega)$ are the frequency domain signals for $s_r(t)$ and $h(t)$, respectively and $\text{IFFT}[\cdot]$ is the inverse FFT.

1.3 PULSE INTEGRATION

In a remote warning surveillance area, the intensity of reflected signals is rather feeble in comparison to the surrounding noise and clutter. Consequently, the signal-to-noise ratio (SNR) is diminished after undergoing matched filtering, since the target signal is overshadowed by the noise. To boost the dynamic detection capability of the radar system and improve the SNR, the radar system utilizes an integration method to preprocess echoes. For short-time integration, the targets are moving at a nearly constant velocity, which enables the coherent integration of many pulse echoes. However, for long-time integration, the echo envelope shifts across range cells, causing the assumption of constant Doppler frequency invalid [3]. This leads to Doppler time-varying echoes across range cells. In this scenario, conventional ways of coherent integration are no longer suitable. Hence, it is imperative to investigate efficient techniques for the long-time integration of radar signals.

Current radar systems can effectively improve target recognition accuracy by collecting several samples across spatial, temporal and frequency domains, therefore mitigating the impact of noise, clutter and interference. The process of integration often involves the repetitive sampling of elements, pulses and frequencies. And the pulse integration can be broadly categorized into two types: coherent integration and non-coherent integration. Non-coherent integration is a technique for accumulating sampled data without taking phase information into account. Nevertheless, the existence of ambient noise impedes the efficiency of integration when SNR is low, along with the threshold phenomenon. High beginning SNR would result in gain of non-coherent integration approaching zero [4].

Coherent integration outperforms non-coherent integration by successfully adjusting for the varying phase across samples, resulting in improved integration performance [5]. Fig. 1.2 depicts the signal processing path for coherent integration achieved via cascading. Firstly, radar utilizes digital beamforming (DBF) technology to accomplish spatial coherent integration between array elements, enabling simultaneous spatial aggregation and energy gathering. Furthermore, the process of temporal coherent integration is accomplished via secondary cascade both intra-pulse and inter-pulse. Intra-pulse integration, or matching filtering, is a highly effective method for detecting echoes of known transmitted waveforms in the presence of background noise that follows a Gaussian distribution. Matched filtering can be employed to accomplish pulse compression for transmitted signal waveforms with high time-bandwidth, leading to substantial enhancement in radar range resolution. The current methods for coherent integration processing involve the cascading of array elements at several spatial-temporal levels, both within and between pulses [5]. A crucial need for

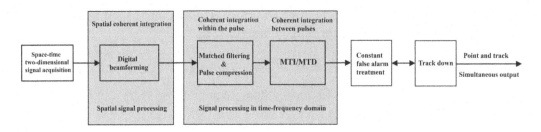

Figure 1.2 Radar signal processing based on cascade mode to realize coherent integration.

the indicated approaches is that when accumulating, the energy of the target is distributed inside a three-dimensional resolution cell known as a "beam-range-Doppler" cell. Otherwise, the unavoidable manifestation of the "three-cross" problem will result in gain loss.

Since coherent integration utilizes phase information from all integrated pulses, it is critical that any phase variation between all integrated pulses be known with a great level of confidence. Consequently, target dynamics should be estimated or computed accurately so that coherent integration can be meaningful. In fact, if a radar coherently integrates pulses from targets without the proper knowledge of the target dynamics, it suffers a loss in SNR rather than the expected SNR build up. Knowledge of target dynamics is not as critical when employing non-coherent integration; nonetheless, target range rate must be estimated so that only the returns from a given target within a specific range bin are integrated.

1.3.1 Non-Coherent Integration

Non-coherent integration refers to the process of adding and accumulating sampled data without taking phase information into account. This method is also known as post-detection integration. Non-coherent integration is a highly practical technique for signal integration in various practical applications. It is advantageous since it does not necessitate coherence in radar systems and may be easily implemented in engineering scenarios. However, the presence of background noise hinders the effectiveness of non-coherent integration at low SNR, resulting in a threshold effect of SNR. When the initial SNR is insufficiently high, the SNR gain of non-coherent integration tends toward zero.

Considering non-coherent integration of N pulses, the SNR can be expressed as

$$I_{S/N \text{ non-coherent}} = 10 \log_{10} \sqrt{N-1}. \tag{1.4}$$

1.3.2 Coherent Integration

In coherent radar receivers, the zero intermediate frequency (IF) preserves the phase information of the signals, ensuring that the signals remain strictly coherence. Coherent integration, often referred to as pre-detection integration or IF integration, is the process of gathering signals at the intermediate frequency. It is a widely employed approach for accumulating radar signals. The variability of radar target echoes has a detrimental impact on the phase coherence, and the operational coherence of radar systems limits the use of coherent integration. Fig. 1.3 displays the schematic diagram of coherent integration.

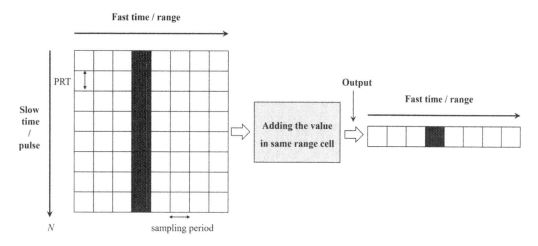

Figure 1.3   Coherent integration diagram in range-pulse slice.

The detailed process of coherent integration is illustrated in Fig. 1.3. Firstly, the received N pulses is sampled along the fast-time domain, with a total number of n sampling points. The position of the ith sampling point of each pulse is kept the same to guarantee their phases are identical. Finally, the values of the same range cell are summed up. And the vertical integration of the received pulses completes the process of coherent integration.

In an ideal scenario, assume the amplitudes of the pulses are identical. After undergoing coherent integration for the coherent pulse series composed of N pulses, the echo signals from pulses are coherently added, resulting in an increase in amplitude by times and an increase in power by times. During the process of coherent integration, the power gain of noise through coherent integration remains N times since the phase of noise is random and the noises among pulses are mutually independent. Therefore, the SNR of radar echoes outputted after coherent integration is improved by N times (or $10 \log_{10} \sqrt{N}$).

Using FFT allows for the rapid construction of a Doppler filter bank which is shown in Fig. 1.4. Different radial velocities correspond to different Doppler frequencies. Therefore, by designing a set of Doppler filter banks with different center frequencies makes it possible to distinguish targets with different velocities. This also serves to suppress various frequencies clutter.

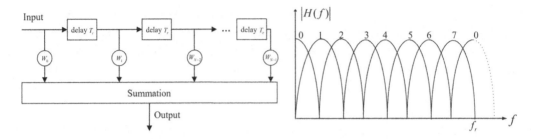

(a) The composition of Doppler filter bank (b) Amplitude-frequency response of Doppler filter bank

Figure 1.4 The composition and amplitude-frequency response of Doppler filter bank.

Perform discrete Fourier transform (DFT) on N-pulse echoes at zero intermediate frequency before detection

$$X(k) = \sum_{n=0}^{N-1} x(n) \exp\left(-j\frac{2\pi}{N}nk\right). \tag{1.5}$$

Assume the Doppler frequency of target is f_d, then $x(n) = \exp\left(-j2\pi f_d nT_r\right) = \exp\left(-j2\pi f_d n/f_r\right)$. For (1.5), if $k = -Nf_d/f_r$, (1.5) could be simplified as

$$X(k) = \left.\sum_{n=0}^{N-1} \exp\left(\frac{-j2\pi f_d n}{f_r}\right) \exp\left(-j\frac{2\pi}{N}nk\right)\right|_{k=\frac{-Nf_d}{f_r}} = N. \tag{1.6}$$

Obviously, the echo signals from each pulse are adjusted to be coherently integrated. Therefore, by utilizing N-point FFT on the echo signals from pulses according to range cells, coherent integration can be achieved. Note that k is understood as the frequency channel number.

If the target is detected in the k-th frequency channel, then the velocity of the target can be represented as

$$v = \frac{k\lambda}{2N_P T_r}. \tag{1.7}$$

Hence, the velocity measurement accuracy of moving target detection (MTD) based on FFT is as follows:

$$\Delta v = \frac{\lambda}{2N_P T_r}. \tag{1.8}$$

According to (1.8), velocity measurement accuracy is inversely proportional to the number of integrated pulses (integration time $T_a = N_p T_r$). When the pulse repetition time is constant, the higher the number of integrated pulses, the higher the velocity measurement accuracy.

1.3.3 Integration Performance Evaluation Indicator

To verify the integration performance of radar echo signals for weak targets adopting non-coherent integration and coherent integration algorithms, the following explanations need to be provided for the involved performance indicators:

1. **SNR** : Suppose the background noise of the multi-pulse echo data follows a complex Gaussian distribution, with both the real and imaginary parts having a variance of σ^2. When only studying the improvement of SNR by pulse accumulation and not considering the improvement of SNR by pulse compression, the SNR after matched filtering can be defined as

$$SNR = 10\log_{10}\frac{|A_k|^2}{2\sigma^2}. \tag{1.9}$$

At each SNR, the amplitude of target signal A_k is constant, and its phase follows uniform distribution on $[0, 2\pi]$.

Since long-time coherent integration is performed on the signal after pulse compression, hence considering the influence of pulse compression on SNR, SNR can be cited as

$$SNR_{\mathrm{OUT}} = SNR_{\mathrm{IN}} + 10\log_{10}\frac{|A_k|^2}{2\sigma^2} + 10\log_{10}(D), \tag{1.10}$$

where SNR_{IN} denotes SNR before pulse compression and SNR_{OUT} denotes SNR after long-time integration.

2. **Integration gain** (*Gain*): Within the range-Doppler cell where the target is located, find the maximum value of the signal's amplitude, denoted as A_c. Replace the data within the range-Doppler cell of with noise data from other locations, removing the target energy. Then, use the entire range-Doppler cells' data to estimate the noise power, denoted as σ_{nc}^2. Finally, the integration gain could be calculated by

$$\mathrm{Gain} = 10\log_{10}\left(\frac{A_c^2}{\sigma_{nc}^2}\right) - (SNR_{\mathrm{OUT}} - SNR_{\mathrm{IN}}). \tag{1.11}$$

According to [5], the theoretical value of integration gain is $10\log_{10}(M)$, where M denotes the pulse number.

3. **Integration efficiency** α: Integration efficiency is related to integration gain, and the relationship can be represented as

$$\mathrm{Gain} = M^\alpha. \tag{1.12}$$

4. **Integration loss** L: The loss of integration is denoted as [6]

$$L = 10\log_{10}(\mathrm{Gain}/M). \tag{1.13}$$

1.4   ORGANIZATION AND OUTLINE OF THE BOOK

The exploitation of coherent integration transform can significantly improve radar detection performance of high-speed target in noise background. Following the growing demands on radar high-speed target detection in noise background, this book provides a comprehensive review and thorough discussions of radar high-speed target detection methods via coherent integration transform, including signal model, the derivations of the transforms and analytical performance. Educating and imparting a holistic understanding of coherent integration transform-based radar high-speed target detection lay a strong foundation for postgraduate students, research scholars and practicing engineers in generating and innovating solutions and products for a broad range of applications. Readers are required to have a quite solid background in statistical signal processing, signal detection and estimation, and radar signal processing.

The remainder of this book is organized as follows:

Chapter 2 introduces the characteristics of high-speed targets and conducts the comprehensive analysis on the advantages of signal integration processing. Then, the challenges of high-speed target signal integration are presented. Last, the recent developments for the radar high-speed target detection via coherent integration transform are discussed.

Chapter 3 considers the coherent integration problem for the radar detection of high-speed targets with a constant velocity. Then, the modified location rotation transform (MLRT)-based coherent accumulation method is given to correct the range migration and achieve the focusing of signal energy.

Chapter 4 discusses the coherent integration problem for the detection of high-speed targets with acceleration. Then, the Radon-Lv's distribution (RLVD)-based coherent integration method is present to remove the range migration and Doppler frequency migration simultaneously, including the transform definition, properties and detailed processing procedure.

Chapter 5 exploits the coupling relationship between target motion and the frequency to address the coherent integration and detection problem for radar high-speed targets with acceleration. The keystone transform and match filtering process (KT-MFP)-based coherent integration method is designed according to the range migration and Doppler frequency migration characteristics.

Chapter 6 (or Chapter 7) addresses the similar problems to those in Chapter 4 (or Chapter 5), respectively. The difference is that Chapter 6 deals with the coherent integration issue with the segmented processing and Chapter 7 obtains the coherent integration without parameters searching, while Chapter 4 (or Chapter 5) deals with the coherent integration issue with parameters searching within the full observation time.

Chapter 8 considers the coherent integration and detection problem for radar high-speed targets with unknown entry time and departure time. Two coherent integration-based detection methods, i.e., window Radon fractional Fourier transform (WRFRFT) and extended generalized Radon Fourier transform (EGRFT) and window fractional Fourier transform (WFRFT), are presented.

Chapter 9 deals with the coherent integration transform-based radar detection problem for high-speed targets with jerk motion. Two fast non-searching methods based on the adjacent cross correlation function (ACCF) are discussed. The advantages of the ACCF-based coherent integration method are also given in detail.

Chapter 10 focuses on the coherent integration problem for detecting a maneuvering weak target with multiple motion models, where the target signal model is established. Then, the short-time generalized Radon Fourier transform (STGRFT)-based coherent integration algorithm is introduced, including the transform definition, property and detailed processing procedure.

Chapter 11 addresses the multi-frame coherent integration (including intra-frame integration and inter-frame integration) problem for the detection of high-speed weak targets. The modified Radon Fourier transform (RFT) domain integration algorithm is presented. Explicit expressions and analysis for various performance measures, such as integration output response, probability of detection, input-output SNR and blind speed sidelobe (BSSL) response are derived.

Chapter 12 further deals with the coherent integration problem for the radar hypersonic target detection with scale effect, where the time-bandwidth product of the radar transmitted signal is increased. The scaled Radon Fourier transform (SCRFT)-based coherent integration method is presented, including the signal modeling, transform definition, computational complexity analysis.

Chapter 13 takes into account the radar high-speed multi-target detection problem via coherent integration transform. A coherent integration method based on keystone transform and dechirp processing (KTGDP) is designed to remove the migrations (RM and Doppler frequency migration [DFM]). Two CLEAN techniques based on sinc-like point spread function and modified point spread function are presented to eliminate the strong target's effect and highlight the weak ones. By this way, the coherent integration of strong targets and weak ones can be achieve iteratively.

Acknowledgments This work was supported in part by the National Natural Science Foundation of China under Grant 62371113, the Young Elite Scientists Sponsorship Program by CAST under Grant YESS20200082, the Natural Science Foundation of Sichuan Province under Grant 2023NSFSC1386 and the Aeronautical Science Foundation under Grant 2023Z017080001.

Bibliography

[1] G. V. Morris, "Airborne pulsed Doppler radar", Norwood, MA, USA: Artech House, 1988.

[2] D. K. Barton, "Modern radar system analysis", Norwood, MA, USA: Artech House, 1988.

[3] B. R. Mahafza, A. Elsherbeni, "MATLAB simulations for radar systems design", New York, NY, USA: Chapman and Hall/CRC, 2008.

[4] V. P. Tuzlukov, "Signal processing in radar systems", Boca Raton, FL, USA: CRC Press/Taylor and Francis, 2013.

[5] D. K. Barton, "Radar system analysis and modeling", Norwood, MA, USA: Artech House, 2005.

[6] L. F. Ding, F. L. Geng, "Principles of radar", Beijing, China: Publishing House of Electronics Industry, 2002.

CHAPTER 2

High-Speed Target Signal Integration Processing

2.1 CHARACTERISTICS OF HIGH-SPEED TARGET

There are many high-speed moving objects around the earth, either man-made or natural [1–3]. For example, 1) near-earth asteroids, which can travel up to Mach 70; 2) space debris, which is not only fast but also with small size, just 1 or 2 m^2; and 3) high-dynamic aircraft in near space, which can fly at speeds up to Mach 20 and perform periodic jump flight. The above targets can be summarized as "far-range, low-observable and high-speed" targets. They share a lot in common that both in time and frequency domain, i.e., the signal-to-noise ratio (SNR) is too low to compete with noise, resulting in poor detection performance. Notably, if these high-speed targets hit the earth, it will cause great harm to the life on the earth. Therefore, how to improve the long-range monitoring ability of a radar on these high-speed targets is of great significance for hazard warning and defense [4–6]. The main characteristics of the high-speed targets can be summarized for modern radars as follows:

High-speed: Targets such as space debris and near-earth asteroids could fly at a speed of 3–25 Mach and can pass through radar space beams and detection units in an instant. Therefore, conventional processing methods are often ineffective due to the limited number of accumulated pulses. At the same time, due to the high-speed motion between the radar and the target, the scale scaling effect may need to be considered; otherwise, it will cause obvious SNR integration loss.

Low radar cross section (RCS): The development of stealth technology (such as the use of stealth materials, stealth shape design, etc.) has made the RCS of high-speed targets drop sharply (compared with traditional non-stealth targets, it can be roughly reduced by two orders of magnitude). Correspondingly, the radar detection performance has dropped to about 30% of that for conventional targets, which seriously affect the effective detection range of a radar.

Strong maneuver: The accelerations of the high-speed targets such as near-space vehicles can reach 3–10 g. In addition, the high-speed moving targets could realize orbital transfer and collision avoidance via the ways of spiral, sinusoidal, jump

DOI: 10.1201/9781003529101-2

and large-corner turns. The maneuverability of high-speed targets will bring about difficulties for the radar signal integration and detection.

Far range: High-speed targets fly fast and are highly threatening. In order to ensure sufficient early warning time, the radar needs to realize the far range detection of these high-speed targets from thousands of kilometers or even farther away so as to provide enough response time for subsequent defensive, interception and other operations.

Complex environment: In addition to the radar echo signals of high-speed targets that we are concerned about, electromagnetic space is full of other interference signals, which also bring challenges to radar high-speed target detection.

2.2 ADVANTAGES OF SIGNAL INTEGRATION PROCESSING

In general, the methods to improve radar detection performance could be mainly divided into two categories. One is to change the parameters of the radar system, such as increasing the transmitting power of the transmitter, reducing the noise figure of the receiver and increasing the antenna aperture. Although this type of method can effectively improve detection performance, it is often limited by engineering implementation and requires a significant increase in system development costs. The other is to increase the SNR of the echo by prolonging the observation time and using advanced signal accumulation processing methods. In comparison, the use of signal integration-based methods can effectively mitigate the limitation of hardware and reduce the cost of system development, which is more flexible and has a broad application prospect [7–9].

The essence of signal integration processing is to exchange energy through time. Fortunately, the development of radar techniques, such as digital beam forming, digital phased array and multi-input multi-output radar technique, makes it possible for the long-time observation and integration of the target [10–12]. Correspondingly, the radar system that uses the signal integration processing method mainly has the following advantages.

Long detection range: By accumulating and processing the multi-frame and multi-pulse echo signals of the radar, the SNR of the echo can be improved, thereby improving the radar detection performance of target and increasing the radar detection range.

Low probability of intercept: In order to cover a larger airspace and obtain a longer observation time, radars often use a wide beam to transmit, where the transmitted signal energy is scattered in multiple directions and will not form an obvious pattern. Since the transmitted signal spreads evenly in space, it is difficult to be intercepted.

High Doppler resolution: Conventional radar requires beam scanning, and the integrated pulse number is very few. However, the radar that uses long-time signal integration processing could stare the interested target by wide-beam transmitting and multi-beam receiving, which could obtain more pulse echo signals for integration and then the Doppler resolution could be improved.

2.3   CHALLENGES OF HIGH-SPEED TARGET SIGNAL INTEGRATION

Due to the motion characteristics of high-speed targets (high velocity, high maneuverability, etc.), the signal integration processing of high-speeds targets faces serious challenges, including:

Range migration: The velocity and acceleration of high-speed target will make the target's envelope vary with time after pulse compression (PC) and across range units during the integration processing time [13]. Range migration (RM) effect would degrade the effectiveness of moving target detection.

Doppler frequency migration (DFM): The acceleration motion of target will cause the target signal to be non-stationary, i.e., target signal's phase changes nonlinearly, and then the Doppler frequency would not be a constant during the coherent processing interval, which leads to energy dispersing in the frequency domain.

Scale effect: For the radar system with large time-bandwidth product, the high-speed targets movement during the intra-pulse time cannot be ignored [14]. Otherwise, the envelope of the echo signal will be scaled and offset, which leads to the mismatch in conventional pulse compression and the failure to accurately obtain the targets information. In this case, the traditional "stop-go" model is invalid.

Variable motion models: The high maneuverability of target means that the target motion during the long observation time is highly irregular. For example, some moving targets, such as near-space high-speed vehicles, are able to make very sharp, evasive maneuvers. Then, the target's motion model during the long observation time will be changeable. Besides, an anunmanned aerial vehicle (UAV) which may exhibit different flight phases (e.g., boost, reentry) within the observation period, then its motion characteristics are significantly changing and must be represented by multiple motion model [15].

Unknown time information: A high-speed moving target may enter a radar coverage area unannounced and leave after an unspecified period, which implies that the target's entry time and departure time are unknown [16]. In the absence of these time information, target signal integration and detection performance will be severely impacted [17].

Interference between multiple targets: When there are multiple targets, if the scattering intensities of different targets differ significantly, the weak ones may be shadowed by the strong targets and then it would be difficult to obtain the parameters estimation and coherent integration for weak targets [18].

Multi-frame joint integration processing: In order to improve the radar's detection performance of high-speed targets as much as possible, in addition to the intra-frame accumulation of signals, the inter-frame accumulation of signals should also be considered to maximize the echo SNR [19]. In this case, the characteristics of intra-frame accumulation output should be considered when designing the inter-frame accumulation method.

High-order motion: The high precision motion estimation of high-speed maneuvering targets attracts much more attention for modern radars due to the increasing demands in applications. As a consequence, the high-order motions, e.g., acceleration, jerk and even the fourth-order motion component, may not be omitted to realize the long-time signal integration [20–22].

2.4 RECENT DEVELOPMENTS

The SNR can be improved through signal accumulation and thus the radar target detection performance could be improved. However, due to the high-speed and maneuvering characteristics of the target, signal integration processing faces some serious challenges such as range migration and Doppler frequency migration, which lead to the failure of traditional coherent accumulation methods. Focusing on the problems above, several scholars have carried out series of studies on signal integration and detection methods for high-speed targets. According to the motion state and maneuvering characteristics of targets, these studies can be divided into three categories: (1) coherent integration methods of uniformly moving target, (2) coherent integration methods of uniformly accelerating moving target and (3) coherent integration methods of moving target signals with variable acceleration.

2.4.1 Coherent Integration Transform for Target with Constant Velocity

Keystone transform (KT), via rescaling the time axis for each frequency, has been employed to correct range walk (RW) and realize the coherent integration for high-speed moving targets [23–25]. KT is capable of eliminating RW for moving targets regardless of the targets' motion information. However, it does not consider Doppler ambiguity. Due to the high speed of targets and low radar pulse repetition frequency, Doppler ambiguity would occur and then the standard KT may suffer from performance loss. The authors [26, 27] then proposed a modified KT method via simultaneous searching of the Doppler ambiguous integers, but it needs repeated implementations of high-complexity KT operators. The axis rotation-moving target detection (AR-MTD) was introduced to correct the range migration via axis rotation and realize the coherent integration by MTD [28]. Unfortunately, the AR-MTD suffers from the peak location shifting and Doppler frequency varying effects, which will result in range estimation errors and target detection performance loss.

Based on the coupling relationship among radial velocity, RW and Doppler frequency of the moving target's echoes, the Radon Fourier transform (RFT) was then introduced to conduct the coherent integration processing for a target with RW [29]. Although RFT can obtain coherent accumulation via jointly searching in the target's motion parameter space, it has two limitations. Firstly, because of discrete pulse sampling, finite range resolution and limited integration time, the blind speed side lobe (BSSL) may inevitably appear in the RFT output, which leads to serious false alarm and detection performance deterioration [30]. A BSSL suppression method, based on the minimum operator of two weighted RFT outputs, was studied in [30]. Nevertheless, it requires to bisect the coherent integration period; thus, the integration performance would be decreased by about 3 dB. Secondly, the RFT method is often computationally prohibitive, since it involves the solution of a two-dimensional searching. Two methods via fast Chirp-Z transform are then proposed for the fast implementation of RFT, i.e., frequency bin Radon-Fourier transform (FBRFT) and sub-band Radon-Fourier transform (SBRFT) [31]. However, these two methods do not consider the BSSL effect and still need searching process.

To realize the coherent integration and detection of high-speed target without parameter-searching process, the time-reversing transform (TRT)-based method is proposed to remove the range migration [32, 33], which could be easily implemented by using complex multiplications (Mc), the fast Fourier transform (FFT) and the inverse FFT (IFFT). Besides, the scaled inverse Fourier transform-based method was presented to eliminate the range migration caused by target's velocity [34], which could estimate the motion parameters and obtain the signal integration by symmetric autocorrelation operation. By employing the parametric symmetric autocorrelation function, the frequency-domain deramp-keystone transform (FDDKT)-based integration method is proposed to deal with the range migration [35]. The FDDKT algorithm is an extension of the scaled inverse Fourier transform (SCIFT)-based algorithm. However, the TRT, SCIFT and FDDKT methods are all of bilinear operations and will confront the cross-term interference in the case of multi-targets.

2.4.2 Coherent Integration Transform for Target with Constant Acceleration

In order to deal with the Doppler frequency migration problem for high-speed targets with constant acceleration, some coherent integration transform-based detection methods, such as Lv's distribution (LVD) [36, 37], fractional Fourier transform (FRFT) [38–40] and polynomial Fourier transform (PFT) [41], are introduced to estimate the target's acceleration and obtain the focused result of target energy. Compared with the PFT and FRFT, LVD can obtain better integration performance and detection ability of weak targets' signal in the centroid frequency chirp rate (CFCR) domain. Unfortunately, these three coherent integration transform-based algorithms can only eliminate the Doppler frequency migration effect and cannot deal with the range migration problem. For the high-speed target with acceleration motion, the range migration and Doppler frequency migration effects will occur simultaneously, and then the methods mentioned above would become invalid.

The Radon transform with a minimum entropy criterion was introduced to remove the range migration and Doppler frequency migration, so as to achieve the coherent integration for a target with acceleration [42]. However, the entropy criterion-based method is not suitable for weak target signal detection since it has a high demand on the input SNR. In addition, Yu et al. [43] introduced a coherent integration method combining the discrete polynomial phase transform (DPT) and KT (DPT-KT) to remove the range migration and Doppler frequency migration effects. Tian et al. [44] proposed an integration and detection method based on the generalized KT and RFT (GKT-RFT). However, the Doppler frequency migration and range migration elimination processing of DPT-KT and GKT-RFT is a step by step process, which means that the Doppler frequency migration elimination performance of these two methods would be influenced by the former range migration calibration procedure.

To eliminate the range migration and Doppler frequency migration effects simultaneously, the keystone transform with fractional Fourier transform (KT-FRFT)-based integration and detection method is proposed by Li et al. [45], which employs the keystone transform and fold factor searching to correct the range migration and applies the FRFT to compensate the Doppler frequency migration. In addition, a coherent integration based on modified axis rotation transform (MART) and Lv's transform (LVT) [46], i.e., MART-LVT, is proposed to achieve the detection and parameter estimation of high-speed maneuvering targets. However, KT-FRFT and MART-LVT could not deal with the range migration induced by the target's acceleration.

In this regard, Chen et al. [47] presented the Radon-fractional Fourier transform (RFRFT) to realize the coherent integration and detection of maneuvering targets. Besides, Li et al. [48] introduced a coherent integration method named Radon-Lv's distribution (RLVD) for detecting weak targets. RLVD and RFRFT could remove the range migration and Doppler frequency migration simultaneously via three-dimensional searching within the parameter space, which makes them of huge computational complexity. In this regard, a coherent integration approach combining the improved axis rotation and fractional Fourier transform (IAR-FRFT) is presented for detecting maneuvering targets [49]. Compared with RLVD and RFRFT, the computational cost of IAR-FRFT could be reduced. Nevertheless, IAR-FRFT method suffers from shift-error problems during the range migration correction process.

2.4.3 Coherent Integration Transform for Target with Varied Acceleration

As for the coherent integration for the detection of high-speed maneuvering targets with jerk motion, a method based on KT and generalized dechirp process (GDP), i.e., KTGDP, is proposed [50]. In this method, KT and fold factor searching are first employed to correct the range migration caused by target's velocity, and then GDP is applied to estimate the target's radial acceleration and jerk motions. With the estimated motion parameters, the Doppler frequency migration can be compensated, and the coherent integration can be achieved via Fourier transform. In addition, at the cost of some performance loss, a fast coherent integration method combing KT and cubic phase function (CPF), i.e., KTCPF, is also introduced to further reduce the computational complexity. After range migration correction via KT, KTCPF applies the cubic phase function to estimate the target's acceleration and jerk. Thereafter, the Doppler frequency migration could be compensated [51]. However, KTGDP and KTCPF methods could deal with the linear RM induced by the target's velocity but could not deal with the SRM and TRM.

A fast coherent integration method based on time reversing transform (TRT), second-order keystone transform (SKT) and LVD, i.e., TRT-SKT-LVD, is proposed [52]. In this method, TRT operation is first presented to remove the TRM and FRM simultaneously. After that, the SKT is employed to correct the remaining SRM, and then the coherent accumulation could be obtained via LVD. The TRT-SKT-LVD method is simple and fast in that it can be easily implemented by using complex multiplications, the FFT and IFFT. Therefore, TRT-SKT-LVD could realize the coherent integration without any searching procedure and can achieve a good balance between the computational cost and detection ability. In addition, a method based on the TRT and special generalized Radon Fourier transform (SGRFT), i.e., TRT-SGRFT, is proposed to achieve the detection and motion parameters estimation [53]. More specifically, the TRT is firstly applied to separate the motion parameters and reduce the order of RM and DFM. Then the SGRFT operation is employed to estimate part of the motion parameters. After that, the residual motion parameters could be achieved via the second SGRFT operation. Since the correction of RM and compensation of DFM are carried out step by step for the TRT-SKT-LVD and TRT-SGRFT methods, their integration and detection performance will be affected to some extent.

To correct and compensate the RM and DFM simultaneously, J. Xu proposed the generalized Radon Fourier transform (GRFT) method [54], which is based on traditional RFT but has added the searching and compensation for the location and phase migration caused by the targets acceleration and jerk. Moreover, L. J. Kong et al. offered the generalized KT and generalized dechirp processing (GKTGDP) method [55], which applies the third-order KT, the sixth-order KT, and the second-order KT in turn and search for the Doppler ambiguity number to correct range migration. After that, the Doppler frequency migration is compensated via the generalized dechirp process.

2.4.4 Problems with Existing Research

Domestic and foreign research institutions and many scholars have conducted in-depth research on radar moving target coherent accumulation signal processing technology and accomplished many achievements in range migration correction and Doppler frequency compensation. However, in terms of long-time coherent accumulation signal processing technology for high-speed targets, it is still in the initial stage of research, and the main problems are as follows.

(1) When the scattering intensities of different targets differ significantly, the weak one would be shadowed by the strong target and makes it difficult to achieve the coherent accumulation for weak targets. How to obtain the signal accumulation for multiple targets when there are large differences in the scattering intensities of targets has become a problem to be solved.

(2) How to obtain the signal integration when the target's entry and departure time are unknown? In some real application scenarios, the time information when a moving target enters/leaves the radar coverage is often unknown, which would lead to severe performance loss for target signal integration and detection.

(3) How to achieve the long time signal integration for high-speed maneuvering targets with multiple motion models? With increasing target's maneuverability and long observation time (needed to detect weak targets), the target's motion model is often changing (i.e., target is of multiple motion models) during the integration time, which would reduce the performance of existing methods that assume the target is of single motion model (i.e., the target's motion model is fixed) during the long integration time.

(4) How to obtain the target signal integration for radar system with large time-width-bandwidth product? the growth of the demand for high-resolution imaging and far-range detection, the radar with large time-bandwidth product, i.e., high average power, is necessary in practical applications. Unfortunately, under the large

time-bandwidth product condition, the scale effect would occur and cause severe SNR loss during pulse accumulation.

(5) How to achieve the long time coherent integration for multi-frame echo signal of radar high-speed targets? Existing inter-frame integration methods are designed without considering the signal characteristics of the target after intra-frame integration. Intuitively, the best way for performance improvement should be to combine both the inter-frame coherent integration and intra-frame coherent integration.

(6) How to achieve the signal integration and detection for high-speed targets with high order motions? In order to achieve refined estimation of micro-motion characteristics (vibration, rotation, precession, nutation, etc.) of high-speed maneuvering targets, the high-order motion components of the target (such as jerk motion) also need to be considered during long-time coherent integration processing.

(7) How to obtain a good compromise between computational cost and integration detection performance? The parametric-searching-based method could obtain superior integration performance, but the computational cost is often high due to multidimensional searching. Meanwhile, the nonparametric-searching-based method can effectively reduce the computational complexity, and the corresponding integration detection performance will suffer a certain loss.

(8) How to achieve the signal integration and detection under lower SNR background? Under the condition of low SNR, how to make full use of the amplitude and phase information of the target signal and increase the difference intensity between the target and noise in the coherent integration transform space is still an issue worthy of further study.

Bibliography

[1] X. L. Li, Z. Sun, W. Yi, G. L. Cui, and L. J. Kong, "Radar detection and parameter estimation of high speed target based on MART-LVT," *IEEE Sensors Journal*, vol. 19, no. 4, pp. 1478–1496, February 2019.

[2] J. Xu, Y. N. Peng, X. G. Xia, and A. Farina, "Focus-before-detection radar signal processing: part i-challenges and methods," *IEEE Aerospace and Electronic Systems Magazine*, vol. 32, no. 9, pp. 48–59, September 2017.

[3] J. Xu, Y. N. Peng, X. G. Xia, T. Long, E. K. Mao, and A. Farina, "Focus-before-detection radar signal processing: part ii-recent developments," *IEEE Aerospace and Electronic Systems Magazine*, vol. 33, no. 1, pp. 34–49, January 2018.

[4] J. B. Zheng, H. W. Liu, J. Liu, X. L. Du, and Q. H. Liu, "Radar high-speed maneuvering target detection based on three-dimensional scaled transform," *IEEE Journal of Selected Topics in Applied Earth Observations and Remote Sensing*, vol. 11, no. 8, pp. 2821–2833, August 2018.

[5] S. Q. Zhu, G. S. Liao, D. Yang, and H. H. Tao, "A new method for radar high-speed maneuvering weak target detection and imaging," *IEEE Geoscience and Remote Sensing Letters*, vol. 11, no. 7, pp. 1175 1179, July 2014

[6] Z. Sun, X. Li, W. Yi, G. Cui, and L. Kong, "A coherent detection and velocity estimation algorithm for the high-speed target based on the modified location rotation transform," *IEEE Journal of Selected Topics in Applied Earth Observations and Remote Sensing*, vol. 11, no. 7, pp. 2346–2361, July 2018

[7] M. Wang, X. Li, L. Gao, Z. Sun, G. Cui, and T. S. Yeo, "Signal Accumulation method for high-speed maneuvering target detection using airborne coherent MIMO radar," *IEEE Transactions on Signal Processing*, vol. 71, pp. 2336–2351, 2023

[8] P. H. Huang, X. G. Xia, G. S. Liao, Z. W. Yang, and Y. H. Zhang, "Long-time coherent integration algorithm for radar maneuvering weak target with acceleration rate," *IEEE Transactions on Geoscience and Remote Sensing*, vol. 57, no. 6, pp. 3528–3542, June 2019.

[9] J. Tian, W. Cui, X. G. Xia, and S. Wu, "A new motion parameter estimation algorithm based on SDFC-LVT," *IEEE Transactions on Aerospace and Electronic Systems*, vol. 52, no. 5, pp. 2331–2346, October 2016.

[10] X. L. Li, Z. Sun, W. Yi, G. L. Gui, L. J. Kong, and X. B. Yang "Computationally efficient coherent detection and parameter estimation algorithm for maneuvering target," *Signal Processing*, vol. 155, pp. 130–142, February 2019.

[11] X. L. Chen, Y. Huang, N. B. Liu, J. Guan, and Y. He, "Radon-fractional ambiguity function-based detection method of low-observable maneuvering target," *IEEE Transactions on Aerospace and Electronic Systems*, vol. 51, no. 2, pp. 815–833, April 2015.

[12] P. H. Huang, G. S. Liao, X. G. Xia, Z. W. Yang, J. J. Zhou, and X. Z. Liu, "Ground moving target refocusing in SAR imagery using scaled GHAF," *IEEE Transactions on Geoscience and Remote Sensing*, vol. 56, no. 2, pp. 1030–1045, February 2018.

[13] X. L. Li, L. J. Kong, G. L. Cui, and W. Yi, "A fast detection method for maneuvering target in coherent radar," *IEEE Sensors Journal*, vol. 15, no. 11, pp. 6722–6729, November 2015.

[14] Z. Sun, X. Li, G. Cui, W. Yi, and L. Kong, 'Hypersonic target detection and velocity estimation in coherent radar system based on scaled Radon Fourier transform," *IEEE Transactions on Vehicular Technology*, vol. 69, no. 6, pp. 6525–6540, June 2020.

[15] X. L. Li, Z. Sun, T. S. Yeo, T. X. Zhang, W. Yi, G. L. Cui, and L. J. Kong, "STGRFT for detection of maneuvering weak target with multiple motion models," *IEEE Transactions on Signal Processing*, vol. 67, no. 7, pp. 1902–1917, April 2019.

[16] X. L. Li, Z. Sun, T. X. Zhang, W. Yi, G. L. Cui, and L. J. Kong, "WRFRFT-based coherent detection and parameter estimation of radar moving target with unknown entry/departure time," *Signal Processing*, vol. 166, 1–14, January 2020.

[17] X. L. Li, Z. Sun, and T. S. Yeo, "Computational efficient refocusing and estimation method for radar moving target with unknown time information," *IEEE Transactions on Computational Imaging*, vol. 6, pp. 544–557, January 2020.

[18] J. Tian, W. Cui, X. L. Lv, S. Wu, J. G. Hou, and S. L. Wu, "Joint estimation algorithm for multi-targets' motion parameters," *IET Radar, Sonar and Navigation*, vol. 8, no. 8, pp. 939–945, October 2014.

[19] X. Li, Y. Yang, Z. Sun, G. Cui, and T. S. Yeo, 'Multi-frame integration method for radar detection of weak moving target," *IEEE Transactions on Vehicular Technology*, vol. 70, no. 4, pp. 3609–3624, April 2021.

[20] X. L. Li, G. L. Cui, W. Yi, and L. J. Kong, "A fast maneuvering target motion parameters estimation algorithm based on ACCF," *IEEE Signal Processing Letters*, vol. 22, no. 3, pp. 270–274, March 2015.

[21] X. L. Li, G. L. Cui, L. J. Kong, and W. Yi, "Fast non-searching method for maneuvering target detection and motion parameters estimation," *IEEE Transactions on Signal Processing*, vol. 64, no. 9, pp. 2232–2244, May 2016.

[22] P. H. Huang, G. S. Liao, Z. W. Yang, X. Xia, J. T. Ma, and J. T. Ma, "Long-time coherent integration for weak maneuvering target detection and high-order motion parameter estimation based on keystone transform," *IEEE Transactions on Signal Processing*, vol. 64, no. 15, pp. 4013–4026, August 2016.

[23] R. P. Perry, R. C. Dipietro, and R. L. Fante, "SAR imaging of moving targets," *IEEE Transactions on Aerospace and Electronic Systems*, vol. 35, no. 1, pp. 188–200, January 1999.

[24] D. Y. Zhu, Y. Li, and Z. D. Zhu, "A keystone transform without interpolation for SAR ground moving-target imaging," *IEEE Geoscience and Remote Sensing Letters*, vol. 4, no. 1, pp. 18–22, January 2007.

[25] G. Li, X. G. Xia, and Y. N. Peng, "Doppler keystone transform: an approach suitable for parallel implementation of SAR moving target imaging," *IEEE Geoscience and Remote Sensing Letters*, vol. 5, no. 4, pp. 573–577, October 2008.

[26] X. L. Li, L. J. Kong, G. L. Cui, and W. Yi, "Detection and RM correction approach for manoeuvring target with complex motions," *IET Signal Processing*, vol. 11, no. 4, pp. 378–386, June 2017.

[27] X. L. Li, G. L. Cui, W. Yi, and L. J. Kong, "Manoeuvring target detection based on keystone transform and Lv's distribution," *IET Radar Sonar and Navigation*, vol. 10, no. 7, pp. 1234 1242, August 2016.

[28] X. Rao, H. H. Tao, J. Su, X. L. Guo, and J. Z. Zhang, "Axis rotation MTD algorithm for weak target detection," *Digital Signal Processing*, vol. 26, pp. 81–86, March 2014.

[29] J. Xu, J. Yu, Y. N. Peng, and X. G. Xia, "Radon-fourier transform (RFT) for radar target detection (I): generalized Doppler filter bank processing," *IEEE Transactions on Aerospace and Electronic Systems*, vol. 47, no. 2, pp. 1186–1202, April 2011.

[30] J. Xu, J. Yu, Y. N. Peng, and X. G. Xia, "Radon-fourier transform (RFT) for radar target detection (II): blind speed sidelobe suppression," *IEEE Transactions on Aerospace and Electronic Systems*, vol. 47, no. 4, pp. 2473–2489, October 2011.

[31] J. Yu, J. Xu, Y. N. Peng, and X. G. Xia, "Radon-fourier transform (RFT) for radar target detection (III): optimality and fast implementations," *IEEE Transactions on Aerospace and Electronic Systems*, vol. 48, no. 2, pp. 991–1004, April 2012.

[32] X. L. Li, G. L. Cui, W. Yi, and L. J. Kong, "Sequence-reversing transform-based coherent integration for high-speed target detection," *IEEE Transactions on Aerospace and Electronic Systems*, vol. 53, no. 3, pp. 1573–1580, June 2017.

[33] X. L. Li, G. L. Cui, W. Yi, and L. J. Kong, "Fast coherent integration for maneuvering target with high-order range migration via TRT-SKT-LVD," *IEEE Transactions on Aerospace and Electronic Systems*, vol. 52, no. 6, pp. 2803–2814, December 2016.

[34] J. B. Zheng, T. Su, W. T. Zhu, X. H. He, and Q. H. Liu, "Radar high-speed target detection based on the scaled inverse Fourier transform," *IEEE Journal of Selected Topics in Applied Earth Observations and Remote Sensing*, vol. 8, no. 3, pp. 1108–1119, March 2015.

[35] J. B. Zheng, T. Su, H. W. Liu, G. S. Liao, Z. Liu, and Q. H. Liu, "Radar high-speed target detection based on the frequency-domain deramp-keystone transform," *IEEE Journal of Selected Topics in Applied Earth Observations and Remote Sensing*, vol. 9, no.1, pp. 285–294, January 2016.

[36] X. L. Lv, G. Bi, C. R. Wan, and M. D. Xing, "Lv's distribution: principle, implementation, properties, and performance," *IEEE Transactions on Signal Processing*, vol. 59, no. 8, pp. 3576–3591, August 2011.

[37] J. Tian, W. Cui, X. L. Lv, S. Wu, and S. L. Wu, "Parameter estimation of manoeuvring targets based on segment integration and Lv's transform," *IET Radar Sonar and Navigation*, vol. 9, no. 5, pp. 600–607, June 2015.

[38] E. Sejdic, I. Djurovic, and L. Stankovic, "Fractional Fourier transform as a signal processing tool: an overview of recent developments," *Signal Processing*, vol. 91, no. 6, pp. 1351–1369, June 2011.

[39] S. C. Pei and J. Ding, "Fractional Fourier transform, Wigner distribution, and filter design for stationary and nonstationary random processes," *IEEE Transactions on Signal Processing*, vol. 58, no. 8, pp. 4079–4092, August 2010.

[40] L. Qi, R. Tao, S. Y. Zhou, and Y. Wang, "Detection and parameter estimation of multicomponent LFM signal based on the fractional Fourier transform," *Science in China Series F: Information Sciences*, vol. 47, no. 2, pp. 184–198, 2004.

[41] X. G. Xia, "Discrete chirp-Fourier transform and its application to chirp rate estimation," *IEEE Transactions on Signal Processing*, vol. 48, no. 11, pp. 3122–3133, November 2010.

[42] M. D. Xing, J. H. Su, G. Y. Wang, and Z. Bao, "New parameter estimation and detection algorithm for high speed small target," *IEEE Transactions on Aerospace and Electronic Systems*, vol. 47, no. 1, pp. 214–224, January 2011.

[43] W. C. Yu, W. M. Su, and H. Gu, "Ground maneuvering target detection based on discrete polynomial-phase transform and Lv's distribution," *Signal Processing*, vol. 144, pp. 364–372, March 2018.

[44] J. Tian, W. Cui, and S. Wu, "A novel method for parameter estimation of space moving targets," *IEEE Geoscience and Remote Sensing Letters*, vol. 11, no. 2, pp. 389–393, February 2014.

[45] X. Li, G. Cui, W. Yi, and L. Kong, "An efficient coherent integration method for maneuvering target detection," *IET International Radar Conference 2015*, pp. 1–6, October 2015.

[46] X. Li, Z. Sun, W. Yi, G. Cui, and L. Kong, "Radar detection and parameter estimation of high-speed target based on MART-LVT," *IEEE Sensors Journal*, vol. 19, no. 4, pp. 1478–1486, February 2019.

[47] X. Chen, J. Guan, N. Liu, and Y. He, "Maneuvering target detection via radon-fractional Fourier transform-based long-time coherent integration," *IEEE Transactions on Signal Processing*, vol. 62, no. 4, pp. 939–953, February 2015.

[48] X. L. Li, G. L. Cui, W. Yi, and L. J. Kong, "Coherent integration for maneuvering target detection based on Radon-Lv's distribution," *IEEE Signal Processing Letters*, vol. 22, no. 9, pp. 1467–1471, September 2015

[49] X. Rao, H. Tao, J. Su, J. Xie, and X. Zhang, "Detection of constant radial acceleration weak target via IAR-FRFT," *IEEE Transactions on Aerospace and Electronic Systems*, vol. 51, no. 4, pp. 3242–3253, October 2015.

[50] X. Li, L. Kong, G. Cui, and W. Yi, "CLEAN-based coherent integration method for high-speed multi-targets detection," *IET Radar, Sonar and Navigation*, vol. 10, no. 9, pp. 1671–1682, December 2016.

[51] X. Li, L. Kong, G. Cui, and W. Yi, "A low complexity coherent integration method for maneuvering target detection," *Digital Signal Processing*, vol. 49, pp. 137–147, February 2016.

[52] X. Li, G. Cui, W. Yi, and L. Kong, "Fast coherent integration for maneuvering target with high-order range migration via TRT-SKT-LVD," *IEEE Transactions on Aerospace and Electronic Systems*, vol. 52, no. 6, pp. 2803–2814, December 2016.

[53] X. L. Li, G. L. Cui, W. Yi, and L. J. Kong, "Radar maneuvering target detection and motion parameter estimation based on TRT-SGRFT," *Signal Processing*, vol. 133, pp. 107–116, April 2017.

[54] J. Xu, X. G. Xia, S. B. Peng, J. Yu, Y. N. Peng, and L. C. Qian, "Radar maneuvering target motion estimation based on generalized radon-Fourier transform," *IEEE Transactions on Signal Processing*, vol. 60, no. 12, pp. 6190–6201, December 2012.

[55] L. J. Kong, X. L. Li, G. L. Cui, W. Yi, and Y. C. Yang, "Coherent integration algorithm for a maneuvering target with high-order range migration," *IEEE Transactions on Signal Processing*, vol. 63, no. 17, pp. 4474–4486, September 2015.

CHAPTER 3

MLRT-Based Coherent Integration Processing

This chapter considers the coherent integration problem for the detection of high-speed targets with a constant velocity, involving range migration within the integration period. We introduce the modified location rotation transform (MLRT)-based accumulation method to correct the range migration and achieve the focusing of signal energy. The signal model, the transform definition and the processing flow are discussed.

3.1 SIGNAL MODEL AND PROBLEM FORMULATION

Assume that the pulse Doppler (PD) radar transmits a narrow-band linear frequency modulated (LFM) signal [1, 2], i.e.,

$$s_{\text{trans}}(t, t_m) = \text{rect}\left(\frac{t}{T_p}\right) \exp\left(j\pi\mu t^2\right) \exp\left[j2\pi f_c(t + t_m)\right], \qquad (3.1)$$

where

$$\text{rect}(u) = \begin{cases} 1 & |u| \leq \frac{1}{2}, \\ 0 & |u| > \frac{1}{2}, \end{cases}$$

$t_m = mT_r$ denotes the slow time, $m = 0, 1, 2, \ldots, M - 1$. M is the total number of integrated pulses, and T_r represents the pulse repetition interval. t denotes the fast time. T_p and f_c denote the pulse duration and the carrier frequency, respectively. μ indicates the frequency modulated rate.

Suppose that the instantaneous slant range from a moving target with a constant speed to the radar, namely $r(t_m)$, satisfies

$$r(t_m) = r_0 + v_0 t_m, \qquad (3.2)$$

where r_0 indicates the initial slant range between radar and the target and v_0 represents target's radial velocity.

DOI: 10.1201/9781003529101-3

After the demodulation, the received signal could be written as [3]

$$s_r(t, t_m) = A_0 \text{rect} \left(\frac{t - 2r(t_m)/c}{T_p} \right) \exp \left(-j \frac{4\pi r(t_m)}{\lambda} \right)$$
$$\times \exp \left[j\pi\mu \left(t - \frac{2r(t_m)}{c} \right)^2 \right],$$
(3.3)

where A_0 is the target reflection coefficient, $\lambda = c/f_c$ represents the radar wavelength and c denotes the light speed.

Note that the pulse compression (PC) should be carried out to increase the signal-to-noise ratio (SNR) gain in the pre-processing of radar signals [4, 5]. The matched filter is utilized to realize PC, which can be given as [6, 7]

$$h(t) = \text{rect} \left(\frac{t}{T_p} \right) \exp \left(j\pi\mu t^2 \right).$$
(3.4)

Here, we perform PC in range frequency domain. Taking respectively the range FT on (3.3) and (3.4), we have

$$S_r(f, t_m) = A_1 \text{rect} \left(\frac{f}{B} \right) \exp \left(-j\pi \frac{f^2}{\mu} \right)$$
$$\times \exp \left(-j \frac{4\pi r(t_m)}{\lambda} \right) \exp \left(-j \frac{4\pi f r(t_m)}{c} \right),$$
(3.5)

and

$$H(f) = \text{rect} \left(\frac{f}{B} \right) \exp \left(j\pi \frac{f^2}{\mu} \right),$$
(3.6)

where A_1 represents the amplitude after FT and $B = \mu T_p$ denotes the signal band-width.

Then the PC result in frequency domain can be expressed as

$$S(f, t_m) = S_r(f, t_m) \times H(f).$$
(3.7)

Substituting (3.5) and (3.6) into (3.7), we can obtain

$$S(f, t_m) = A_1 \text{rect} \left(\frac{f}{B} \right) \exp \left(-j \frac{4\pi r(t_m)}{\lambda} \right)$$
$$\times \exp \left(-j \frac{4\pi f r(t_m)}{c} \right).$$
(3.8)

Next, applying inverse Fourier transform (IFT) along the range direction and the compressed signal in $(t_m - t)$ plane can be stated as [8]

$$s(t, t_m) = A_2 \text{sinc} \left[B \left(t - \frac{2r(t_m)}{c} \right) \right]$$
$$\times \exp \left(-j \frac{4\pi r(t_m)}{\lambda} \right),$$
(3.9)

where $\mathrm{sinc}(x) = \sin \pi x/(\pi x)$ is the sinc function and A_2 is echo's compressed amplitude. Substituting (3.2) into (3.9) yields

$$
\begin{aligned}
s(t, t_m) =& A_2 \mathrm{sinc} \left\{ B \left[t - \frac{2\,(r_0 + v_0 t_m)}{c} \right] \right\} \\
& \times \exp \left[-j \frac{4\pi\,(r_0 + v_0 t_m)}{\lambda} \right].
\end{aligned}
\tag{3.10}
$$

Let $t = 2r/c$, then (3.10) can be expressed as

$$
\begin{aligned}
s(r, t_m) =& A_2 \mathrm{sinc} \left\{ \frac{2B}{c} \left[r - (r_0 + v_0 t_m) \right] \right\} \\
& \times \exp \left[-j \frac{4\pi\,(r_0 + v_0 t_m)}{\lambda} \right],
\end{aligned}
\tag{3.11}
$$

where r represents the slant range, which corresponds to t.

When the sampling frequency is $f_s = eB$, we have $r = \rho n$ and $r_0 = \rho n_{r_0}$, where e (we use $e = 2$ in this chapter) is the ratio between the sampling frequency and the bandwidth; $\rho = c/(2f_s)$ denotes the range cell, n is the number of range cell for r and n_{r_0} indicates the range cell number of r_0. Then, (3.11) can be rewritten in the $(m-n)$ domain, i.e.,

$$
\begin{aligned}
s(n, m) =& A_2 \mathrm{sinc} \left[\frac{1}{e} \left(n - n_{r_0} - \frac{v_0 mT}{\rho} \right) \right] \\
& \times \exp \left(-j \frac{4\pi n_{r_0} \rho}{\lambda} \right) \exp \left(-j \frac{4\pi v_0 mT}{\lambda} \right).
\end{aligned}
\tag{3.12}
$$

According to (3.12), we can see that the peak location of $s(n, m)$ changes with the new slow time m. For the target with high speed, the rang walk (RW) effect would occur and will make it hard to obtain the coherent integration (CI). Thus, it is necessary to remove the RW before long-time CI.

3.2 COHERENT DETECTION VIA MLRT

In this section, the location rotation transform (LRT) based integration algorithm in [9] is introduced at first. Then, the MLRT-based integration algorithm for a single target is discussed in detail. Subsequently, the coherent detection via the MLRT for multi-target and comparison with axis rotation moving target detection (AR-MTD) is analyzed [10]. Besides, the computational complexity analysis and consideration of target's acceleration are also given. Finally, the main procedures of MLRT are summarized.

3.2.1 Brief Introduction of LRT

In [9], the RW can be corrected via the location rotation as

$$
\begin{bmatrix} m \\ n \end{bmatrix} = \begin{bmatrix} \cos \theta' & -\sin \theta' \\ \sin \theta' & \cos \theta' \end{bmatrix} \times \begin{bmatrix} m_1' \\ n_1' \end{bmatrix},
\tag{3.13}
$$

where (m_1', n_1') represents the new location after performing LRT. Moreover, $\theta' \in (-\pi/2, \pi/2)$ denotes the rotation angle.

Substituting (3.13) into (3.12) yields

$$
\begin{aligned}
&s(n_1', m_1'; \theta') \\
&= A_1 \mathrm{sinc} \left\{ \frac{1}{e} \left[n_1' \left(\cos\theta' + \frac{v_0 T \sin\theta'}{\rho} \right) - n_{r_0} + Q \right] \right\} \\
&\quad \times \exp \left[-j \frac{4\pi \left(n_{r_0}\rho - v_0 n_1' T \sin\theta' \right)}{\lambda} \right] \\
&\quad \times \exp \left(-j \frac{4\pi v_0 m_1' T \cos\theta'}{\lambda} \right),
\end{aligned}
\tag{3.14}
$$

where

$$
Q = m_1' \left(\sin\theta' - \frac{v_0 T \cos\theta'}{\rho} \right).
\tag{3.15}
$$

When $Q = 0$, the RW effect is eliminated and the estimated parameters $\hat{\theta}_1$ and $\hat{v}_1$ can be obtained. Thus, we have

$$
\begin{aligned}
&s(n_1', m_1') \\
&= A_1 \mathrm{sinc} \left[\frac{1}{e} \left(n_1' - n_{r_0} \cos\hat{\theta}_1 \right) \left(\tan\hat{\theta}_1 \sin\hat{\theta}_1 + \cos\hat{\theta}_1 \right) \right] \\
&\quad \times \exp \left[-j \frac{4\pi \left(n_{r_0}\rho - \hat{v}_1 n_1' T \sin\hat{\theta}_1 \right)}{\lambda} \right] \\
&\quad \times \exp \left(-j \frac{4\pi \hat{v}_1 m_1' T \cos\hat{\theta}_1}{\lambda} \right).
\end{aligned}
\tag{3.16}
$$

Based on (3.13)–(3.16), we can get the sketch map of LRT exhibited in Fig. 3.1. Before the RW correction, the target's energy is spread along the oblique line in Fig.

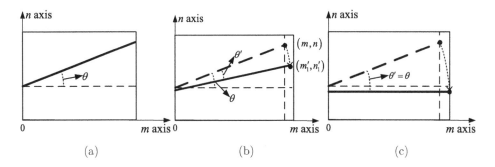

Figure 3.1 Sketch map of LRT. (a) RW effect: target energy is distributed along an oblique line (θ is corresponding to the slope value of the oblique line). (b) LRT result ($\theta' \neq \theta$): the RW still exists (θ' denotes the rotation angle). (c) LRT result ($\theta' = \theta$): the RW has been eliminated.

3.1(a). Assume the initial location before the LRT is (n, m), then the new location is (m'_1, n'_1), as shown in Fig. 3.1(b). After the RW correction, the target energy is distributed along the horizontal line in Fig. 3.1(c). Therefore, we can see that the LRT has range estimation error and Doppler frequency cell extension, which will affect the accuracy of parametric estimation.

3.2.2   MLRT for a Single Target

Because of the RW effect, the target's energy is dispersed into several range cells, which corresponds to an oblique line, as shown in Fig. 3.2(a). To correct the RW, a coherent integration method via location rotation is introduced. Suppose that the initial location before MLRT is (n, m), then after the MLRT with a rotation angle, the corresponding new location is (n', m'), as displayed in Fig. 3.2(b). The relationship between (n, m) and (n', m') can be expressed as

$$\begin{bmatrix} m \\ n \end{bmatrix} = \begin{bmatrix} 1 & 0 \\ \tan \theta' & 1 \end{bmatrix} \times \begin{bmatrix} m' \\ n' \end{bmatrix}. \tag{3.17}$$

Substituting (3.17) into (3.16) yields

$$s(n', m'; \theta') = A_1 \text{sinc} \left[\frac{1}{e} \left(n' - n_{r_0} + P \right) \right]$$
$$\times \exp \left(-j \frac{4\pi n_{r_0} \rho}{\lambda} \right) \exp \left(-j \frac{4\pi v_0 m' T}{\lambda} \right), \tag{3.18}$$

where

$$P = m' \left(\tan \theta' - \frac{v_0 T}{\rho} \right). \tag{3.19}$$

From (3.18) and (3.19), we can see that as long as $P = 0$, i.e., $\theta' = \arctan(v_0 T / \rho)$, the RW effect will be corrected (as shown in Fig. 3.2(c)) and the accumulation of

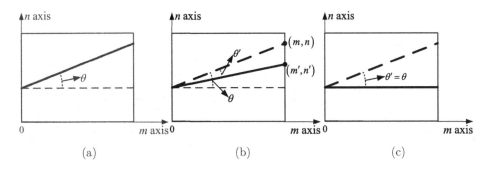

Figure 3.2   Sketch map of MLRT. (a) RW effect: target energy is distributed along an oblique line (θ is corresponding to the slope value of the oblique line). (b) MLRT result ($\theta' \neq \theta$): the RW still exists (θ' denotes the rotation angle). (c) MLRT result ($\theta' = \theta$): the RW has been eliminated.

target's energy via FT along the slow time can reach its maximum value. Besides, the corresponding searching rotation angle is equal to the initial angle θ. Therefore, θ and v_0 could be estimated as follows

$$\hat{\theta} = \arg \max_{\theta'} \left| \left\{ \mathop{\mathrm{FT}}_{m'} [s(n', m'; \theta')] \right\} \right|, \tag{3.20}$$

$$\hat{v} = \frac{\rho}{T} \tan \hat{\theta}, \tag{3.21}$$

where $\mathop{\mathrm{FT}}_{m'}(\cdot)$ indicates FT with respect to m'.

After the RW correction via MLRT, equation (3.18) can be rewritten as

$$\begin{aligned} s(n', m') = & A_1 \mathrm{sinc} \left[\frac{1}{e} (n' - n_{r_0}) \right] \exp \left(-j \frac{4\pi n_{r_0} \rho}{\lambda} \right) \\ & \times \exp \left(-j \frac{4\pi \hat{v}_0 m' T}{\lambda} \right). \end{aligned} \tag{3.22}$$

Next, perform the slow time FT for (3.22) with respect to m', and we could realize the CI as follow:

$$\begin{aligned} s_{\mathrm{int}}(n', f_{m'}) = & A_2 \mathrm{sinc} \left[\frac{1}{e} (n' - n_{r_0}) \right] \\ & \times \mathrm{sinc} \left[T_{sum} \left(f_{m'} + \frac{2\hat{v}_0 T}{\lambda} \right) \right], \end{aligned} \tag{3.23}$$

where $A_2 = A_1 G_s \exp \left(-j \frac{4\pi n_{r_0} \rho}{\lambda} \right)$ indicates the CI amplitude and G_s is the compression gain after the slow time FT. Besides, $T_{sum} = MT$ denotes the coherent processing interval and $f_{m'}$ indicates the Doppler frequency corresponding to m'.

According to (3.23), it is easy to find that the CI for a single target is obtained. Then, when the peak of (3.23) is larger than the given threshold means, it means the target detection is accomplished.

3.2.3 MLRT for Multiple Targets

Consider that K targets with a constant radial velocity move in the scene. For simplicity, the derivation for the kth target based on the MLRT is provided in what follows. The transient slant range of the Kth target satisfies

$$r_k(t_m) = r_{0,k} + v_{0,k} t_m, \tag{3.24}$$

where $r_{0,k}$ represents the initial slant range of the kth target and $v_{0,k}$ is the kth target's radial velocity.

Thus, the compressed signal after PC can be written as

$$\begin{aligned} s_{\mathrm{mc}}(t, t_m) = & \sum_{k=1}^{K} A_{1,k} \mathrm{sinc} \left[B \left(t - \frac{2r_{0,k}(t_m)}{c} \right) \right] \\ & \times \exp \left(-j \frac{4\pi r_{0,k}(t_m)}{\lambda} \right), \end{aligned} \tag{3.25}$$

where $A_{1,k} = A_{0,k} \times \sqrt{D}$ is the complex amplitude of the compressed signal and $A_{0,k}$ denotes the kth target's reflectivity.

Similarly, still let $t = 2r/c$. Thus, when $f_s = nB$, we can get $r = \rho n$ and $r_{0,k} = \rho n_{r_{0,k}}$. Therefore, equation (3.25) can be recast as

$$
\begin{aligned}
s_{\text{mc}}(n, m) = \sum_{k=1}^{K} A_{1,k} \text{sinc}\left[\frac{1}{e}\left(n - n_{r_{0,k}} - \frac{v_{0,k} mT}{\rho} \right) \right] \\
\times \exp\left(-j\frac{4\pi n_{r_{0,k}}\rho}{\lambda} \right) \exp\left(-j\frac{4\pi v_{0,k} mT}{\lambda} \right).
\end{aligned}
\tag{3.26}
$$

Assume that the initial angle of the kth target is equal to θ and the rotation angle of the kth target is θ'. Thus, the relationship between (n, m) and (n', m') is the same with equation (3.17).

Thus, substituting (3.17) into (3.26) yields

$$
\begin{aligned}
s_{\text{mc}}(n', m'; \theta') = \sum_{k=1}^{K} A_{1,k} \text{sinc}\left[\frac{1}{e}\left(n' - n_{r_{0,k}} + P_k \right) \right] \\
\times \exp\left(-j\frac{4\pi n_{r_{0,k}}\rho}{\lambda} \right) \exp\left(-j\frac{4\pi v_{0,k} m'T}{\lambda} \right),
\end{aligned}
\tag{3.27}
$$

where

$$
P_k = m'\left(\tan\theta' - \frac{v_{0,k} T}{\rho} \right).
\tag{3.28}
$$

As long as $P_k = 0$, i.e., $\theta' = \arctan(v_{0,k} T/\rho)$, the RW effect of the kth target will be removed. At this moment, the kth target's rotation angle is equivalent to its initial angle, namely, $\theta' = \theta$. Then, after the RW correction by the MLRT operation, we have

$$
\begin{aligned}
s_{\text{mc}}(n', m') = A_{1,k} \text{sinc}\left[\frac{1}{e}\left(n' - n_{r_{0,k}} \right) \right] \exp\left(-j\frac{4\pi n_{r_{0,k}}\rho}{\lambda} \right) \\
\times \exp\left(-j\frac{4\pi \hat{v}_k m'T}{\lambda} \right) + s_{\text{other}}(n', m'; \theta'),
\end{aligned}
\tag{3.29}
$$

where

$$
\begin{aligned}
s_{\text{other}}(n', m'; \theta') = \sum_{l=1, l\neq k}^{K} A_{1,l} \text{sinc}\left\{ \frac{1}{e}\left[n' - n_{r_{0l}} \right.\right. \\
\left.\left. + m'\left(\tan\theta' - \frac{v_{0,l} T}{\rho} \right) \right] \right\} \\
\times \exp\left(-j\frac{4\pi n_{r_{0l}}\rho}{\lambda} \right) \\
\times \exp\left(-j\frac{4\pi \hat{v}_{0,l} m'T}{\lambda} \right).
\end{aligned}
\tag{3.30}
$$

Next, perform the slow time FT for (3.29) with respect to m' and we can get

$$
\begin{aligned}
s_{\text{mint}}(n', f_{m'}) =& A_{2,k}\text{sinc}\left[\frac{1}{e}\left(n' - n_{r_{0,k}}\right)\right] \\
& \times \text{sinc}\left[T_{sum}\left(f_{m'} + \frac{2\hat{v}_{0,k}T}{\lambda}\right)\right] \\
& + s_{\text{other}}(n', f_{m'}; \theta'),
\end{aligned}
\tag{3.31}
$$

where

$$
s_{\text{other}}(n', f_{m'}; \theta') = \underset{m'}{\text{FT}}(s_{\text{other}}(n', m'; \theta')),
\tag{3.32}
$$

$A_{2,k} = A_{1,k}G_{sk}\exp\left(-j\frac{4\pi n_{r_{0,k}}\rho}{\lambda}\right)$ is the integration amplitude and G_{sk} denotes the compression gain after the slow time FT.

3.2.4 Comparison between AR-MTD and MLRT

In this section, we consider one of the I targets to compare the differences between the AR-MTD algorithm and the MLRT algorithm, which mainly include three parts, i.e., the peak location shifting, the Doppler frequency varying and the computational complexity.

1) The peak location shifting: In [10], after the AR-MTD, using the parameters in this chapter, one can achieve the coherent integration as follows:

$$
\begin{aligned}
s_{\text{int}}(n', f_{m'}) =& A_2\text{sinc}\left[\frac{1}{e}\left(\frac{n'}{\cos\hat{\theta}} - n_{r_0}\right)\right] \\
& \times \text{sinc}\left[T_{sum}\left(f_{m'}\cos\hat{\theta} + \frac{2\hat{v}_0 T}{\lambda}\right)\right].
\end{aligned}
\tag{3.33}
$$

From equation (3.33), one can see that the target's energy is concentrated in $n_{r_0}\cos\hat{\theta}$ for the AR-MTD. That is to say, the peak location of the target's energy has been shifted after the AR-MTD operation. For MLRT-based method, it can be seen that the energy is situated in n_{r_0} via (3.23). Thus, the MLRT algorithm has no peak location shifting effect compared with the AR-MTD.

2) The Doppler frequency varying: According to (3.33), the Doppler frequency after the AR-MTD operation is $f_{m'} = -2\hat{v}_0 T/(\lambda\cos\hat{\theta})$, i.e., $f_{m'}\cos\hat{\theta} = f_m$ (f_m denotes the initial Doppler frequency). Hence, the Doppler frequency of the AR-MTD will change with the rotation angle. In fact, the Doppler frequency after the MLRT is $f_{m'} = -2\hat{v}_0 T/\lambda$ via equation (3.23), which is equal to the initial Doppler frequency. Therefore, the MLRT algorithm's Doppler frequency will not vary in comparison with the AR-MTD.

Next, an example will be given to evaluate the RW correction and coherent integration results of the MLRT and AR-MTD. Besides, this example will also validate how the peak location shifts and the Doppler frequency varies after the AR-MTD.

Example: A high-speed target with the radar parameters listed in Table 3.1 is considered in this example, and its motion parameters are given as follows: initial

TABLE 3.1 Simulation Parameters of Radar

Parameters	Values
Carrier frequency f_c	1.5 GHz
Bandwidth B	5 MHz
Sample frequency f_s	10 MHz
Pulse repetition frequency PRF	500 Hz
Pulse duration T_p	20 μs
Number of pulses M	256

range cell $n_{r_0} = 200$ (the corresponding slant range is 300 km), radial velocity $v_0 = 2500$ m/s (the corresponding rotation angle is 18.43°). Suppose that there is no noise added in the echo data. The simulation results is shown in Fig. 3.3.

In particular, Fig. 3.3(a) gives the PC result, where 85 range cell migrations appear due to the target's high speed. Fig. 3.3(b) gives the RW correction result via the MLRT. The target's energy is located in the range cell 200, and the total Doppler frequency cell is still 256. The RW correction result through the AR-MTD is shown in Fig. 3.3(c), from which we could find that the target's energy is situated in the range cell 190 and that the total Doppler frequency cell changes into 269 after the AR-MTD operation. It should be pointed out that the $200 \times \cos(18.43) = 190$, which satisfies the peak shifting relationship, i.e., $n_{r_0} \cos \hat{\theta}$ ($\hat{\theta} = \arctan(\hat{v}T/\rho)$).

Besides, Fig. 3.3(d) and Fig. 3.3(e) give the CI results of MLRT and AR-MTD, respectively. One could see that the integrated energy of the MLRT is in the Doppler frequency cell 129 (which is equal to the initial Doppler frequency cell of f_m), while the target's energy varies in the Doppler frequency cell 135 after the AR-MTD. Note that $135 \times \cos(18.43) = 129$, which meets the relationship, namely, $f_{m'} \cos \hat{\theta} = f_m$.

Furthermore, because of peak location shifting and Doppler frequency varying, the RW correction result of MLRT is better than the AR-MTD's (i.e., the target's energy of MLRT is distributed more evenly than that of the AR-MTD), according to Fig. 3.3(b) and Fig. 3.3(c). Thus, the coherent integration result of the MLRT (its integration amplitude is 510.8) is better than the AR-MTD's (the integration amplitude is 431.8), as exhibited in Fig. 3.3(d) and Fig. 3.3(e).

3) Comparison of computational complexity: We define the number of integrated pulses and range cells as M and N, respectively. Through [10], for each rotation angle, the AR-MTD needs $4MN$ multiplications and $2MN$ additions for rotating the axis. Next, it requires $MN \log_2 N/2$ multiplications and $MN \log_2 N$ additions for the MTD operation. For the MLRT, when we rotate the echo data at each angle, MN multiplications and MN additions are necessary. Then, the integration of the target's energy demands $MN \log_2 N/2$ multiplications and $MN \log_2 N$ additions via slow time FT. Fig. 3.4 shows that the computational complexity ratio of AR-MTD and MLRT is higher than 1.55, which means the computational burden of MLRT is less than the AR-MTD's.

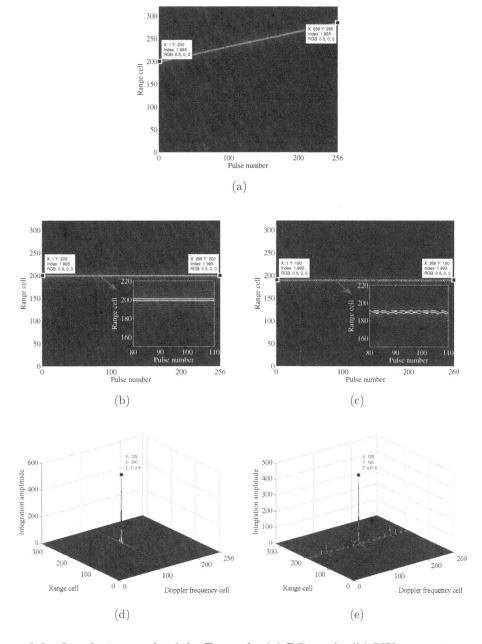

Figure 3.3 Simulation result of the Example. (a) PC result. (b) RW correction result after MLRT. (c) RW correction result after AR-MTD. (d) CI of MLRT. (e) CI of AR-MTD.

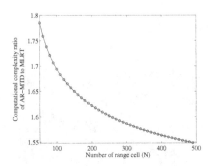

Figure 3.4   Computational complexity ratio of AR-MTD to MLRT.

3.2.5   Consideration of Target's Acceleration

The MLRT algorithm aims at correcting the RW and realizing CI for the target with constant radial velocity. Nevertheless, many targets might have radial acceleration in actual situation, which will result in both the range curvature (RC) and Doppler frequency migration (DFM). Therefore, the CI gain of the MLRT algorithm may be decreased because of the RC and DFM effects.

Suppose that the target moves with the acceleration a and the target's instantaneous range has changed from (3.2) into

$$r_T(t_m) = r_0 + v_0 t_m + a_0 t_m^2, t_m \in \left[-\frac{T_{sum}}{2}, \frac{T_{sum}}{2} \right]. \tag{3.34}$$

The acceleration will lead to the RC as (3.34). Since the MLRT could not compensate the nonlinear RC, the RC should be restricted as [14]

$$a_{\max} t_m^2 \Big|_{t_m = \pm \frac{T_{sum}}{2}} = \frac{a_0 T_{sum}^2}{4} < \rho, \tag{3.35}$$

namely

$$a_{\max} < \frac{4\rho}{T_{sum}^2}. \tag{3.36}$$

Furthermore, the instantaneous Doppler frequency can be given as [14]

$$f_d(t_m) = -\frac{dr_T(t_m)}{dt_m} \frac{2}{\lambda} = -\frac{2(v_0 + 2a_0 t_m)}{\lambda},$$
$$t_m \in \left[-\frac{T_{sum}}{2}, \frac{T_{sum}}{2} \right]. \tag{3.37}$$

To guarantee the CI performance of the MLRT, the DFM should be limited within the Doppler resolution $\rho_d = 1/T_{sum}$, namely

$$\triangle f_d(t_m) \Big|_{t_m \in \left[-\frac{T_{sum}}{2}, \frac{T_{sum}}{2} \right]} = \frac{4a_0 T_{sum}}{\lambda} < \rho_d. \tag{3.38}$$

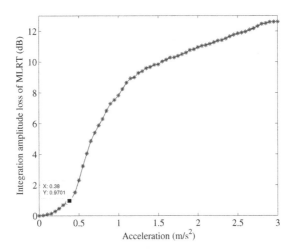

Figure 3.5 The relationship between the integration amplitude loss of MLRT and the target's acceleration.

Thus, the CI time should be limited into

$$a_{\max} < \frac{\lambda}{4T_{sum}{}^2}. \tag{3.39}$$

Usually, radar's range resolution satisfies $\rho_r \gg \lambda$, so it is easy to see that (3.39) is a more strict condition than (3.36).

Generally, in order to obtain good CI result via MLRT, the target's acceleration should satisfy (3.39). Fig. 3.5 gives the relationship between the CI amplitude loss and target's acceleration for the MLRT. According to (3.36), we can work out that the acceleration should satisfy $a_{\max} < \lambda/\left(4T_{sum}{}^2\right) \approx 0.19$ m/s^2 if the RC and DFM don't exceed the resolution cell. From Fig. 3.5, we can see that the integration amplitude loss of MLRT is less than 1 dB when target's acceleration is 0.19 m/s^2, which means that the MLRT can work well when the acceleration satisfies (3.39). Otherwise, if the acceleration value is high than 0.19 m/s^2, we need to consider other methods [15, 19, 21] to remove the RC and DFM effects before the CI.

3.2.6 Primary Procedures of MLRT

In this section, the flow chart of the MLRT is given in Fig. 3.6, and the summarized procedures could be described as:

Procedure 1: Transmit an LFM signal $s_{\mathrm{trans}}(t, t_m)$ and obtain the demodulated echo signal $s_{\mathrm{r}}(t, t_m)$;

Procedure 2: Perform PC (along the range direction) on target's echo $s_{\mathrm{r}}(t, t_m)$ and get the compressed signal $s(t, t_m)$;

Procedure 3: Express $s(t, t_m)$ as $s(n, m)$ via the variable substitution according to (3.10)–(3.12) and initialize the searching region and angle;

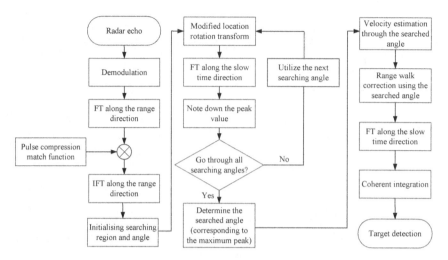

Figure 3.6 Flow chart of MLRT-based coherent integration processing.

Procedure 4: Go through all the searching angles and determine the searched angle via MLRT, which is corresponding to the maximum peak value, as given in (3.20);

Procedure 5: Utilize the searched angle to estimate target's velocity via (3.21) and correct the RW. After the RW correction, we can obtain $s(n', m')$ in (3.23);

Procedure 6: Perform FT with respect to m' and realize the CI $s_{\text{int}}(n', f_{m'})$, as shown in (3.23).

Procedure 7: Compare the peak amplitude of the CI and the threshold value determined by the constant false alarm rate (CFAR) technique. If the peak value of the CI is larger than the given threshold, there exists a target and detection is accomplished, i.e.,

$$| s_{\text{int}}(n', f_{m'}) | \geq \text{TH}, \qquad (3.40)$$

where TH represents the CFAR threshold [3, 15, 16].

3.3 SIMULATION PROCESSING RESULTS

To assess the effectiveness of the presented coherent detection algorithm, several simulations, which include CI for a single target, CI for a multi-target, detection performance analysis, initial range error propagation analysis of AR-MTD, range estimation performance and velocity estimation performance, are provided in this section. It should be pointed out that we still utilize the radar parameters of the example in Table 3.1. Besides, several algorithms, i.e., MTD, Hough transform (HT) [22–24], scaled inverse Fourier transform (SCIFT) [3], frequency-domain deramp-keystone transform (FDDKT) [16], AR-MTD [10], KT [11–13] and RFT [14], are also presented to contrast hereinbelow.

3.3.1 CI for a Single Target

To assess CI performance for a single target via MLRT, we provide a simulation in Fig. 3.7, and the target's motion parameters are the same as those of the example in Section 3.2.4. The SNR after PC is given at 6 dB. In particular, Fig. 3.7(a) shows the PC result, which indicates that there appears severe RW. Besides, Fig. 3.7(b) gives the rotation angle searching result, where the peak denotes the estimated rotation angle $\hat{\theta}$. Using the estimated angle $\hat{\theta}$ ($\hat{\theta}=18.43°$), one could calculate the estimated velocity $\hat{v}$ is equivalent to 2499.3 m/s based on (3.12). Fig. 3.7(c) gives the RW correction result utilizing the estimated rotation angle after the MLRT operation, and Fig. 3.7(d) shows the CI result by employing the slow time FT, from which one can see the energy is integrated well (the integration amplitude is 521).

To make a comparison, the CI results of four methods, i.e., MTD, AR-MTD, KT and RFT, are shown in Fig. 3.8. Fig. 3.8(a)–Fig. 3.8(d), respectively. Give the CI results of the four algorithms aforementioned. Particularly, Fig. 3.8(a) gives the CI result of MTD, and this method becomes invalid due to the RW induced by the target's high speed. The CI result of AR-MTD is shown in Fig. 3.8(b), and its integration amplitude is 442.2. So, the CI performance of AR-MTD is inferior to the MLRT's. Fig. 3.8(c) gives the CI result of KT. One can find that though the target's energy is still integrated, the peak of the KT output (the integration amplitude is 449.2) is also lower than the MLRT's because of the sinc-like interpolation loss.

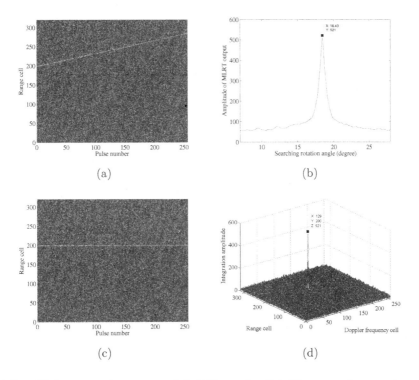

Figure 3.7 CI for a single target. (a) PC result. (b) The rotation angle searching result via MLRT. (c) RW correction result after MLRT. (d) CI result of MLRT.

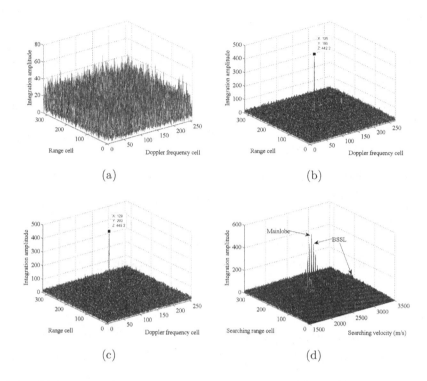

Figure 3.8 CI for a single target via several other methods. (a) CI result of MTD. (b) CI result of AR-MTD. (c) CI result of KT. (d) CI result of RFT.

Fig. 3.8(d) is the CI result of RFT. Despite the RFT could also obtain the CI, it might suffer the BSSL effect, which may lead to serious false alarm and decrease multi-target detection performance.

3.3.2 CI for Multi-Target

In this section, the result of CI for multi-target via the MLRT-based algorithm and RFT is shown in Fig. 3.9. Table 3.2 gives the motion parameters of targets A and B. Particularly, Fig. 3.9(a) gives the PC result, and then Fig. 3.9(b) shows the rotation angles searching result, from which one could find that the estimated rotation angles are 18.43° (for target A) and 11.31° (for target B). With the estimated angles, one can easily work out the estimated velocities are 2499.3 m/s (for target A) and 1500 m/s (for target B). Fig. 3.9(c) and Fig. 3.9(d) give the CI results of targets A and B via MLRT, respectively. On account of the differences in SNR and velocity, the peak of the integration outputs for targets A and B is unequal. Finally, the CI result of these two targets via the RFT is shown in Fig. 3.9(e). In this figure, we can see the BSSL effect appears in the integration output both for Target A and B, which would affect the multi-target detection.

In this section, we also discuss whether multi-targets are separated in different experiment scenes. Here, we add four targets (i.e., Target C, D, E and F) into the scenes with the motion parameters given in Table 3.3. Note that the SNR after PC in

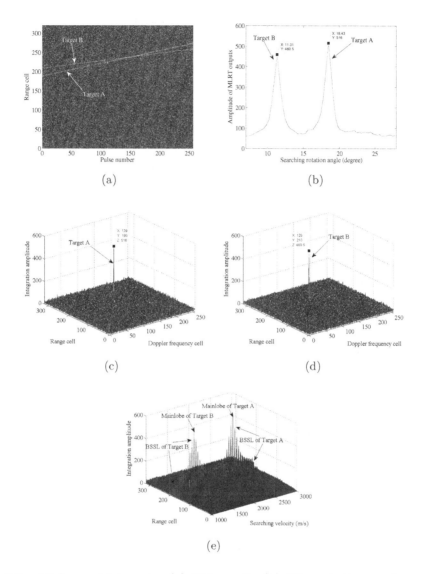

Figure 3.9 CI for multi-targets. (a) PC result. (b) The rotation angles searching result. (c) CI result for target A. (d) CI result for target B. (e) CI results for targets A and B via the RFT.

TABLE 3.2 Motion Parameters of Two Moving Targets

Parameters	Target A	Target B
Initial range cell	190	210
Radial velocity (m/s)	2500	1500
SNR (after PC) (dB)	6	5

TABLE 3.3   Motion Parameters of Multi-Target

Targets	Initial range cell	Radial velocity (m/s)
C	200	2500
D	198	2020
E	100	2490
F	197	2486

each scene is 6 dB. Fig. 3.10 shows the results of each scene. Particularly, Fig. 3.10(a) gives the result of the first scene; namely, targets C and D are in this scene, which have near distance and much different velocities. Fig. 3.10(b) gives the separating result of targets C and D. Then, Fig. 3.10(c) and Fig. 3.10(d) show the scene including targets C and F as well as the separating result of these two targets. Finally, we consider the scene with very close distance and velocities (i.e., targets C and F) in Fig. 3.10(e). Then, Fig. 3.10(f) gives the distinguishing result of these two targets. According to the simulation results above, we can see that the MLRT method could be separated clearly in different scenes. If multiple targets have identical initial range cells and velocities, we still need to utilize the "CLEAN" technique or other methods to process the targets one by one [17].

3.3.3   Detection Performance Analysis

In this experiment, the detection performance of HT, SCIFT, FDDKT, AR-MTD, KT, RFT and MLRT is analyzed by performing Monte Carlo trials under the false alarm probability $P_{f_a} = 10^{-4}$ [18], as shown in Fig. 3.11. The SNRs change in the region of $[-20,20]$ dB. While after 500 times of Monte Carlo experiments, the detection probability of the MLRT exceeds the HT's. Because of having no interpolation loss, MLRT's detection performance also surpasses that of the KT. Besides, the MLRT has better detection performance than the SCIFT and FDDKT. Moreover, the detection probability of the MLRT is better than the AR-MTD's and close to the RFT's.

3.3.4   Initial Range Error Propagation Analysis of AR-MTD

In this simulation, we analyze the initial range error propagation of AR-MTD, which is shown in Fig. 3.12. It is worth pointing out that we focus on the variation of the first range cell, i.e., the initial range cell. Assume that the initial range cell after the AR-MTD operation is r_0', then the initial range error of AR-MTD (namely Δr_0) satisfies $\Delta r_0 = |r_0' - r_0|$ (where r_0 is the initial range before AR-MTD). Fig. 3.12(a) shows the initial range error varying with the initial slant range, and initial range error varying with the target's initial velocity is shown in Fig. 3.12(b). It is observed that the initial range error of the AR-MTD will increase with the target's initial slant range or velocity.

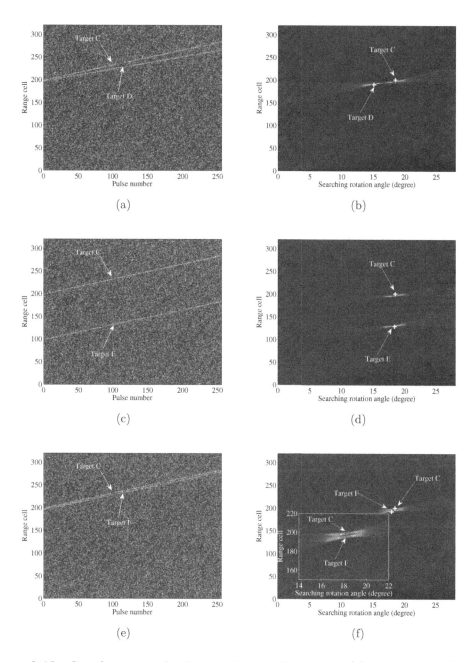

Figure 3.10 Simulation result of separating multi-targets. (a) Result after PC with targets C and D. (b) Separating result of targets C and D. (c) Result after PC with targets C and E. (d) Separating result of targets C and E. (e) Result after PC with targets C and F. (f) Separating result of targets C and F.

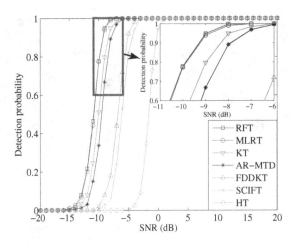

Figure 3.11 Detection probability of HT, SCIFT, FDDKT, AR-MTD, KT, RFT and MLRT.

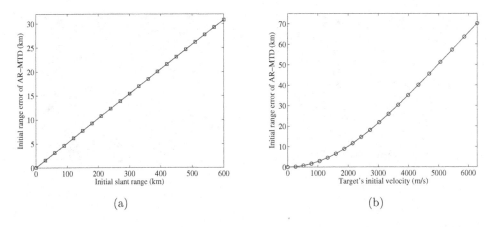

(a)   (b)

Figure 3.12 Initial range error propagation of AR-MTD. (a) Initial range error varying with the initial slant range. (b) Initial range error varying with the target's initial velocity.

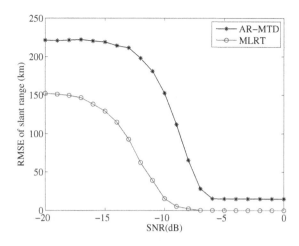

Figure 3.13 Range estimation performance of AR-MTD and MLRT.

3.3.5 Range Estimation Performance

We also assess the range estimation performance of AR-MTD and MLRT by Monte Carlo trials, as shown in Fig. 3.13. Note that the initial slant range is 300 km and that the SNRs after the PC vary from −20 to 0 dB. We measure the root-mean-square errors (RMSEs) for the slant range via performing 500 times simulations [19]. In Fig. 3.13, the RMSEs for the slant range of MLRT are lower than the AR-MTD's. Besides, the errors of MLRT are equivalent to 0 when the SNRs exceed −7 dB and the AR-MTD's estimation errors are approximately equal to 15 km when the SNRs are bigger than −6 dB. That is to say, the MLRT method can achieve more accurate range estimation value than the AR-MTD and avoid the range estimation error.

3.3.6 Velocity Estimation Performance

In this experiment, we evaluate the estimation performance of velocity for SCIFT, FDDKT, AR-MTD, KT, RFT and the MLRT method through Monte Carlo trials in Fig. 3.14, from which the SNRs vary in [−20, 0 dB]. A total of 500 times simulations are performed to measure the RMSEs for target velocity [20]. According to Fig. 3.14, we can see that the RMSEs for velocity of the MLRT are less than those of the AR-MTD and KT but close to the RFT method's at the SNRs from −20 to −7 dB. It should be pointed out that the symmetric autocorrelation operation of SCIFT and FDDKT will reduce the antinoise and parameter estimation performance. Therefore, the MLRT could achieve better velocity estimation performance than SCIFT and FDDKT. Besides, the errors are equal to 0 when the SNRs are more than −7dB, which means we can obtain an accurate velocity estimation value.

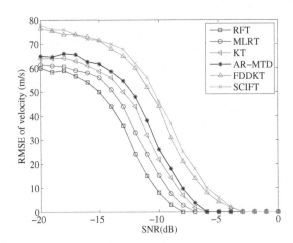

Figure 3.14 Velocity estimation performance of SCIFT, FDDKT, AR-MTD, KT, RFT, and MLRT.

3.4 SUMMARY

In this chapter, we have discussed radar high-speed target detection based on the MLRT, which could correct the range migration induced by target's velocity, via the location rotation. Compared with AR-MTD, the MLRT could achieve better CI and detection performance but avoid the range estimation error. Besides, in comparison with RFT, the MLRT can avert the BSSL and obtain close detection and estimation performance without increasing the computational complexity.

Bibliography

[1] M. D. Xing, J. H. Su, G. Y. Wang, and Z. Bao, "New parameter estimation and detection algorithm for high speed small target," *IEEE Transactions on Aerospace and Electronic Systems*, vol. 47, no. 1, pp. 214–224, January 2011.

[2] J. Tian, W. Cui, X. G. Xia, and S. L. Wu, "Parameter estimation of ground moving targets based on SKT-DLVT processing," *IEEE Transactions on Computational Imaging*, vol. 2, no. 1, pp. 13–26, March 2016.

[3] J. B. Zheng, T. Su, W. T. Zhu, X. H. He, and Q. H. Liu, "Radar high-speed target detection based on the scaled inverse Fourier transform," *IEEE Journal of Selected Topics in Applied Earth Observation and Remote Sensing*, vol. 8, no. 3, pp. 1108–1119, March 2015.

[4] J. C. Zhang, T. Su, J. B. Zheng, and X. H. He, "Novel fast coherent detection algorithm for radar maneuvering target with jerk motion," *IEEE Journal of Selected Topics in Applied Earth Observation and Remote Sensing*, vol. 10, no. 5, pp. 1792–1803, May 2017.

[5] P. H. Huang, G. S. Liao, Z. W. Yang, and X. G. Xia, "Long-time coherent integration for weak maneuvering target detection and high-order motion parameter estimation based on keystone transform," *IEEE Transactions on Signal Processing*, vol. 64, no. 15, pp. 4013–4026, August 2016.

[6] Z. Sun, X. L. Li, W. Yi, G. L. Cui, and L. J. Kong, "Detection of weak maneuvering target based on keystone transform and matched filtering process," *Signal Processing*, vol. 140, pp. 127–138, May 2017.

[7] X. L. Li, G. L. Cui, W. Yi, and L. J. Kong, "Manoeuvring target detection based on keystone transform and Lv's distribution," *IET Radar Sonar and Navigation*, vol. 10, no. 7, pp. 1234–1242, August 2016.

[8] S. Q. Zhu, G. S. Liao, D. Yang, and H. H. Tao, "A new method for radar high-speed maneuvering weak target detection and imaging," *IEEE Geoscience and Remote Sensing Letters*, vol. 11, no. 7, pp. 1175–1179, July 2014.

[9] Z. Sun, X. L. Li, W. Yi, G. L. Cui, and L. J. Kong, "Range walk correction and velocity estimation for high-speed target detection," in *Proceedings of 2017 IEEE Radar Conference*, pp. 1478–1482, June 2017.

[10] X. Rao, H. H. Tao, J. Su, X. L. Guo, and J. Z. Zhang, "Axis rotation MTD algorithm for weak target detection," *Digital Signal Processing*, vol. 26, pp. 81–86, March 2014.

[11] R. P. Perry, R. C. Dipietro, and R. L. Fante, "SAR imaging of moving targets," *IEEE Transactions on Aerospace and Electronic Systems*, vol. 35, no. 1, pp. 188–200, January 1999.

[12] D. Y. Zhu, Y. Li, and Z. D. Zhu, "A keystone transform without interpolation for SAR ground moving-target imaging," *IEEE Geoscience and Remote Sensing Letters*, vol. 4, no. 1, pp. 18–22, 2007.

[13] G. Li, X. G. Xia, and Y. N. Peng, "Doppler keystone transform: an approach suitable for parallel implementation of SAR moving target imaging," *IEEE Geoscience and Remote Sensing Letters*, vol. 5, no. 4, pp. 573–577, October 2008.

[14] J. Xu, J. Yu, Y. N. Peng, and X. G. Xia, "Radon-Fourier transform (RFT) for radar target detection (I): generalized Doppler filter bank processing," *IEEE Transactions on Aerospace and Electronic Systems*, vol. 47, no. 2, pp. 1186–1202, April 2011.

[15] X. L. Li, G. L. Cui, L. J. Kong, and W. Yi, "Fast non-searching method for maneuvering target detection and motion parameters estimation," *IEEE Transactions on Signal Processing*, vol. 64, no. 9, pp. 2232–2244, January 2016.

[16] J. B. Zheng, T. Su, H. W. Liu, G. S. Liao, Z. Liu, and Q. H. Liu, "Radar high-speed target detection based on the frequency-domain deramp-keystone transform," *IEEE Journal of Selected Topics in Applied Earth Observation and Remote Sensing*, vol. 9, no. 1, pp. 285–294, January 2016.

[17] X. L. Li, L. J. Kong, G. L. Cui, and W. Yi, "CLEAN-based coherent integration method for high-speed multi-targets detection," *IET Radar Sonar and Navigation*, vol. 10, no. 9, pp. 1671–1682, December 2016.

[18] L. J. Kong, X. L. Li, G. L. Cui, W. Yi, and Y. C. Yang, "Coherent integration algorithm for a maneuvering target with high-order range migration," *IEEE Transactions on Signal Processing*, vol. 63, no. 17, pp. 4474–4486, September 2015.

[19] X. L. Chen, J. Guan, N. B. Liu, and Y. He, "Maneuvering target detection via Radon-Fractional Fourier transform-based long-time coherent integration," *IEEE Transactions on Signal Processing*, vol. 62, no. 4, pp. 939–953, February 2014.

[20] X. L. Li, G. L. Cui, W. Yi, and L. J. Kong, "Coherent integration for maneuvering target detection based on Radon-Lv's distribution," *IEEE Signal Processing Letters*, vol. 22, no. 9, pp. 1467–1471, September 2015.

[21] L. B. Almeida, "The fractional Fourier transform and time-frequency represnetations," *IEEE Transactions on Signal Processing*, vol. 42, no. 11, pp. 3084–3091, November 1994.

[22] B. D. Carlson, E. D. Evans, and S. L. Wilson, "Search radar detection and track with the Hough transform(I): system concept" *IEEE Transactions on Aerospace and Electronic Systems*, vol. 30, no. 1, pp. 102–108, January 1994.

[23] B. D. Carlson, E. D. Evans, and S. L. Wilson, "Search radar detection and track with the Hough transform(II): detection statistics" *IEEE Transactions on Aerospace and Electronic Systems*, vol. 30, no. 1, pp. 109–115, January 1994.

[24] B. D. Carlson, E. D. Evans, and S. L. Wilson, "Search radar detection and track with the Hough transform(III): detection performance with binary integration" *IEEE Transactions on Aerospace and Electronic Systems*, vol. 30, no. 1, pp. 116–125, January 1994.

CHAPTER 4

RLVD-Based Coherent Integration Processing

This chapter addresses the coherent integration problem for the detection of high-speed targets with acceleration, involving range migration (RM) and Doppler frequency migration (DFM) within the integration time. The Radon-Lv's distribution (RLVD)-based coherent integration method is discussed in this chapter, including the transform definition, properties and detailed processing procedure. Several simulations are provided to demonstrate the effectiveness. The results show that for detection ability, the RLVD-based method is superior to the moving target detection (MTD), Radon-Fourier transform (RFT) and Radon-fractional Fourier transform (RFRFT) under low signal-to-noise ratio (SNR) environment.

4.1 SIGNAL MODEL

Suppose that the radar transmits a linear frequency modulated (LFM) signal, i.e.,

$$s_{trans}(t, t_m) = \text{rect}\left(\frac{t}{T_p}\right) \exp\left(j\pi\mu t^2\right) \exp\left[j2\pi f_c\left(t + t_m\right)\right], \tag{4.1}$$

where $\text{rect}(u) = \begin{cases} 1, & |u| \leq \frac{1}{2} \\ 0, & |u| > \frac{1}{2} \end{cases}$, T_p is the pulsewidth, μ is the frequency modulated rate, f_c is the carrier frequency, $t_m = mT_r$ $(m = 0, 1, \cdots, M-1)$ is the slow time, T_r denotes the pulse repetition time, M is the number of coherent integrated pulses and t is the fast time.

Assume that there is a moving target with slant range r_0 at $t_m = 0$. Neglecting the high-order components, the instantaneous range between radar and the target satisfies

$$r(t_m) = r_0 + v_0 t_m + a_0 t_m^2, t_m \in [0, T_{sum}] \tag{4.2}$$

where $T_{sum} = MT_r$ is the total integration time and v_0 and a_0 denote, respectively, the target's radial velocity and acceleration.

DOI: 10.1201/9781003529101-4

The received baseband echoes can be stated as [1]

$$s_r(t, t_m) = A_0 \text{rect}\left(\frac{t - 2r(t_m)/c}{T_p}\right) \exp\left[j\pi\mu\left(t - \frac{2r(t_m)}{c}\right)^2\right]$$
$$\times \exp\left(-j\frac{4\pi r(t_m)}{\lambda}\right), \tag{4.3}$$

where A_0 is the target reflectivity, c is the light speed and $\lambda = c/f_c$ denotes the wavelength.

After pulse compression (PC), the compressed signal can be expressed as

$$s(t, t_m) = A_1 \text{sinc}\left[\pi B\left(t - \frac{2r(t_m)}{c}\right)\right] \exp\left(-j\frac{4\pi f_c r(t_m)}{c}\right), \tag{4.4}$$

where B is the bandwidth of the transmitted signal.

Equation (4.4) shows that the target's envelope changes with the slow time after PC. When the offset exceeds the range resolution, i.e, $\rho_r = c/2B$, the RM effect would occur. Additionally, the phase of (4.4) is a quadratic phase function of slow time due to the target's radial acceleration, which would lead to DFM and make the target energy defocused. Both RM and DFM will create difficulties during the coherent integration processing. In the following, RLVD is presented to remove the migrations (RM and DFM) and achieve the coherent integration.

4.2 RLVD

In this section, the RLVD is proposed based on the standard RFT and LVD. Hence, we give a brief introduction of the RFT and LVD first, and then the RLVD is proposed.

4.2.1 Introduction of RFT

Based on the coupling relationship among radial velocity, range migration and Doppler frequency of the moving target, RFT is presented to obtain the coherent integration for a target with range migration. The definition of standard RFT is as follows [1]

$$R(r, v) = \int_0^T f(t, r + vt_m) \exp\left(-j\frac{4\pi vt_m}{\lambda}\right) dt_m, \tag{4.5}$$

where $f(t, r_s) \in C$ is a two-dimensional complex function defined in the (t, r_s) plane and $r_s = r + vt_m$. RFT can extract the observation values in the range-slow time plane according to the searching motion parameters and finally integrate the target's energy as a peak by accumulating these observations with Fourier transform (FT). Unfortunately, RFT is only suitable for the target with constant radial velocity and would become invalid in the case of DFM.

4.2.2 Introduction of LVD

Let us consider an LFM signal expressed as

$$x(t_m) = A_2 \exp\left(j2\pi f_0 t_m + j\pi\gamma_0 t_m^2\right). \tag{4.6}$$

where A_2, f_0 and γ_0 denote, respectively, the constant amplitude, centroid frequency and chirp rate of the LFM signal.

Its parametric symmetric instantaneous autocorrelation function (PSIAF) is defined as [2, 3]

$$
\begin{aligned}
R_x^c(t_m, \tau) &= x\left(t_m + \frac{\tau + b}{2}\right) x^*\left(t_m - \frac{\tau + b}{2}\right) \\
&= A_2^2 \exp[j2\pi f_0(\tau + b) + j2\pi\gamma_0(\tau + b)t_m],
\end{aligned}
\tag{4.7}
$$

where b denotes a constant time-delay related to a scaling operator. It can be seen from (4.7) that the time variable t_m and lag variable τ couple with each other in the exponential phase term. To remove the coupling, conduct a variable transform as follow

$$
t_m = \frac{t_n}{h(\tau + b)},
\tag{4.8}
$$

where h is a scaling factor. Then, inserting (4.8) into (4.7), we have

$$
R_x^c(t_n, \tau) = A_2^2 \exp\left[j2\pi f_0(\tau + b) + j2\pi\frac{\gamma_0}{h}t_n\right].
\tag{4.9}
$$

According to [2, 3], we generally use the parameters $b = 1$ and $h = 1$ for obtaining a desirable CFCR representation. Then performing two-dimensional (2D) FT on (4.9) with respect to t_n and τ, we obtain the LVD as follows:

$$
L(f_{ce}, \gamma) = A_3 \exp(j2\pi f_{ce})\text{sinc}(f_{ce} - f_0)\text{sinc}(\gamma - \gamma_0).
\tag{4.10}
$$

From (4.10), it can be seen that the LVD is able to accumulate the energy of a LFM signal as a obvious peak in the CFCR domain.

4.2.3 Definition of RLVD

We borrow the idea of RFT and LVD and propose a novel transform known as the RLVD to achieve the coherent integration for the maneuvering target. Without loss of generality, the definition of RLVD is given as follows. Suppose that $f(t_m, r) \in C$ is a two-dimensional complex function defined in the (t_m, r) plane and the line equation $r = r_0' + vt_m + at_m^2$ representing accelerated motion is used for searching lines in the plane. Then, the RLVD of the compressed signal $s(t, t_m)$ shown in (4.4) is defined as

$$
\begin{aligned}
RLVD(f_{ce}, \gamma) &= LVD[s(2r/c, t_m)] \\
&= LVD[s(2(r_0' + vt_m + at_m^2)/c, t_m)],
\end{aligned}
\tag{4.11}
$$

where $LVD(\cdot)$ denotes the LVD operator. By (4.11), the coherent integration outputs of $RLVD(f_{ce}, \gamma)$ can be obtained with respect to different searching pairs. Only when the searching initial slant range, searching radial velocity and searching radial acceleration are respectively equal to r_0, v_0 and a_0, $|RLVD(f_{ce}, \gamma)|$ can reach its maximum value. Then based on the peak location, the target can be detected and the motion parameters of target can be estimated.

For comparison, the RFRFT introduced in [4] is also given

$$G_r(\alpha, u) = F^\alpha[s(2(r_0' + vt_m + at_m^2)/c, t_m)](u), \qquad (4.12)$$

where F^α denotes the operator corresponding to the RFRFT of angle α.

Interestingly, the RLVD has similar parts as RFRFT, and they both use the signal along the target's trajectory. The main difference lies in the way of accumulation, using LVD or FRFT. Therefore, the proposed method is named as RLVD. Compared with MTD, RFT and RFRFT, the advantages and differences of RLVD are as follows:

1) The RLVD combines the merits of RFT and LVD, so it not only has the same integration time as RFT but also works as a useful tool for nonstationary and time-varying signal processing. Besides, since RLVD can deal with RM and DFM for maneuvering targets, the integration time of RLVD is much longer than MTD and RFT. Thus, the integration gain and detection performance will be further improved.

2) Although the RLVD is superior to the MTD and RFT for coherent integration of a maneuvering target, the improved performance comes at a cost in computational complexity. In addition, the accumulation way of RLVD is LVD, which can obtain a better performance on signal concentration and detection than FRFT without requiring more computational complexity [3, 5]. Hence, compared with RFRFT, RLVD can acquire a better detection ability with similar computational cost.

3) Similar to RFT and RFRFT, RLVD can also be used to achieve the coherent integration for multiple targets due to the excellent cross term suppression ability of LVD [2, 3, 5]. Furthermore, if the scattering intensities of different targets differ significantly, the CLEAN technique could be employed to eliminate the effect of the strong target [6–8]. In this way, the coherent integration of strong moving targets and weak ones can be achieved iteratively.

4.2.4 Properties of RLVD

As a new kind coherent integration method, RLVD satisfies some important properties. They are listed as follows.

4.2.4.1 *Asymptotic Linearity*

Assume that $z(t_m) = c_1 x(t_m) + c_2 y(t_m)$, where c_1 and c_2 are constant coefficients; then, we have

$$\begin{aligned}
RLVD_{z(t_m)}(f_{ce}, \gamma) = &|c_1|^2 RLVD_{x(t_m)}(f_{ce}, \gamma) \\
&+ |c_2|^2 RLVD_{y(t_m)}(f_{ce}, \gamma) + R_{cross,1} + R_{cross,2},
\end{aligned} \qquad (4.13)$$

where $R_{cross,1}$ and $R_{cross,2}$ denote the cross terms.

From (4.13), we can see that the RLVD of $z(t_m)$ contains both auto terms and cross terms. Fortunately, the cross terms in (4.13) can be ignored relative to the auto terms because of the excellent cross-terms suppression ability of LVD [2, 3]. Then, (4.13) can be approximated as

$$\begin{aligned}
RLVD_{z(t_m)}(f_{ce}, \gamma) \approx &|c_1|^2 RLVD_{x(t_m)}(f_{ce}, \gamma) \\
&+ |c_2|^2 RLVD_{y(t_m)}(f_{ce}, \gamma).
\end{aligned} \qquad (4.14)$$

Hence, RLVD satisfies the asymptotic linearity, which is helpful to the detection of multiple LFM signals.

4.2.4.2 *Scaling*

For a nonzero real number c_3, if $g(t_m) = x(c_3 t_m)$, applying the scaling property of LVD [2] easily leads to

$$RLVD_{g(t_m)}(f_{ce}, \gamma) = \frac{1}{c_3^2} RLVD_{x(t_m)} \left(\frac{1}{c_3} f_{ce}, \frac{1}{c_3} \gamma \right). \tag{4.15}$$

4.2.4.3 *Frequency and Chirp Rate Shift*

For any real numbers t_1, f_1 and γ_1, if

$$g(t_m) = x(t_m - t_1) \exp(j2\pi f_1 t_m) \exp(j2\pi \gamma_1 t_m^2), \tag{4.16}$$

then

$$RLVD_{g(t_m)}(f_{ce}, \gamma) = RLVD_{x(t_m)}(f_{ce} + \gamma t_1 - f_1, \gamma - \gamma_1) \\ \times \exp(-j2\pi\gamma t_1) \exp(j2\pi f_1), \tag{4.17}$$

which indicates that the two-order modulation of $x(t_m)$ will cause the chirp rate shift and that the one-order modulation or translation will cause the frequency shift.

4.2.5 Procedure of the Coherent Integration Algorithm via RLVD

To summarize, the procedure of the proposed integration algorithm based on RLVD can be described as follows:

- Input: raw data $s_r(t, t_m)$, searching area of initial slant range $[r_1, r_2]$, searching area of radial velocity $[-v_{max}, v_{max}]$ and searching area of radial acceleration $[-a_{max}, a_{max}]$.

- Process:

  Step 1) Perform pulse compression on $s_r(t, t_m)$ and obtain the compressed signal $s(t, t_m)$.

  Step 2) Determine the searching interval of initial slant range, radial velocity and radial acceleration, respectively, i.e., $\Delta r = \frac{c}{2B}$, $\Delta v = \frac{\lambda}{2T}$ and $\Delta a = \frac{\lambda}{2T^2}$ [4]. Then the searching number of initial slant range, radial velocity and radial acceleration is $N_r = \text{round}(\frac{r_2 - r_1}{\Delta r})$, $N_v = \text{round}(\frac{2v_{max}}{\Delta v})$ and $N_a = \text{round}(\frac{2a_{max}}{\Delta a})$, respectively. Where round$(\cdot)$ denotes the integer operator.

  Step 3) Determine the the moving trajectory of the target to be searched for according to the searching parameters pair (r_i, v_p, a_q), i.e.,

$$r(t_m) = r_i + v_p t_m + a_q t_m^2, \tag{4.18}$$

where $r_i \in [r_1, r_2]$, $i = 1, 2, \cdots, N_r$; $v_p \in [-v_{\max}, v_{\max}]$, $p = 1, 2, \cdots, N_v$; $a_q \in [-a_{\max}, a_{\max}]$, $q = 1, 2, \cdots, N_a$.

Step 4) Perform the RLVD to achieve the coherent integration based on the searching trajectory.

Step 5) Go through all the searching parameters and obtain the integration outputs in the RLVD domain.

Step 6) Make a detection decision and estimate the motion parameters based on the peak location of RLVD.

- Output: detection result and motion parameters estimation of target.

4.3 NUMERICAL RESULTS

This section is devoted to evaluating the performance of the proposed method via computer simulations, and the parameters of radar are shown in Table 4.1. For the sake of comparison, we will also simulate the MTD, RFT and RFRFT.

4.3.1 Coherent Integration for a Weak Target

We first evaluate the coherent integration performance for a weak target via MTD, RFT, RFRFT and RLVD in Fig. 4.1, where the motion parameters of a maneuvering target are $r_0 = 100.15$ km, $v_0 = 100$ m/s and $a_0 = 20$ m/s^2. Fig. 4.1(a) shows the result after pulse compression, which indicates that the target energy is totally buried in the noise with SNR $= -13$ dB. To show the target's trajectory clearly, the pulse compression result without noise is also given in Fig. 4.1(b). It can be seen that serious RM occurs. Fig. 4.1(c) and Fig. 4.1(d) show, respectively, the integration results of MTD and RFT. Because of the RM and DFM, MTD and RFT become ineffectiveness. Moreover, Fig. 4.1(e) and Fig. 4.1(f) show the integration results of RFRFT and RLVD, respectively. We can see that the RLVD obtains a much larger noise margin than that obtained from the RFRFT, which implies that a better detection ability can be achieved.

4.3.2 Coherent Integration for Multiple Targets

We also analyze the coherent integration performance of RLVD for multiple targets in Fig. 4.2, where the motion parameters of two maneuvering targets are listed in

TABLE 4.1 Simulation Parameters of Radar

Parameters	Values
Carrier frequency	0.15 GHz
Bandwidth	20 MHz
Sample frequency	100 MHz
Pulse repetition frequency	500 Hz
Pulse duration	5 μs
Pulse number	512

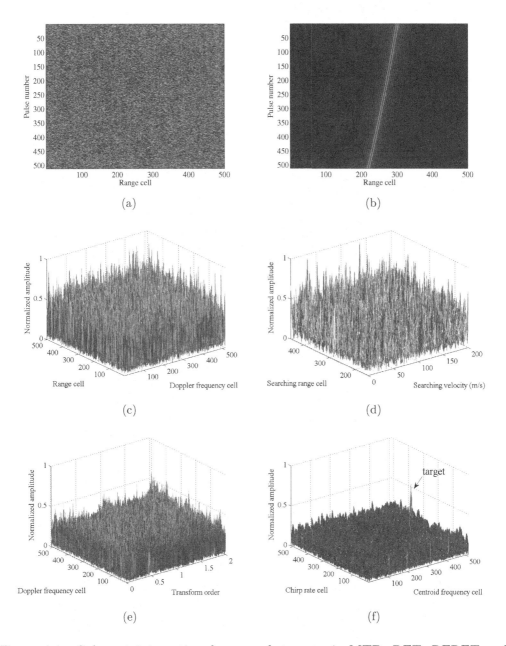

Figure 4.1 Coherent integration for a weak target via MTD, RFT, RFRFT and RLVD. (a) Result after pulse compression. (b) Result after pulse compression without noise. (c) MTD. (d) RFT. (e) RFRFT. (f) RLVD.

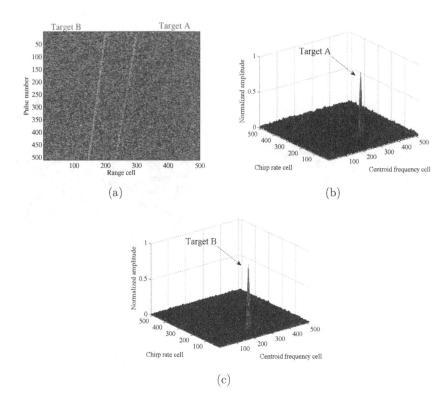

Figure 4.2 Coherent integration for multiple targets via RLVD. (a) Result after pulse compression. (b) Coherent integration result of target A ($r_0 = 100.15$ km). (c) Coherent integration result of target B ($r_0 = 100$ km).

Table 4.2. Fig. 4.2(a) shows the result after pulse compression. Fig. 4.2(b) and Fig. 4.2(c) shows, respectively, the integration result of targets A and B based on RLVD. It can be seen that the targets' energy is accumulated as two obvious peaks in the corresponding RLVD domain, which is helpful to the target detection.

4.3.3 Coherent Integration Detection Ability

The detection performances of MTD, RFT, RFRFT and RLVD are further investigated by Monte Carlo trials. For simplicity, we only consider target A in the scene. We combine the constant false alarm rate (CFAR) detector and the four methods as corresponding detectors. The Gaussian noises are added to the target echoes, and the false alarm ratio is set as $P_{fa} = 10^{-2}$. Fig. 4.3 shows the detection probability of the four detectors versus different SNR levels, and in each case, 1000 times of Monte

TABLE 4.2 Motion Parameters of Targets

Motion parameters	Target A	Target B
Initial slant range (kM)	100.15	100
Radial velocity (m/s)	100	80
Radial acceleration (m/s^2)	20	10
SNR (after pulse compression) (dB)	1	4

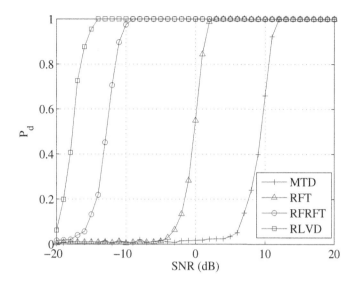

Figure 4.3 Detection probability of MTD, RFT, RFRFT and RLVD.

Carlo simulations are done. The simulation results show that the probability of the detector based on RLVD is superior to the others, thanks to its ability to deal with the RM and DFM as well as the better performance on signal concentration.

4.4 SUMMARY

In this chapter, we have addressed the coherent integration problem for the detection of high-speed targets with constant acceleration, involving RM and DFM with the coherent integration time. The RLVD-based coherent integration method, which accomplishes the data extraction on the pulse-compressed signal with a three-dimensional searching along range, radial velocity and radial acceleration directions, is discussed. Several simulations are provided to demonstrate the effectiveness.

Bibliography

[1] J. Xu, J. Yu, Y. N. Peng, and X. G. Xia, "Radon-Fourier transform (RFT) for radar target detection (I): generalized Doppler filter bank processing," *IEEE Transactions on Aerospace and Electronic Systems*, vol. 47, no. 2, pp. 1186–1202, April 2011.

[2] X. L. Lv, G. Bi, C. R. Wan, and M. D. Xing, "Lv's distribution: principle, implementation, properties, and performance," *IEEE Transactions on Signal Processing*, vol. 59, no. 8, pp. 3576–3591, August 2011.

[3] S. Luo, G. Bi, X. L. Lv, and F. Y. Hu, "Performance analysis on Lv distribution and its applications," *Digital Signal Processing*, vol. 23, no. 3, pp. 797–807, May 2013.

[4] X. L. Chen, J. Guan, N. B. Liu, and Y. He, "Maneuvering target detection via Radon-Fractional Fourier transform-based long-time coherent integration," *IEEE Transactions on Signal Processing*, vol. 62, no. 4, pp. 939–953, February 2014.

[5] S. Luo, X. Lv, and G. Bi, "Lv's distribution for time-frequency analysis," in *Proceedings of 2011 International Conference on Circuits, Systems, Control, Signals*, 2011, pp. 110–115.

[6] J. Tsao and B. D. Steinberg, "Reduction of sidelobe and speckle artifacts in microwave imaging: the CLEAN technique," *IEEE Transactions on Antennas and Propagation*, vol. 36, no. 4, pp. 543–556, April 1998.

[7] H. Deng, "Effective CLEAN algorithms for performance-enhanced detection of binary coding radar signals," *IEEE Transactions on Signal Processing*, vol. 52, no. 1, pp. 72–78, January 2004.

[8] Q. Wang, M. D. Xing, G. Y. Lu, and Z. Bao, "SRMF-CLEAN imaging algorithm for space debris," *IEEE Transactions on Antennas and Propagation*, vol. 55, no. 12, pp. 3524–3533, December 2007.

CHAPTER 5

KT-MFP-Based Coherent Integration Processing

The Radon-Lv's distribution (RLVD)-based coherent integration method in Chapter 3 realizes the radar high-speed target signal integration and detection via the three-dimensional searching in the time domain. In this chapter, we discuss the coherent integration problem for radar high-speed target detection in the frequency domain, while the keystone transform and match filtering process (KT-MFP)-based coherent integration method is presented. In particular, KT-MFP corrects the linear range migration (RM) caused by the unambiguous velocity via KT, and then the match filtering process (MFP) is carried out to remove the residual range migration and compensate the Doppler frequency migration (DFM). At last, the coherent integration of the target's energy is achieved through the slow time Fourier transform (FT).

5.1 SIGNAL MODEL AND PROBLEM FORMULATION

Consider that the pulse Doppler (PD) radar transmits a narrow-band linear frequency modulated (LFM) signal with the mathematical model as follows [1, 2]:

$$s_{trans}(t, t_m) = \text{rect}\left(\frac{t}{T_p}\right) \exp\left(j\pi\mu t^2\right) \exp\left[j2\pi f_c(t + t_m)\right], \qquad (5.1)$$

where

$$\text{rect}\left(\frac{t}{T_p}\right) = \begin{cases} 1 & |t| \leq T_p/2, \\ 0 & |t| > T_p/2, \end{cases}$$

$\exp(\cdot)$ is the exponential function [3–5]. t denotes the fast time, and $t_m = mT_r$ is the slow time, $m = 0, 1, 2, \ldots, M - 1$. M represents the total integration pulse number of radar, and T_r indicates the pulse repetition interval. T_p and f_c denote the pulse duration and the carrier frequency, respectively. $\mu = \frac{B}{T_p}$ is the frequency modulated rate, and B is the signal bandwidth.

DOI: 10.1201/9781003529101-5

Ignoring the high-order components and supposing that the instantaneous slant range between the radar and a moving target with an acceleration satisfies [6–11]

$$r(t_m) = r_0 + v_0 t_m + a_0 t_m^2, \tag{5.2}$$

where r_0 is the initial slant range between the radar and the target and v_0 and a_0 denote the radial velocity and acceleration, respectively.

Assume $|v_0| \ll c$ (c denotes the speed of light), and then the received baseband echo signal after the demodulation can be expressed as [12, 13]

$$
\begin{aligned}
s_r(t, t_m) =& A_0 \text{rect}\left(\frac{t - \beta}{T_p}\right) \exp\left(-j \frac{4\pi r(t_m)}{\lambda}\right) \\
& \times \exp\left[j\pi\mu (t - \beta)^2\right] \exp\left(-j \frac{4\pi v_0 t_m}{\lambda}\right),
\end{aligned}
\tag{5.3}
$$

where A_0 is the target reflectance, $\lambda = \frac{c}{f_c}$ represents the radar wavelength and $\beta = \frac{2r(t_m)}{c}$ is the delay time.

Then, we can use the matched filter, i.e., $h(t) = \text{rect}\left(\frac{t}{T_p}\right) \exp(j\pi\mu t^2)$, to realize the pulse compression (PC), and the compressed signal in the range frequency-slow time $(f - t_m)$ domain may be written as [14]

$$
\begin{aligned}
S(f, t_m) =& A_1 \text{rect}\left(\frac{f + f_d/2}{B - f_d}\right) \exp\left(-j \frac{\pi f_d^2}{\mu}\right) \\
& \times \exp\left[-j \frac{4\pi (f + f_c + f_d) r(t_m)}{c}\right],
\end{aligned}
\tag{5.4}
$$

where $A_1 = A_0 \times \sqrt{D}$ denotes the complex amplitude of the signal echoes after PC and $D = B \times T_p$ is the pulse compression ratio [15]. $f_d = \frac{2v_0}{\lambda}$ is the Doppler frequency.

Because of the target's high-speed and low pulse repetition frequency (PRF) of radar, the velocity may exceed the value of PRF, which will bring about the velocity ambiguity effect [16]. So, the velocity v_0 satisfies

$$v_0 = l v_{amb} + v_{unamb}, \tag{5.5}$$

where $v_{amb} = \frac{\lambda PRF}{2}$ is the blind velocity, l is the fold factor and v_{unamb} denotes the unambiguous velocity with the range of $[-v_{amb}/2, v_{amb}/2]$ [17].

Substituting (5.2) and (5.5) into (5.4) yields

$$
\begin{aligned}
S(f, t_m) =& A_2 \exp\left[-j \frac{4\pi (f + f_c + f_d) r_0}{c}\right] \\
& \times \exp\left[-j \frac{4\pi (f + f_c + f_d) v_{unamb} t_m}{c}\right] \\
& \times \exp\left[-j \frac{4\pi (f + f_c + f_d) a_0 t_m^2}{c}\right] \\
& \times \exp\left[-j \frac{4\pi (f + f_c + f_d) l v_{amb} t_m}{c}\right],
\end{aligned}
\tag{5.6}
$$

where $A_2 = A_1 \text{rect}\left(\frac{f+f_d/2}{B-f_d}\right)\exp\left(-j\frac{\pi f_d^2}{\mu}\right)$. Ignoring the effect of f_d, (5.6) can be expressed as follow [14, 18]:

$$
\begin{aligned}
S(f,t_m) =& A_2 \exp\left[-j\frac{4\pi(f+f_c)r_0}{c}\right] \\
&\times \exp\left[-j\frac{4\pi(f+f_c)v_{unamb}t_m}{c}\right] \\
&\times \exp\left[-j\frac{4\pi(f+f_c)a_0 t_m^2}{c}\right] \\
&\times \exp\left(-j\frac{4\pi f l v_{amb} t_m}{c}\right) \\
&\times \exp\left(-j2\pi PRF l t_m\right).
\end{aligned} \tag{5.7}
$$

It should be pointed out that $\exp\left(-j2\pi PRF l t_m\right) = 1$ [14, 19]. So, (5.7) can be rewritten as

$$
\begin{aligned}
S(f,t_m) =& A_2 \exp\left[-j\frac{4\pi(f+f_c)r_0}{c}\right] \\
&\times \exp\left[-j\frac{4\pi(f+f_c)v_{unamb}t_m}{c}\right] \\
&\times \exp\left[-j\frac{4\pi(f+f_c)a_0 t_m^2}{c}\right] \\
&\times \exp\left(-j\frac{4\pi f l v_{amb} t_m}{c}\right).
\end{aligned} \tag{5.8}
$$

Clearly, the last three exponential terms of (5.8) are all coupling terms with respect to slow-time t_m and range-frequency f. In particular,

- $\exp\left[-j\frac{4\pi(f+f_c)v_{unamb}t_m}{c}\right]$ denotes the term of unambiguous velocity, which will cause the linear RM.

- $\exp\left(-j\frac{4\pi f l v_{amb} t_m}{c}\right)$ represents the term of blind velocity, which would also result in the linear RM.

- $\exp\left[-j\frac{4\pi(f+f_c)a_0 t_m^2}{c}\right]$ may bring about the quadratic RM and DFM, which can make the integration energy dispersed.

In order to solve the problems of RM and DFM and obtain the coherent accumulation, KT-MFP is proposed in the following section.

5.2 COHERENT INTEGRATION VIA KT-MFP

In this section, the detailed processes of the KT-MFP-based coherent integration method are introduced. Furthermore, the coherent integration via the KT-MFP for the multi-target is also analyzed. Finally, the KT-MFP method is also compared with generalized Radon Fourier transform (GRFT) in terms of the computational complexity.

5.2.1 Keystone Transform

Firstly, apply KT, which is performed as $t_m = \frac{f_c}{f+f_c} t_n$ (t_n is the new slow-time variable), to eliminate the linear range migration [20–22]. Then (5.8) can be represented in the $f - t_n$ domain as

$$
\begin{aligned}
S_{\text{KT}}(f, t_n) = & A_2 \exp\left[-j\frac{4\pi(f + f_c)r_0}{c}\right] \\
& \times \exp\left(-j\frac{4\pi v_{unamb}t_n}{\lambda}\right) \\
& \times \exp\left[-j\frac{4\pi l v_{amb}t_n f}{\lambda(f + f_c)}\right] \\
& \times \exp\left[-j\frac{4\pi a_0 t_n^2 f_c}{\lambda(f + f_c)}\right].
\end{aligned}
\tag{5.9}
$$

Consider that the radar transmits a narrowband signal and the range-frequency variable f satisfies $f \ll f_c$, so $\frac{f_c}{f+f_c} \approx 1 - \frac{f}{f_c}$ and $f + f_c \approx \frac{f_c^2}{f_c-f}$ [23]. Thus (5.9) can be recast as

$$
\begin{aligned}
S_{\text{KT}}(f, t_n) = & A_2 \exp\left[-j\frac{4\pi(f + f_c)r_0}{c}\right] \\
& \times \exp\left(-j\frac{4\pi v_{unamb}t_n}{\lambda}\right) \\
& \times \exp\left[-j\frac{4\pi l v_{amb}t_n f(f_c - f)}{c f_c}\right] \\
& \times \exp\left[-j\frac{4\pi a_0 t_n^2(f_c - f)}{c}\right].
\end{aligned}
\tag{5.10}
$$

Through the second exponential term in (5.10), it can be seen that the linear RM resulted from unambiguous velocity v_{unamb} has been corrected. However, the last two exponential terms indicate that the linear RM caused by the blind velocity and the quadratic RM induced by acceleration still exist.

5.2.2 Matched Filtering Process

To eliminate the residual RM and the DFM in (5.10), we define the matched filtering function as

$$
\begin{aligned}
H_{\text{m}}(f, t_n; l', a_0') = & \exp\left[j\frac{4\pi l' v_{amb}t_n f(f_c - f)}{c f_c}\right] \\
& \times \exp\left[j\frac{4\pi a_0' t_n^2(f_c - f)}{c}\right],
\end{aligned}
\tag{5.11}
$$

where l' is the searching fold factor and a_0' denotes the searching acceleration. Multiplying (5.11) by (5.10) yields

$$
\begin{aligned}
S_{\mathrm{KT}}(f, t_n; l', a_0') =& A_2 \exp\left[-j\frac{4\pi\,(f + f_c)\,r_0}{c}\right] \\
&\times \exp\left(-j\frac{4\pi v_{unamb} t_n}{\lambda}\right) \\
&\times \exp\left[j\frac{4\pi(l' - l)v_{amb} t_n f\,(f_c - f)}{c f_c}\right] \\
&\times \exp\left[j\frac{4\pi(a_0' - a_0)t_n^2\,(f_c - f)}{c}\right].
\end{aligned}
$$
(5.12)

When $l' = l$ and $a_0' = a_0$, the searching fold factor and the acceleration are matched with the target's fold factor and acceleration, respectively. Therefore, (5.12) can be rewritten as

$$
\begin{aligned}
S_{\mathrm{match}}(f, t_n) =& A_2 \exp\left[-j\frac{4\pi\,(f + f_c)\,r_0}{c}\right] \\
&\times \exp\left(-j\frac{4\pi v_{unamb} t_n}{\lambda}\right).
\end{aligned}
$$
(5.13)

From (5.13), one can see that the residual RM caused by the blind velocity and the acceleration can be removed. Besides, the DFM effect resulted from the acceleration could also be compensated. After the RM and DFM compensation via KT and MFP, applying the range inverse Fourier transform (IFT) to (5.13), we can get

$$
\begin{aligned}
s_{\mathrm{match}}(t, t_n) =& A_3 \mathrm{sinc}\left[B\left(t - \frac{2r_0}{c}\right)\right]\exp\left(-j\frac{4\pi r_0}{\lambda}\right) \\
&\times \exp\left(-j\frac{4\pi v_{unamb} t_n}{\lambda}\right),
\end{aligned}
$$
(5.14)

where $\mathrm{sinc}(x) = \frac{\sin\pi x}{\pi x}$ is the sinc function [5, 24]. In addition, $A_3 = A_2 G_r$ denotes the complex amplitude (after the range IFT), and G_r is the compression gain for range IFT. Finally, performing the slow time FT on (5.14), we can achieve the coherent integration, i.e.,

$$
\begin{aligned}
s_{\mathrm{int}}(t, f_{t_n}) =& A_5 \mathrm{sinc}\left[B\left(t - \frac{2r_0}{c}\right)\right] \\
&\times \mathrm{sinc}\left[T_{sum}\left(f_{t_n} + \frac{2v_{unamb}}{\lambda}\right)\right],
\end{aligned}
$$
(5.15)

where $A_5 = A_4 \exp\left(-j\frac{4\pi r_0}{\lambda}\right)$. Furthermore, $A_4 = A_3 G_s$ is the integration amplitude (after the slow time FT), and G_s indicates the compression gain of the slow time FT. Additionally, $T_{sum} = M T_r$ represents the coherent processing interval, while f_{t_n} is the Doppler frequency with respect to t_n.

From (5.14) and (5.15), we can see that the coherent integration reaches its maximum value when the searching fold factor l' and acceleration a_0' are precisely matched.

Thus, the estimated function of l and a_0, i.e., $(\hat{l}, \hat{a}_0)$, can be achieved as follows:

$$
\begin{aligned}
(\hat{l}, \hat{a}_0) = \underset{(l', a_0')}{\arg\max} \, | \underset{t_n}{\mathrm{FT}} \{ \underset{f}{\mathrm{IFT}} [S_{\mathrm{KT}}(f, t_n) \\
\times H_{\mathrm{m}}(f, t_n; l', a_0')] \} |,
\end{aligned}
\tag{5.16}
$$

where $\underset{f}{\mathrm{IFT}}(\cdot)$ and $\underset{t_n}{\mathrm{FT}}(\cdot)$ indicate, respectively, the IFT over the range frequency variable f and the FT with respect to the slow time variable t_n.

According to the peak location of (5.15), we could also obtain the value of the Doppler frequency f_{t_n}, which has the relationship with unambiguous velocity as $v_{unamb} = -\frac{\lambda f_{t_n}}{2}$. Then, the estimation value of $\hat{v}_{unamb}$ can be written as

$$
\begin{aligned}
\hat{v}_{unamb} = -\frac{\lambda}{2} \underset{f_{t_n}}{\arg\max} \, | \underset{t_n}{\mathrm{FT}} \{ \underset{f}{\mathrm{IFT}} [S_{\mathrm{KT}}(f, t_n) \\
\times H_{\mathrm{m}}(f, t_n; l', a_0')] \} |.
\end{aligned}
\tag{5.17}
$$

Thereby, we can achieve the value of $\hat{v}_0$, which indicates the estimated velocity, as follows:

$$
\hat{v}_0 = \hat{v}_{unamb} + \hat{l} v_{amb}.
\tag{5.18}
$$

One can see that the coherent integration of target's energy can be obtained from (5.15). On condition that the peak value of (5.15) exceeds the given threshold, the target could be detected. It should be noted that the searching scopes of l' and a_0' are $[l'_{\min}, l'_{\max}]$ and $[a_{0\min}', a_{0\max}']$, where the searching intervals are $\triangle l' = f_c/(M f_s)$ (f_s denotes the sampling rate) [14] and $\triangle a_0' = \lambda/(2T_{sum}^2)$ [24], respectively. Additionally, the summarized framework of the KT-MFP-based method is given in Algorithm 1, while the flowchart is shown in Fig. 5.1.

5.2.3 KT-MFP for Multi-Target

Assume that there are K targets in the scene. Without loss of generality, the derivation of KT-MFP for the kth target is given in the following. The instantaneous slant range of the kth target with complex motions satisfies [5–7]

$$
r_k(t_m) = r_{0,k} + v_{0,k} t_m + a_{0,k} t_m^2,
\tag{5.19}
$$

where $r_{0,k}$ is the initial slant range from the radar to the kth target and $v_{0,k}$ and $a_{0,k}$, respectively, indicate the kth target's radial velocity and acceleration.

Note that the velocity of the kth target could be written as follows:

$$
v_{0,k} = l_k v_{amb} + v_{unamb,k},
\tag{5.20}
$$

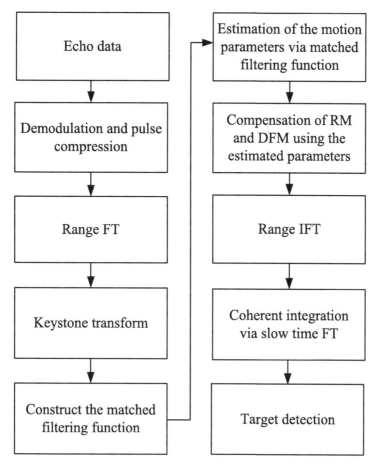

Figure 5.1 Flow chart of the KT-MFP-based coherent integration method.

where l_k and $v_{unamb,k}$ represent the fold factor and the unambiguous velocity of the kth target, respectively.

Furthermore, the received signal after PC could be expressed as [25, 26]

$$S_{\mathrm{m}}(f, t_m) = \sum_{k=1}^{K} A_{1,k} \mathrm{rect}\left(\frac{f + f_{d,k}/2}{B - f_{d,k}}\right) \exp\left(-j\frac{\pi f_{d,k}^2}{\mu}\right)$$
$$\times \exp\left[-j\frac{4\pi\left(f + f_c + f_{d,k}\right) r_k\left(t_m\right)}{c}\right], \qquad (5.21)$$

where $A_{1,k}$ is the complex amplitude of the kth target after PC and $f_{d,k}$ denotes the kth target's Doppler frequency.

Ignoring the effect of $f_{d,k}$, then substituting (5.19) and (5.20) into (5.21) yields

$$
\begin{aligned}
S_{\mathrm{m}}(f, t_m) = \sum_{k=1}^{K} A_{2,k} & \exp\left[-j\frac{4\pi(f+f_c)r_{0,k}}{c}\right] \\
& \times \exp\left[-j\frac{4\pi(f+f_c)v_{unamb,k}t_m}{c}\right] \\
& \times \exp\left[-j\frac{4\pi(f+f_c)a_{0,k}t_m^2}{c}\right] \\
& \times \exp\left(-j\frac{4\pi f l_k v_{amb} t_m}{c}\right),
\end{aligned}
\tag{5.22}
$$

where $A_{2,k} = A_{1,k}\mathrm{rect}\left(\frac{f+f_{d,k}/2}{B-f_{d,k}}\right)\exp\left(-j\frac{\pi f_{d,k}^2}{\mu}\right)$. It should also be noted that $\exp(-j2\pi PRF l_k t_m) = 1$ in the above equation.

Then, performing KT on (5.22), we have

$$
\begin{aligned}
S_{\mathrm{mKT}}(f, t_n) = \sum_{k=1}^{K} A_{2,k} & \exp\left[-j\frac{4\pi(f+f_c)r_{0,k}}{c}\right] \\
& \times \exp\left(-j\frac{4\pi v_{unamb,k}t_n}{\lambda}\right) \\
& \times \exp\left[-j\frac{4\pi l_k v_{amb} t_n f}{\lambda(f+f_c)}\right] \\
& \times \exp\left[-j\frac{4\pi a_{0,k}t_n^2 f_c}{\lambda(f+f_c)}\right].
\end{aligned}
\tag{5.23}
$$

Similarly, since the radar transmits a narrowband signal ($f \ll f_c$), we still have $\frac{f_c}{f+f_c} \approx 1 - \frac{f}{f_c}$ and $f + f_c \approx \frac{f_c^2}{f_c-f}$. So, (5.23) could be reconstructed as

$$
\begin{aligned}
S_{\mathrm{mKT}}(f, t_n) = \sum_{k=1}^{K} A_{2,k} & \exp\left[-j\frac{4\pi(f+f_c)r_{0,k}}{c}\right] \\
& \times \exp\left(-j\frac{4\pi v_{unamb,k}t_n}{\lambda}\right) \\
& \times \exp\left[-j\frac{4\pi l_k v_{amb} t_n f(f_c-f)}{c f_c}\right] \\
& \times \exp\left[-j\frac{4\pi a_{0,k}t_n^2(f_c-f)}{c}\right].
\end{aligned}
\tag{5.24}
$$

Next, we define the matched filtering function for the kth target as

$$
\begin{aligned}
H_{\mathrm{mul}}(f, t_n; l_k', a_{0,k}') = & \exp\left[j\frac{4\pi l_k' v_{amb} t_n f(f_c-f)}{c f_c}\right] \\
& \times \exp\left[j\frac{4\pi a_{0,k}'t_n^2(f_c-f)}{c}\right],
\end{aligned}
\tag{5.25}
$$

where l'_k represents the searching fold factor of the kth target and $a'_{0,k}$ is the kth target's searching acceleration. Multiplying (5.25) by (5.24), we can get

$$
\begin{aligned}
S_{\mathrm{mKT}}(f, t_n; l'_k, a'_{0,k}) = \sum_{k=1}^{K} & A_{2,k} \exp\left[-j\frac{4\pi\left(f + f_c\right)r_{0,k}}{c}\right] \\
& \times \exp\left(-j\frac{4\pi v_{unamb,k}t_n}{\lambda}\right) \\
& \times \exp\left[j\frac{4\pi(l'_k - l_k)v_{amb}t_n f\left(f_c - f\right)}{cf_c}\right] \\
& \times \exp\left[j\frac{4\pi(a'_{0,k} - a_{0,k})t_n^2\left(f_c - f\right)}{c}\right].
\end{aligned}
\tag{5.26}
$$

When $l'_k = l_k$ and $a'_{0,k} = a_{0,k}$, i.e., the searching fold factor and searching acceleration are equal to the kth target's fold factor and acceleration, respectively. Then (5.26) could be recast as

$$
\begin{aligned}
S_{\mathrm{mul}}(f, t_n) = & A_{2,k} \exp\left[-j\frac{4\pi\left(f + f_c\right)r_{0,k}}{c}\right] \\
& \times \exp\left(-j\frac{4\pi v_{unamb,k}t_n}{\lambda}\right) + S_{\mathrm{other}}(f, t_n),
\end{aligned}
\tag{5.27}
$$

where

$$
\begin{aligned}
S_{\mathrm{other}}(f, t_n) = \sum_{g=1,g\neq k}^{K} & A_{2,g} \exp\left[-j\frac{4\pi\left(f + f_c\right)r_{0,g}}{c}\right] \\
& \times \exp\left(-j\frac{4\pi v_{unamb,g}t_n}{\lambda}\right) \\
& \times \exp\left[j\frac{4\pi(l'_g - l_k)v_{amb}t_n f\left(f_c - f\right)}{cf_c}\right] \\
& \times \exp\left[j\frac{4\pi(a'_{0,g} - a_{0,k})t_n^2\left(f_c - f\right)}{c}\right].
\end{aligned}
\tag{5.28}
$$

From (5.27), we can see that the RM and DFM of the kth target are both corrected. By conducting the range IFT and the slow time FT on (5.27), we have

$$
\begin{aligned}
s_{\mathrm{mint}}(t, f_{t_n}) = & A_{5,k}\mathrm{sinc}\left[B\left(t - \frac{2r_{0,k}}{c}\right)\right] \\
& \times \mathrm{sinc}\left[T_{sum}\left(f_{t_n} + \frac{2v_{unamb,k}}{\lambda}\right)\right] + s_{\mathrm{other}}(t, f_{t_n}),
\end{aligned}
\tag{5.29}
$$

where

$$
s_{\mathrm{other}}(t, f_{t_n}) = \mathrm{FT}_{t_n}[\mathrm{IFT}_{f}(S_{\mathrm{other}}(f, t_n))],
\tag{5.30}
$$

TABLE 5.1   Computational Complexity of GRFT and KT-MFP

Methods	Multiplications	Additions
GRFT	$MNN_{v_0}N_{a_0}$	$(M-1)NN_{v_0}N_{a_0}$
KT-MFP	$MN\log_2 N/2$	$MN\log_2 N$
	$+(M^2+2M)N$	$+(M+2N)(M-1)$
	$+N_lN_{a_0}MN$	$+N_lN_{a_0}MN$
	$\times\log_2 M/2$	$\times\log_2 M$

$A_{5,k} = A_{2,k}G_{r,k}G_{s,k}\exp\left(-j\frac{4\pi r_{0,k}}{\lambda}\right)$ denotes the integration amplitude. Moreover, $G_{r,k}$ is the compression gain for the range IFT, and $G_{s,k}$ indicates the compression gain of the slow time FT.

Remark: Note that the KT is a linear transform and could simultaneously correct the effect of the linear RM induced by the unambiguous velocity. In addition, the MFP is also a linear procedure. Therefore, there is no cross-talk among the multi-target.

5.2.4   Computational Complexity

In what follows, we analyze the computational complexity of the major steps in GRFT [5] and the KT-MFP based method. For the arithmetic types, the complex multiplication and addition operations are presented. We define the number of searching range cells, echo pulses, searching velocity, searching acceleration and searching fold factor as N, M, N_{v_0}, N_{a_0} and N_l, respectively. In terms of the KT-MFP method, the KT operation needs $MN\log_2 N/2 + (M^2 + 2M)N$ multiplications and $MN\log_2 N + (M+2N)(M-1)$ additions to correct the linear RM caused by the unambiguous velocity. When we perform the MFP to eliminate the residual RM and the DFM, $N_lN_{a_0}MN\log_2 M/2$ and $N_lN_{a_0}MN\log_2 M$ are necessary. The GRFT method needs $MNN_{v_0}N_{a_0}$ multiplications and $(M-1)NN_{v_0}N_{a_0}$ additions [5]. The detailed computational complexity of GRFT and KT-MFP is listed in Table 5.1. Then, using the parameters of the radar in Table 5.2, we can achieve the ration of the computational complexity between GRFT and KT-MFP (which vary with the pulse number), as shown in Fig. 5.2.

TABLE 5.2   Simulation Parameters of Radar

Carrier frequency f_c	10 GHz
Bandwidth B	1 MHz
Sample frequency f_s	5 MHz
Pulse repetition frequency PRF	200 Hz
Pulse duration T_p	100 μs
Number of pulses M	201

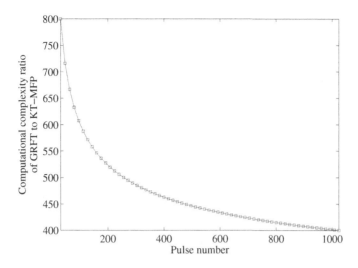

Figure 5.2 Computational complexity ratio of GRFT to KT-MFP versus the pulse number.

5.3 NUMERICAL RESULTS

To evaluate the efficiency of the KT-MFP-based coherent integration method, numerical simulations are provided in this section. In order to carry out the comparison experiment for a single target with high speed, several popular methods, i.e., RFT, IAR-FRFT [6], KT-FRFT [27] and GRFT [5], are presented. Moreover, the detection performance and the coherent integration for multi-targets are also evaluated hereinbelow.

5.3.1 Coherent Integration Performance

In the following simulation, we assess the coherent integration for a high-speed target via the KT-MFP method in Fig. 5.3. The simulation parameters of radar in this section are listed in Table 5.2. The motion parameters of the high-speed target are $r_0 = 500$ km, $v_0 = 3000.8$ m/s and $a_0 = 310$ m/s^2. Here, we also set the SNR after PC as 6 dB. According to Fig. 5.3(a), we could see that serious RM happens after the PC. Fig. 5.3(b) shows the jointly searching result of the fold factor and acceleration after applying KT and MFP. The peak value denotes that the estimated acceleration is equal to 310 m/s^2 and the estimated fold factor is 1000. Fig. 5.3(c) gives the unambiguous velocity estimation result, which is 0.81 m/s. So, the estimated velocity is equal to 3000.81 m/s. Fig. 5.3(d) shows that RM has been corrected using the estimated motion parameters. After eliminating the effect of RM, the coherent integration result through KT-MFP is given in Fig. 5.3(e) where the energy of the high-speed target is also integrated well.

   In order to make a comparison, the coherent integration performance of other popular methods, i.e., RFT, IAR-FRFT, KT-FRFT and GRFT, is also shown in Fig. 5.4.

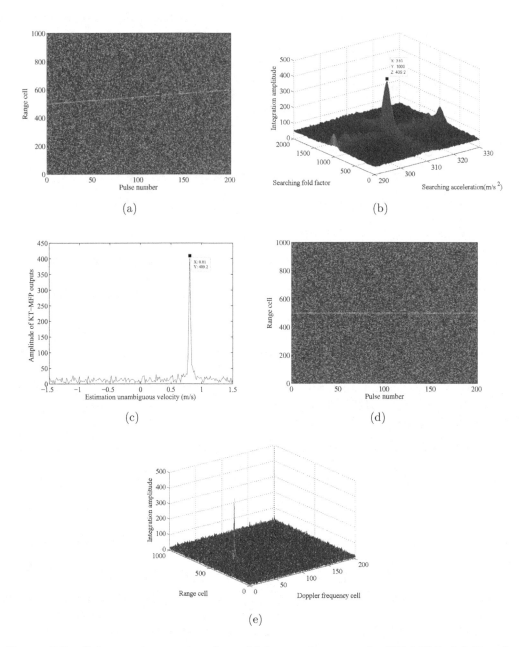

Figure 5.3 Coherent integration for a high-speed target via KT-MFP. (a) Result after PC. (b) Jointly searching result of the fold factor and acceleration. (c) Estimation result of unambiguous velocity. (d) RM correction result. (e) Integration result of KT-MFP.

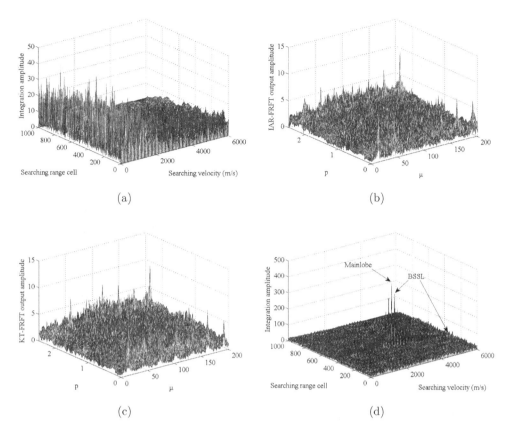

Figure 5.4 Coherent integration for a high-speed target via RFT, IAR-FRFT, KT-FRFT and GRFT. (a) Integration result of RFT. (b) Integration result of IAR-FRFT. (c) Integration result of KT-FRFT. (d) Integration result of GRFT.

Fig. 5.4(a)–Fig. 5.4(d) show the integration results of the aforementioned methods. Specifically, Fig. 5.4(a) shows that RFT cannot accumulate the energy mainly because of the DFM effect caused by target's acceleration. Moreover, Fig. 5.4(b)–Fig. 5.4(c) show the integration results of IAR-FRFT and KT-FRFT, respectively. Because of the target's high speed and high radar carrier frequency, the undersampling along the slow time dimension would occur and then the FRFT-based integration methods (i.e., IAR-FRFT and KT-FRFT) will become invalid. Furthermore, Fig. 5.4(d) gives the integration result of the GRFT method. Although GRFT also could obtain the coherent integration, there appears the blind speed sidelobe (BSSL) effect in the GRFT output, which might bring about severe false alarm and reduce the multi-target detection ability.

5.3.2 Detection Performance

We compare the detection performance of RFT, IAR-FRFT, KT-FRFT, GRFT and KT-MFP for a high-speed target in this simulation, where the Monte Carlo trials are

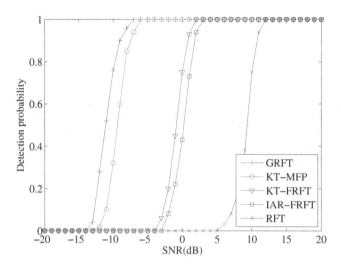

Figure 5.5 Detection probability of RFT, IAR-FRFT, KT-FRFT, GRFT and KT-MFP for a high-speed target.

performed. One should note that we still use the radar parameters in Table 5.2 and the motion parameters are identical to those in Fig. 5.3. The false alarm probability is set as $P_{f_a} = 10^{-4}$, and the Gaussian noises are added to the target's echoes. Fig. 5.5 shows the detection probabilities of the KT-MFP method with other four methods in different SNR cases. The SNRs also vary in $[-20, 20 \text{ dB}]$. One could clearly see that the detection performance of KT-MFP is far better than that of KT-FRFT, IAR-FRFT and RFT. In fact, the undersampling of the slow time might appear on account of the target's high speed and radar's high carrier frequency, so IAR-FRFT and KT-FRFT would become invalid and their detection performance would decrease a lot. Although the detection performance of the proposed method declines slightly (still because of the sinc-like interpolation of KT) compared with GRFT, the KT-MFP method has lower computational cost without the BSSL effect.

5.3.3 Coherent Integration for Multiple Targets

In this section, we evaluate the coherent integration ability of KT-MFP for multiple targets, and the motion parameters of Target A and B are listed in Table 5.3. The SNRs (after PC) of the two targets are both 6 dB. The simulation results of KT-MFP

TABLE 5.3 Motion Parameters of Two Moving Targets

Motion parameters	Target A	Target B
Initial slant range (km)	180	220
Radial velocity (m/s)	80	-100
Radial acceleration (m/s^2)	20	-15
SNR(after PC) (dB)	6	6

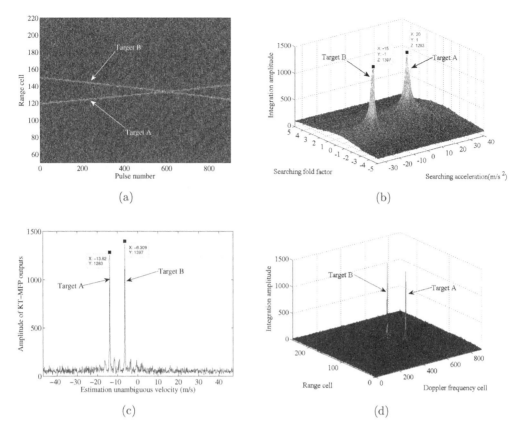

Figure 5.6 Coherent integration for multiple targets via the KT-MFP method. (a) Result after PC. (b) Jointly searching result of the fold factors and accelerations. (c) Estimation result of unambiguous velocities. (d) Integration result.

for multiple targets are given in Fig. 5.6. Particularly, Fig. 5.6(a) shows the result after the PC, while Fig. 5.6(b) gives the jointly searching result of fold factors and accelerations. Specifically, the estimated fold factors are 1 and −1 for the two targets, respectively. Besides, according to the peak values, the estimated accelerations are 20 and −15 m/s^2. Fig. 5.6(c) gives the estimation result of unambiguous velocities. As for target A, the estimated unambiguous velocity is −13.82 m/s, and then we can work out that the estimation of target A's velocity is 79.93 m/s. In terms of target B, the estimated unambiguous velocity is −6.31 m/s, and the estimated velocity is equal to −100.06 m/s. Furthermore, Fig. 5.6(d) gives the coherent integration result via the KT-MFP method, and we can see that the energy of the two targets is well focused.

5.4 SUMMARY

In this chapter, we have discussed the radar high-speed target detection via the KT-MFP-based coherent integration method. The KT-MFP method can both eliminate the RM and DFM effects with the keystone transform and match filtering process in the frequency domain. Detailed simulation experiments show that the KT-MFP method is able to perform well for the high-speed target and could avoid the BSSL effect but requires lower computational complexity compared with GRFT. In addition, the detection performance of KT-MFP outperforms that of RFT, IAR-FRFT and KT-FRFT.

Bibliography

[1] X. L. Chen, J. Guan, Y. Huang, N. B. Liu, and Y. He, "Radon-linear canonical ambiguity function-based detection and estimation method for marine target with micromotion," *IEEE Transactions on Geoscience and Remote Sensing*, vol. 53, no. 4, pp. 2225–2240, April 2015.

[2] P. H. Huang, G. S. Liao, Z. W. Yang, and X. G. Xia, "A fast SAR imaging method for ground moving target using a second-order WVD transform," *IEEE Transactions on Geoscience and Remote Sensing*, vol. 54, no. 4, pp. 1940–1956, April 2016.

[3] J. Xu, J. Yu, Y. N. Peng, and X. G. Xia, "Radon-Fourier transform (RFT) for radar target detection (I): generalized Doppler filter bank processing," *IEEE Transactions on Aerospace and Electronic Systems*, vol. 47, no. 2, pp. 1186–1202, April 2011.

[4] J. Xu, J. Yu, Y. N. Peng, and X. G. Xia, "Radon-Fourier transform (RFT) for radar target detection (II): blind speed sidelobe suppression," *IEEE Transactions on Aerospace and Electronic Systems*, vol. 47, no. 4, pp. 2473–2489, October 2011.

[5] J. Xu, X. G. Xia, S. B. Peng, J. Yu, Y. N. Peng, and L. C. Qian, "Radar maneuvering target motion estimation based on Generalized Radon-Fourier transform," *IEEE Transactions on Signal Processing*, vol. 60, no. 12, pp. 6190–6201, December 2012.

[6] X. Rao, H. H. Tao, J. Su, J. Xie, and X. Y. Zhang, "Detection of constant radial acceleration weak target via IAR-FRFT," *IEEE Transactions on Aerospace and Electronic Systems*, vol. 51, no. 4, pp. 3242–3253, October 2015.

[7] Y. Wang, and Y. C. Jiang, "ISAR imaging of maneuvering target based on the L-Class of fourth-order complex-lag PWVD," *IEEE Transactions on Geoscience and Remote Sensing*, vol. 48, no. 3, pp. 1518–1527, March 2010.

[8] Y. Wang, and Y. C. Jiang, "Inverse synthetic aperture radar imaging of maneuvering target based on the product generalized cubic phase function," *IEEE Geoscience and Remote Sensing Letters*, vol. 8, no. 5, pp. 958–962, September 2011.

[9] Y. C. Li, M. D. Xing, Y. H. Quan, and Z. Bao, "A new algorithm of ISAR imaging for maneuvering targets with low SNR," *IEEE Transactions on Aerospace and Electronic Systems*, vol. 49, no. 1, pp. 543–557, January 2013.

[10] J. B. Zheng, T. Su, W. T. Zhu, and Q. H. Liu, "ISAR imaging of targets with complex motions based on the keystone time-chirp rate distribution," *IEEE Geoscience and Remote Sensing Letters*, vol. 11, no. 7, pp. 1275–1279, July 2014.

[11] J. B. Zheng, T. Su, L. Zhang, W. T. Zhu, and Q. H. Liu, "ISAR imaging of targets with complex motion based on the chirp rate-quadratic chirp rate distribution," *IEEE Transactions on Geoscience and Remote Sensing*, vol. 52, no. 11, pp. 7276–7289, November 2014.

[12] M. R. Sharif, S. S. Abeysekera, "Efficient wideband signal parameter estimation using a single slice of Radon-ambiguity transform", in *Proceedings of IEEE International Conference on Accoustics, Speech and Signal Processing*, vol. 5, pp. 605–608, March 2005.

[13] M. R. Sharif, S. S. Abeysekera, "Efficient wideband signal parameter estimation using a Radon-ambiguity transform slice", *IEEE Transactions on Aerospace and Electronic Systems*, vol. 43, no. 2, pp. 673–688, April 2007.

[14] M. D. Xing, J. H. Su, G. Y. Wang, and Z. Bao, "New parameter estimation and detection algorithm for high speed small target," *IEEE Transactions on Aerospace and Electronic Systems*, vol. 47, no. 1, pp. 214–224, January 2011.

[15] X. Rao, H. T. Huang, J. Xie, J. Su, and W. P. Li, "Long-time coherent integration detection of weak manoeuvring target via integration algorithm, improved axis rotation discrete chirp-Fourier transform," *IET Radar Sonar and Navigation*, vol. 9, no. 7, pp. 917–926, February 2015.

[16] G. C. Sun, M. D. Xing, X. G. Xia, Y. R. Wu, and Z. Bao, "Robust ground moving-target imaging using Deramp-Keystone processing," *IEEE Transactions on Geoscience and Remote Sensing*, vol. 51, no. 2, pp. 966–982, February 2013.

[17] J. Tian, W. Cui, X. G. Xia, and S. L. Wu, "Parameter estimation of ground moving targets based on SKT-DLVT processing," *IEEE Transactions on Computational Imaging*, vol. 2, no. 1, pp. 13–26, March 2016.

[18] J. Tian, W. Cui, and S. L. Wu, "A novel method for parameter estimation of space moving targets," *IEEE Geoscience and Remote Sensing Letters*, vol. 11, no. 2, pp. 389–393, February 2014.

[19] X. L. Li, L. J. Kong, G. L. Cui, and W. Yi, "A low complexity coherent integration method for manueuvering target detection," *Digital Signal Processing*, vol 49, pp. 137–147, February 2016.

[20] R. P. Perry, R. C. Dipietro, and R. L. Fante, "SAR imaging of moving targets," *IEEE Transactions on Aerospace and Electronic Systems*, vol. 35, no. 1, pp. 188–200, January 1999.

[21] D. Y. Zhu, Y. Li, and Z. D. Zhu, "A keystone transform without interpolation for SAR ground moving-target imaging," *IEEE Geoscience and Remote Sensing Letters*, vol. 4, no. 1, pp. 18–22, 2007.

[22] G. Li, X. G. Xia, and Y. N. Peng, "Doppler keystone transform: an approach suitable for parallel implementation of SAR moving target imaging," *IEEE Geoscience and Remote Sensing Letters*, vol. 5, no. 4, pp. 573–577, October 2008.

[23] P. H. Huang, G. S. Liao, Z. W. Yang, and X. G. Xia, "Long-time coherent integration for weak maneuvering target detection and high-order motion parameter estimation based on keystone transform," *IEEE Transactions on Signal Processing*, vol. 64, no. 15, pp. 4013–4026, August 2016.

[24] X. L. Chen, J. Guan, N. B. Liu, and Y. He, "Maneuvering target detection via Radon-Fractional Fourier transform-based long-time coherent integration," *IEEE Transactions on Signal Processing*, vol. 62, no. 4, pp. 939–953, February 2014.

[25] J. Xu, Y. N. Peng, X. G. Xia, and A. Farina, "Focus-before-detection radar signal processing: part i—challenges and methods," in IEEE Aerospace and Electronic Systems Magazine, vol. 32, no. 9, pp. 48–59, September 2017.

[26] X. L. Li, L. J. Kong, G. L. Cui, and W. Yi, "CLEAN-based coherent integration method for high-speed multi-targets detection," *IET Radar Sonar and Navigation*, vol. 10, no. 9, pp. 1671–1682, December 2016.

[27] X. L. Li, G. L. Cui, W. Yi, and L. J. Kong, "An efficient coherent integration method for maneuvering target detection," *IET International Radar Conference*, Hangzhou: IET 2015: 1–6.

CHAPTER 6

Segmented Processing-Based Coherent Integration

The previous two chapters discussed the radar high-speed target detection via coherent integration transform in a full observation time, which may make the computational burden become huge. This chapter presents a computational efficient segmented processing-based coherent integration method for radar high-speed target detection, which could reduce the computational cost with slight integration performance loss. In particular, the segmentation criteria, intra-segment integration processing, inter-segment integration processing, integration gain and computational cost analysis are discussed. In addition, the signal accumulation and detection performance of the segmented-processing coherent integration method is analyzed through simulation experiments.

6.1 SIGNAL MODELING

Supposing that the ground-based pulse Doppler radar sends narrow-band linear frequency modulation (LFM) waves, its baseband waveform can be expressed as [1, 2]:

$$s_{trans}(t) = \text{rect}\left(\frac{t}{T_P}\right)\exp\left(j\pi\mu t^2\right), \tag{6.1}$$

where $\text{rect}(x) = \begin{cases} 1, & |x| \leq \frac{1}{2} \\ 0, & |x| > \frac{1}{2} \end{cases}$, t is the fast time, T_P is the pulse duration, $\mu = B/T_P$ is the frequency modulated rate and B is the signal bandwidth.

Considering a high velocity maneuvering target in the radar surveillance area, its instantaneous radial distance can be expressed as [3]:

$$R(t_m) = r_0 + v_0 t_m + a_0 t_m^2, \tag{6.2}$$

where $t_m = (m-1)T_r$ $(m = 1, 2, \ldots, M)$ is the slow time; T_r is the pulse repetition time; M is the pulse number; r_0, v_0 and a_0 represent, respectively, the target's initial radial distance, initial radial velocity and initial radial acceleration. Then, the

DOI: 10.1201/9781003529101-6

baseband echo signal of target can be written as:

$$s_r(t, t_m) = A_0 \text{rect}\left(\frac{t - \tau(t_m)}{T_m}\right) \exp\left[j\pi\mu(t - \tau(t_m))^2\right]$$
$$\times \exp\left(-j4\pi\frac{R(t_m)}{\lambda}\right), \tag{6.3}$$

where A_0 denotes the received echo's complex amplitude, t is the fast time, $\tau(t_m) = 2R(t_m)/c$ is the echo time delay, $\lambda = c/f_c$ is the wavelength, c is the light velocity and f_c is the radar carrier frequency.

After pulse compression, the compressed signal can be represented as [4]:

$$s_{pc}(t, t_m) = A_1 \text{sinc}\left[\pi B(t - \tau(t_m))\right]$$
$$\times \exp\left(-j4\pi\frac{R(t_m)}{\lambda}\right), \tag{6.4}$$

where A_1 is the echo's complex amplitude after pulse compression. Let $r = ct/2$, then equation (6.4) could be recast as:

$$s_{pc}(r, t_m) = A_1 \text{sinc}\left[\pi\frac{r - R(t_m)}{\rho_r}\right]$$
$$\times \exp\left(-j4\pi\frac{R(t_m)}{\lambda}\right), \tag{6.5}$$

where $\rho_r = c/2B$ is the range resolution and r is the range corresponding to the fast time t.

As shown in (6.5), the envelope and phase of the target signal will be affected by the target's motions (i.e., velocity and acceleration). Within the whole integration time of M pulses, the target's velocity will cause range migration (RM) (i.e., first-order RM), and the acceleration will also cause RM (i.e., second-order RM) and Doppler spread (DS).

6.2 SEGMENTED INTEGRATION METHOD

The traditional generalized Radon Fourier transform (GRFT) method removes the first-order RM, second-order RM and Doppler spread via multi-dimensional search, but the calculation is large [6]. In this section, we present a computational efficient segmented coherent accumulation method to reduce the computational burden. The sketch map of the flow chart for the segmented processing-based coherent integration method is given in Fig. 6.1. In essence, the segmented integration method contains three main steps: the first is the segmented processing and the second is the integration within the sub-segment, while the last is the integration across different sub-segments.

6.2.1 Segmented Processing

First, divides the whole M pulses into N_c sub-segments, where each sub-segment has $N_{sub} = M/N_c$ pulses. Then, the instantaneous radial distance, radial velocity and

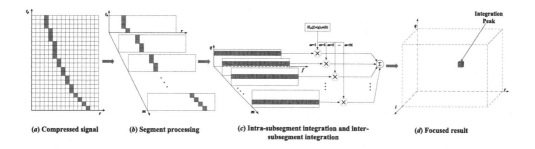

(a) Compressed signal   (b) Segment processing   (c) Intra-subsegment integration and inter-subsegment integration   (d) Focused result

Figure 6.1   Sketch map of the flow chart for the segmented processing-based method.

radial acceleration of target in the lth sub-segments can be represented as:

$$R_l(t_n) = r_0^l + v_0^l t_n + a_0^l t_n^2, \tag{6.6}$$

$$v_l(t_n) = v_0^l + 2a_0^l t_n, \tag{6.7}$$

$$a_l(t_n) = a_0^l, \tag{6.8}$$

where $t_n = (n-1)T_r$, $n \in [1, 2, \ldots, N_{sub}]$ is the pulse index of the lth sub-segment; $l \in [1, 2, \ldots, N_c]$ is the index of the sub-segment; and r_0^l, v_0^l and a_0^l denote, respectively, the initial radial distance, velocity and acceleration of target in the lth sub-segment.

Note that the last pulse of the former sub-segment (i.e., $l-1$) is adjacent to the first pulse of the later sub-segment (i.e., l). Correspondingly, we have the following relationship

$$r_0^l = r_0^{l-1} + v_0^{l-1}T_s + a_0^{l-1}T_s^2, \tag{6.9}$$

$$v_0^l = v_0^{l-1} + 2a_0^l T_s, \tag{6.10}$$

$$a_0^l = a_0^{l-1} = a_0, \tag{6.11}$$

where $T_s = N_{sub}T_r$ is the duration of each sub-segment.

Besides, the pulse compression signal within the lth sub-segment after segmentation can be expressed as:

$$s_l(r, t_n) = A_1 \text{sinc}\left[\pi \frac{r - R_l(t_n)}{\rho_r}\right] \\ \times \exp\left(-j4\pi \frac{R_l(t_n)}{\lambda}\right). \tag{6.12}$$

During the segmented processing, in order to ensure the RM and DS induced by the acceleration are negligible (i.e., the RM caused by acceleration within the sub-segment cannot overstep half of one range unit, and the DS caused by acceleration in the sub-segment is less than half of one Doppler frequency unit), the constraints on the time of each sub-segment should satisfy the following equations:

$$a_{\max}T_s^2 \leq \frac{c}{4f_s}, \tag{6.13}$$

$$\frac{4a_{\max}T_s}{\lambda} \leq \frac{1}{2T_s}, \tag{6.14}$$

where $a_{\max}$ is the possible maximum acceleration and f_s is the fast time sampling rate. By solving equations (6.13) and (6.14), we can get the following formula:

$$T_s \leq \sqrt{\frac{c}{4a_{\max}f_s}}, \tag{6.15}$$

$$T_s \leq \sqrt{\frac{\lambda}{8a_{\max}}}, \tag{6.16}$$

So, the time of sub-segment should satisfy:

$$T_s \leq \min\left(\sqrt{\frac{c}{4a_{\max}f_s}}, \sqrt{\frac{\lambda}{8a_{\max}}}\right). \tag{6.17}$$

With the segmented processing, the RM and DS caused by target's acceleration could be neglected. As a result, the compressed signal show in (6.12) could be approximated as:

$$
\begin{aligned}
s_l(r, t_n) &\approx A_1 \mathrm{sinc}\left[\pi \frac{r - (r_0^l + v_0^l t_n)}{\rho_r}\right] \\
&\times \exp\left(-j4\pi \frac{r_0^l + v_0^l t_n}{\lambda}\right).
\end{aligned}
\tag{6.18}
$$

6.2.2 Coherent Integration within the Sub-Segment

Performing FFT operation on (6.18) with respect to the range, we could obtain its frequency format as:

$$
\begin{aligned}
S_l(f, t_n) &\approx A_2 \mathrm{rect}\left[\frac{f}{B}\right] \exp\left(-j4\pi f \frac{r_0^l + v_0^l t_n}{c}\right) \\
&\times \exp\left(-j4\pi f_c \frac{r_0^l + v_0^l t_n}{c}\right) \\
&= A_2 \mathrm{rect}\left[\frac{f}{B}\right] \exp\left(-j4\pi \frac{f + f_c}{c} r_0^l\right) \\
&\times \exp\left(-j4\pi \frac{f + f_c}{c} v_0^l t_n\right),
\end{aligned}
\tag{6.19}
$$

where A_2 is the complex amplitude after FFT and f is the range frequency.

The frequency bin Radon Fourier transform (FBRFT) [5] is then applied to realize the coherent integration within each sub-segment, and its expression could be represented as:

$$
\begin{aligned}
G_l(f, v(q)) &= \sum_{n=1}^{N_{sub}} S_m(f, t_n) \\
&\times \exp\left(j4\pi \frac{f + f_c}{c} v(q) t_n\right),
\end{aligned}
\tag{6.20}
$$

where $v(q)$ is the velocity searching sequence:

$$v(q) = q\Delta v, q = 1, 2, \ldots, N_v, \tag{6.21}$$

$\Delta v = \lambda/(2T)$, $T = N_c T_s$ is the total integration time, N_v is the total number of the searching velocity.

Substitute (6.19) into equation (6.20) yields:

$$
\begin{aligned}
G_l(f, v(q)) \\
&= A_2 \mathrm{rect}\left[\frac{f}{B}\right] \exp\left(-j4\pi \frac{f + f_c}{c} r_0^l\right) \\
&\times \sum_{n=1}^{N_{sub}} \exp\left(j4\pi \frac{f + f_c}{c}(v(q) - v_0^l)(n - 1)T_r\right).
\end{aligned}
\tag{6.22}
$$

Due to the summation formula of the geometric sequence, (6.22) can be recast as:

$$
\begin{aligned}
G_l(f, v(q)) \\
&= A_2 \mathrm{rect}\left[\frac{f}{B}\right] \exp\left(-j4\pi \frac{f + f_c}{c} r_0^l\right) \\
&\times \frac{1 - \exp\left(j4\pi \frac{f+f_c}{c}(v(q) - v_0^l)N_{sub}T_r\right)}{1 - \exp\left(j4\pi \frac{f+f_c}{c}(v(q) - v_0^l)T_r\right)}.
\end{aligned}
\tag{6.23}
$$

Simplifying, (6.23) becomes:

$$
\begin{aligned}
G_l(f, v(q)) \\
&= A_2 \mathrm{rect}\left[\frac{f}{B}\right] \exp\left(-j4\pi \frac{f + f_c}{c} r_0^l\right) \\
&\times \frac{\exp\left(j2\pi \frac{f+f_c}{c}(v(q) - v_0^l)N_{sub}T_r\right)}{\exp\left(j2\pi \frac{f+f_c}{c}(v(q) - v_0^l)T_r\right)} \\
&\times \frac{\sin\left(2\pi \frac{f+f_c}{c}(v(q) - v_0^l)N_{sub}T_r\right)}{\sin\left(2\pi \frac{f+f_c}{c}(v(q) - v_0^l)T_r\right)}.
\end{aligned}
\tag{6.24}
$$

Due to $2\pi \frac{f+f_c}{c}(v(q) - v_m)T_r \ll 1$:

$$
\begin{aligned}
G_l(f, v(q)) \\
&\approx A_2 N_{sub} \mathrm{rect}\left[\frac{f}{B}\right] \exp\left(-j4\pi \frac{f + f_c}{c} r_0^l\right) \\
&\times \exp\left(j2\pi\left(\frac{f + f_c}{c}(v(q) - v_0^l)(N_{sub} - 1)T_r\right)\right) \\
&\times \frac{\sin\left(2\pi \frac{f+f_c}{c}(v(q) - v_0^l)N_{sub}T_r\right)}{2\pi \frac{f+f_c}{c}(v(q) - v_0^l)N_{sub}T_r} N_{Sub}.
\end{aligned}
\tag{6.25}
$$

Through mathematical operations and approximate transformation, the result of (6.22) could be further expressed as:

$$
\begin{aligned}
G_l&(f, v(q)) \\
&\approx A_2 N_{sub} \text{rect}\left[\frac{f}{B}\right] \exp\left(-j4\pi \frac{f + f_c}{c} r_0^l\right) \\
&\times \exp\left(j2\pi\left(\frac{f + f_c}{c}(v(q) - v_0^l)(N_{sub} - 1)T_r\right)\right) \\
&\times \text{sinc}\left(2\pi \frac{f + f_c}{c}((v(q) - v_0^l)T_s)\right).
\end{aligned}
\tag{6.26}
$$

When $v(q) = v_0^l$, (6.26) can be rewritten as:

$$
\begin{aligned}
G_l(f, v_0^l) &= A_2 N_{sub} \text{rect}\left[\frac{f}{B}\right] \\
&\times \exp\left(-j4\pi \frac{f + f_c}{c} r_0^l\right).
\end{aligned}
\tag{6.27}
$$

The time domain of equation (6.27) could be achieved via inverse fast Fourier transform (IFFT) operation along the f dimension:

$$
\begin{aligned}
G_l(r, v_0^l) &= A_1 N_{sub} \text{sinc}\left[\pi B\left(\frac{2r}{c} - \frac{2r_0^l}{c}\right)\right] \\
&\times \exp\left(-j4\pi \frac{r_0^l}{\lambda}\right).
\end{aligned}
\tag{6.28}
$$

Combing (6.27) and (6.28), it could be found that the peak of the FBRFT output would be located at (r_0^l, v_0^l) in time domain.

Notably, equation (6.20) can be fast implemented through Chirp Z-transform (CZT). In particular, let $C = \exp\left(j4\pi \Delta v T_r (f + f_c)/c\right)$, then (6.20) can be recast as:

$$
\begin{aligned}
G_l&(f, v(q)) \\
&= \sum_{n=1}^{N_{sub}} S_l(f, t_n) C^{q(n-1)} \\
&= \sum_{n=1}^{N_{sub}} S_l(f, t_n) C^{\frac{1}{2}[q^2 + (n-1)^2 - (n-1-q)^2]} \\
&= C^{\frac{1}{2}q^2} \sum_{n=1}^{N_{sub}} C^{-\frac{1}{2}(n-1-q)^2}\left(C^{\frac{1}{2}(n-1)^2} S_l(f, t_n)\right) \\
&= C^{\frac{1}{2}q^2}\left(C^{-\frac{1}{2}(n-1)^2} \otimes \left(C^{\frac{1}{2}(n-1)^2} S_l(f, t_n)\right)\right),
\end{aligned}
\tag{6.29}
$$

where $\otimes$ is the convolution operation.

6.2.3 Coherent Integration across Sub-Segments

After the accumulation of multiple pulses in each sub-segment, we also need to complete the integration of the peak energy of FBRFT output across different sub-segments in order to further improve the signal-to-noise ratio (SNR). Next, we firstly analyze the property of the FBRFT output. Then, the integration of the FBRFT output is discussed.

(1) **Property analysis of the FBRFT output**

Substitute (6.9) into (6.27):

$$
\begin{aligned}
G_l(f, v_0^l) = {} & A_2 N_{sub} \mathrm{rect} \left[\frac{f}{B} \right] \\
& \times \exp \left(-j4\pi \frac{f + f_c}{c} r_0^{l-1} \right) \\
& \times \exp \left(-j4\pi \frac{f + f_c}{c} v_0^{l-1} T_s \right) \\
& \times \exp \left(-j4\pi \frac{f + f_c}{c} a_0 T_s^2 \right) \\
= {} & G_{l-1}(f, v_0^{l-1}) \\
& \times \exp \left(-j4\pi \frac{f + f_c}{c} v_0^{l-1} T_s \right) \\
& \times \exp \left(-j4\pi \frac{f + f_c}{c} a_0 T_s^2 \right),
\end{aligned}
\tag{6.30}
$$

where

$$
\begin{aligned}
G_{l-1}(f, v_0^{l-1}) = {} & A_2 N_{sub} \mathrm{rect} \left[\frac{f}{B} \right] \\
& \times \exp \left(-j4\pi \frac{f + f_c}{c} r_0^{l-1} \right).
\end{aligned}
\tag{6.31}
$$

It should be noted that $G_l(f, v_0^l)$ and $G_{l-1}(f, v_0^{l-1})$ are, respectively, the FBRFT outputs of the lth sub-segment and $l - 1$th sub-segment.

From (6.30), it could be found that there are two differences between the integration outputs $G_l(f, v_0^l)$ and $G_{l-1}(f, v_0^{l-1})$. First, the peak locations along the searching velocity axis are different, that is, v_0^l for $G_l(f, v_0^{l-1})$ and v_0^{l-1} for $G_{l-1}(f, v_0^{l-1})$. Note that the relationship of v_m and v_{m-1} is as shown in (6.10). Second, the phases of $G_l(f, v_0^l)$ and $G_{l-1}(f, v_0^{l-1})$ are different, where the phase difference is determined by the following exponential term: $\exp \left(-j4\pi \frac{f+f_c}{c} v_0^{l-1} T_s \right) \exp \left(-j4\pi \frac{f+f_c}{c} a_0 T_s^2 \right)$

Using (6.30) to iterate, we have

$$
\begin{aligned}
G_l(f, v_0^l) \\
= G_1(f, v_0^1) \exp\left(-j4\pi \frac{f + f_c}{c} v_0^1 (l-1) T_s\right) \\
\times \exp\left(-j4\pi \frac{f + f_c}{c} a_0 ((l-1) T_s)^2\right).
\end{aligned}
\tag{6.32}
$$

Taking the FBRFT result of the first sub-segment as a reference, the positional relationship between the $G_l(f, v_0^l)$ and $G_1(f, v_0^1)$ along the searching velocity dimension satisfies:

$$
v_0^l = v_0^1 + 2a_0(l-1)T_s.
\tag{6.33}
$$

Besides, the exponential term difference (corresponding to the phase difference) between $G_l(f, v_0^l)$ and $G_1(f, v_0^1)$ is

$$
\begin{aligned}
\varphi_m = \exp\left(-j4\pi \frac{f + f_c}{c} v_0^1 (l-1) T_s\right) \\
\times \exp\left(-j4\pi \frac{f + f_c}{c} a_0 ((l-1) T_s)^2\right).
\end{aligned}
\tag{6.34}
$$

(2) Integration of the FBRFT output

According to the above analysis, we could find that the phase difference between the $G_l(f, v_0^l)$ and $G_1(f, v_0^1)$ (i.e., FBRFT output of the lth sub-segment and the FBRFT output of the first sub-segment) is determined by (6.34). In addition, the location difference along the searching velocity axis for the peak of $G_l(f, v_0^l)$ and $G_1(f, v_0^1)$ is determined by (6.33). In order to realize further accumulation of the FBRFT peaks across multiple sub-segments, we need to compensate for the phase difference and align them in the velocity direction. To do this, the following operation could be employed:

$$
\begin{aligned}
G_{\mathrm{pro}}(f, v(q), a(i)) \\
= \sum_{l=1}^{N_c} G_l\left(f, v(q) + 2a(i)(l-1)T_s\right) \\
\times H_l(f, v(q), a(i)),
\end{aligned}
\tag{6.35}
$$

where $a(i) = i\Delta a, i = 1, 2, \ldots, N_a$ is the acceleration searching sequence, $\Delta a = \lambda/(2N_c T_s^2)$ is the searching interval of acceleration and N_a is the number of the searching intervals. $H_l(f, v(q), a(i))$ is the phase compensation function:

$$
\begin{aligned}
H_l(f, v(q), a(i)) \\
= \exp\left(j4\pi \frac{f + f_c}{c} v(q)(l-1)T_s\right) \\
\times \exp\left(j4\pi \frac{f + f_c}{c} a(i)((l-1)T_s)^2\right).
\end{aligned}
\tag{6.36}
$$

Substituting equations (6.26) and (6.36) into (6.35), we have:

$$
\begin{aligned}
G_{\mathrm{pro}}(f, v(q), a(i)) = {} & A_2 N_{sub}\mathrm{rect}\left[\frac{f}{B}\right]\exp\left(-j4\pi\frac{f+f_c}{c}r_0\right) \\
& \times \sum_{l=1}^{N_c}\exp\left[j4\pi\frac{f+f_c}{c}(v(q)-v_0)(l-1)T_s\right] \\
& \times \exp\left[j4\pi\frac{f+f_c}{c}(a(i)-a_0)((l-1)T_s)^2\right] \\
& \times \exp\left(j2\pi\frac{f+f_c}{c}((v(q)-v_0)+2(a(i)-a_0)(l-1)T_s)(N_{sub}-1)T_r\right) \\
& \times \mathrm{sinc}\left(2\pi\frac{f+f_c}{c}((v(q)-v_0)+2(a(i)-a_0)(l-1)T_s)T_s\right).
\end{aligned}
\tag{6.37}
$$

When the searching parameters are equal to the target's real motion parameters (i.e., $v(q)=v_0$ and $a(i)=a_0$), (6.37) can be recast as:

$$
\begin{aligned}
G_{\mathrm{pro}}(f, v_0, a_0) = {} & A_2 N_{sub}N_c\mathrm{rect}\left[\frac{f}{B}\right] \\
& \times \exp\left[-j4\pi\frac{f+f_c}{c}r_0\right].
\end{aligned}
\tag{6.38}
$$

Then, the integration result in the range domain could be achieved using IFFT operation on (6.38) with respect to the variable f:

$$
\begin{aligned}
G_{\mathrm{pro}}(r, v_1, a_1) = {} & A_1 N_{sub}N_c\mathrm{sinc}\left[\pi\frac{r-r_0}{\rho_r}\right] \\
& \times \exp\left[-j4\pi\frac{r_0}{\lambda}\right].
\end{aligned}
\tag{6.39}
$$

From (6.39), it could be found that the integration of signal energy across different sub-segments has be obtained.

6.2.4 Procedure of the Segmented Integration

In order to illustrate the detection steps of the segmented processing-based coherent integration method in detail, the flow chart is given in Fig. 6.2. We can see that the execution of the segmented processing-based coherent integration method is divided into the following steps:

Step 1) Perform pulse compression on the radar raw echo $s_r(t, t_m)$, and then we could obtain the compressed signal $s_{pc}(t, t_m)$.

Step 2) According to the segmentation criterion shown in (6.16), divide the compressed signal $s_{pc}(t, t_m)$ into N_c sub-segments, and each sub-segment (e.g., $s_l(r, t_n)$) contains N_{sub} pulses.

Step 3) Apply fast Fourier transform (FFT) operation on $s_l(r, t_n)$ along the range axis, and we could obtain the corresponding signal in the frequency-slow time domain, i.e., $s_l(f, t_n)$.

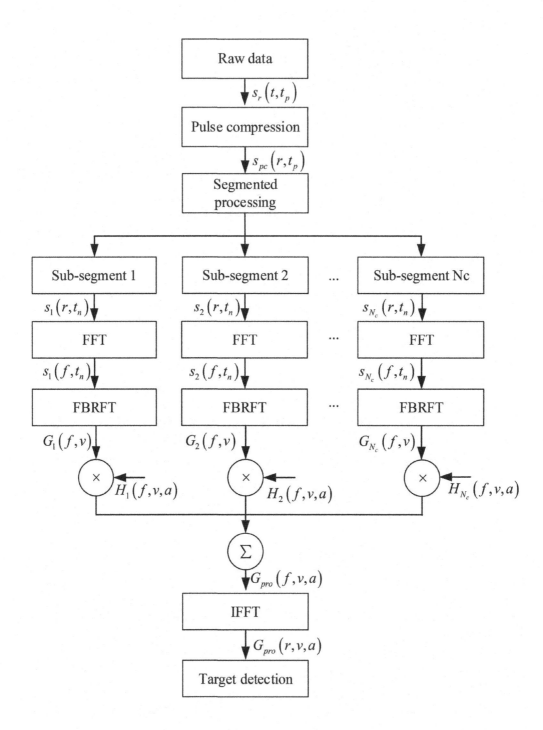

Figure 6.2   Detailed processing flowchart of the segmented processing-based method.

Step 4) Perform FBRFT processing on $s_l(f, t_n)$, as shown in (6.20), and the accumulation result of each sub-segment via FBRFT can be obtained. In addition, the IFFT operation could be performed on $G_l(f, v_0^l)$ along the f direction, and then the spectrum of the FBRFT integration output can be obtained.

Step 5) Based on the property of the FBRFT output, apply the inter-segment integration operation (as shown in (6.35)) to realize the integration across multiple sub-segments.

Step 6) Carry out the IFFT operation on (6.38) with respect to the variable f. Then, the final integration result in the range-domain could be achieved, i.e., $G_{\text{pro}}(r, v, a)$.

Step 7) Take the amplitude after integration as test statistic, and perform the constant false alarm ratio (CFAR) [9] detection to confirm a target, i.e.,

$$|G_{\text{pro}}(r, v, a)| \underset{H_0}{\overset{H_1}{\underset{<}{\gtrless}}} \widehat{T}, \tag{6.40}$$

where $\widehat{T}$ denotes the adaptive threshold for a given false alarm probability.

6.3   SNR GAIN AND COMPUTATIONAL BURDEN

6.3.1   SNR Gain Analysis

During the segmented processing, we use the constraints shown in (6.13) and (6.14), which assumed that the range migration and Doppler frequency migration caused by the target acceleration can be ignored within the integration time of each sub-segment. Ideally, if the assumption is strictly satisfied, the SNR gain of the proposed integration method is $10 \log_{10}(N_{sub}N_c)$, which is benefit from the accumulation of $N_{sub}N_c$ pulses. However, the assumption is approximately satisfied, and the approximation will bring slight performance loss. Notably, the approximate error of DS has a greater impact on the integration performance than that of RM. Therefore, in the following, we analyze the SNR gain of the segmented processing-based integration method which only considers the DS effect but neglects the RM within the integration period of each sub-segment.

Combining (6.12) and (6.9), the compressed signal of the lth sub-segment can be rewritten as:

$$s_l(r, t_n) = A_1 \text{sinc} \left[\pi \frac{r - (r_0^l + v_0^l t_n + a_0^l t_n^2)}{\rho_r} \right]$$
$$\times \exp \left(-j4\pi \frac{r_0^l + v_0^l t_n + a_0^l t_n^2}{\lambda} \right). \tag{6.41}$$

Neglecting the RM induced by the target's acceleration, (6.41) could be approximated as

$$s'_l(r, t_n) \approx A_1 \text{sinc} \left[\pi \frac{r - (r^l_0 + v^l_0 t_n)}{\rho_r} \right]$$

$$\times \exp \left(-j4\pi \frac{r^l_0 + v^l_0 t_n + a^l_0 t^2_n}{\lambda} \right). \tag{6.42}$$

And its expression in the range frequency domain could be written as:

$$s'_l(f, t_n) = A_2 \text{rect} \left[\frac{f}{B} \right] \exp \left(-j4\pi \frac{f + f_c}{c} r^l_0 \right)$$

$$\times \exp \left(-j4\pi \frac{f + f_c}{c} v^l_0 t_n \right) \exp \left(-j4\pi \frac{a^l_0 t^2_n}{\lambda} \right). \tag{6.43}$$

With the FBRFT operation, the integration result of (6.43) is

$$G'_l(f, v(q))$$

$$= A'_{pc} \text{rect} \left[\frac{f}{B} \right] \exp \left(-j4\pi r^l_0 \frac{f + f_c}{c} \right)$$

$$\times \sum_{n=1}^{N_{sub}} \exp \left(j4\pi \frac{f + f_c}{c} (v(q) - v^l_0)(n-1)T_r \right) \tag{6.44}$$

$$\times \exp \left(-j4\pi \frac{a^l_0 ((n-1)T_r)^2}{\lambda} \right).$$

When the search parameters matched the actual parameters, the output of integration across sub-segments can be written as:

$$G'_{\text{Pro}}(f, v^l_0, a_0)$$

$$\approx A_2 N_c \text{rect} \left[\frac{f}{B} \right] \exp \left(-j4\pi r^l_0 \frac{f + f_c}{c} \right)$$

$$\times \sum_{n=1}^{N_{sub}} \exp \left(-j4\pi \frac{a_0 ((n-1)T_r)^2}{\lambda} \right). \tag{6.45}$$

And its time domain signal is:

$$G'_{\text{Pro}}(r, v_m, a_1)$$

$$\approx A_{pc} N_c A'_{sub} \text{sinc} \left[\pi \frac{r - r^l_0}{\rho_r} \right] \exp \left(-j4\pi \frac{r^l_0}{\lambda} \right), \tag{6.46}$$

where

$$A'_{sub} = \sum_{n=1}^{N_{sub}} \exp \left(-j4\pi \frac{a_0 ((n-1)T_r)^2}{\lambda} \right). \tag{6.47}$$

Suppose that the additive white Gaussian noise after pulse compression is with variance σ^2. Then, the input SNR of the compressed echo is

$$SNR_1 = 10 \log_{10} \left(\frac{A^2_1}{\sigma^2} \right). \tag{6.48}$$

It should be noted that the segmented processing-based coherent integration method (including integration within sub-segment and integration across different sub-segments) is a linear operation. Thus, the noise after integration is still Gaussian distribution, and the variance is $N_p\sigma^2$. Correspondingly, the output SNR of the segmented processing-based coherent integration method could be calculated as:

$$SNR_2 = 10\log_{10}\left(\frac{|A_1 N_c A'_{sub}|^2}{M\sigma^2}\right). \tag{6.49}$$

Therefore, the SNR gain of the segmented processing-based integration method can be represented as:

$$SNR_{\text{gain}} = SNR_2 - SNR_1$$
$$= 10\log_{10}\left(\frac{|N_c A'_{sub}|^2}{M}\right). \tag{6.50}$$

The theoretical analytical value of A'_{sub} shown in (6.47) is difficult to calculate. However, we could obtain its approximate value in the following way:

$$A'_{sub} = \sum_{n=1}^{N_{sub}} \exp\left(-j4\pi\frac{a_0((n-1)T_r)^2}{\lambda}\right)$$
$$\approx \frac{1}{T_r}\int_0^{T_{sub}} \exp\left(-j4\pi\frac{a_0 t^2}{\lambda}\right) dt. \tag{6.51}$$

The integral of $\exp(ax^2 + bx + c)$ can be expressed as:

$$\int \exp\left(ax^2 + bx + c\right) dx$$
$$= \frac{1}{2}\sqrt{\frac{\pi}{2}}\exp\left(\frac{ac - b^2}{a}\right)\text{erfi}\left(\sqrt{a}x + \frac{b}{\sqrt{a}}\right), \tag{6.52}$$

where $\text{erfi}(x) = -i\text{erf}(ix) = \frac{2}{\sqrt{\pi}}\int_0^x e^{t^2} dt$ [10]; thus, (6.51) can be recast as:

$$A'_{sub} \approx \frac{1}{2T_r}\sqrt{\frac{j\lambda}{4a_0}}\text{erfi}\left(\sqrt{-j\frac{4\pi a_0}{\lambda}}T_{sub}\right). \tag{6.53}$$

Substitute (6.53) into (6.50), we could obtain the SNR gain as:

$$SNR_{\text{gain}}$$
$$\approx 10\log_{10}\left(\frac{N_c^2\lambda\left|\text{erfi}\left(\sqrt{-j\frac{4\pi a_0}{\lambda}}T_{sub}\right)\right|^2}{16Ma_0 T_r^2}\right). \tag{6.54}$$

From (6.54), it could be seen that the SNR gain of the segmented processing-based integration method will get larger with N_c increasing. The reason is that when M

is fixed and N_c is increased, N_{sub} decreases. Then, the duration of each sub-segment becomes shorter, and the approximation error induced by target acceleration would be smaller and the corresponding integration performance loss would become smaller. As a consequence, the SNR gain of the segmented processing-based integration method will increase.

On the other hand, the existing hybrid integration methods (such as hybrid integration [HI] method [7] and subspace hybrid integration [SHI] method [8]) integrate the target signal coherently within each sub-segment and then accumulate the integration results of multiple sub-segments non-coherently. Ideally (i.e., without considering the approximate error in the segmentation process), the SNR gain of HI/SHI could be expressed as [7, 8]:

$$SNR'_{\text{gain}} = 10\log_{10}N_{sub}\sqrt{N_c}. \tag{6.55}$$

Since $N_{sub} = M/N_c$, (6.55) can be further rewritten as:

$$
\begin{aligned}
SNR'_{\text{gain}} &= 10\log_{10}\frac{M}{N_c}\sqrt{N_c} \\
&= 10\log_{10}\frac{M}{\sqrt{N_c}}.
\end{aligned}
\tag{6.56}
$$

From (6.56), we can see that the integration gain of HI/SHI method decreases as N_c increases. Correspondingly, the integration and detection performance of HI/SHI method suffers losses as N_c increases.

6.3.2 Computational Complexity Comparison

This subsection analyzes the computational complexity of the segmented processing-based method in detail, and for comparison we also give the computational complexity of GRFT, HI and SHI.

Segmented processing-based method: The segmented processing-based method performs M groups of FFT operation, and each group contains N_{r1} points. The computational complexity of N_{r1} points FFT can be represent as:

$$I_{\text{FFT}}(N_{r1}) = \frac{N_{r1}}{2}\log_2(N_{r1})I_{cm} + N_r\log_2(N_{r1})I_{ca}, \tag{6.57}$$

where I_{cm} and I_{ca} represent the computational complexity of complex multiplication and complex addition, respectively. So, the computational complexity of M groups of N_{r1} points FFT is $MI_{\text{FFT}}(N_{r1})$. Subsequently, according to (6.17), M pulses are divided into N_{c1} sub-segments, and each sub-segment has N_{sub1} pulses. Then, the coherent integration in each sub-segment is achieved by CZT, and the computational complexity of CZT can be expressed as:

$$I_{\text{CZT}}(J) = (2J + J\log_2(J))I_{cm} + 2N_r\log_2(J)I_{ca}, \tag{6.58}$$

where $J = N_{v1} \mid N_{sub1}$. Note that $N_{r1}N_{c1}$ times of CZT operations were used in the sub-segment integration, and thus its computational complexity is $N_{r1}N_{c1}I_{\text{CZT}}(J)$.

For the phase compensation and range compensation among sub-segments in the frequency domain, it needs $N_{r1}N_{v1}N_{a1}N_{c1}$ times complex multiplication and $N_{r1}N_{v1}N_{a1}(N_{c1}-1)$ times complex addition, i.e., the computational complexity is $N_{r1}N_{v1}N_{a1}N_{c1}I_{cm} + N_{r1}N_{v1}N_{a1}(N_{c1}-1)I_{ca}$. Finally, we get the time domain signal through IFFT, for which computational complexity is $N_{v1}N_{a1}I_{FFT}(N_{r1})$. Thus, the total computational complexity of the segmented processing-based method can be written as:

$$I_{\text{Proposed}} = (N_p + N_{v1}N_{a1})I_{\text{FFT}}(N_{r1}) + N_{r1}N_{c1}I_{\text{CZT}}(J) \\ + N_{r1}N_{v1}N_{a1}N_{c1}I_{cm} + N_{r1}N_{v1}N_{a1}(N_{c1}-1)I_{ca}, \tag{6.59}$$

where N_{r1}, N_{v1} and N_{a1} separately denote the number of searching range cell, searching velocity unit and searching acceleration unit for the segmented processing-based method.

GRFT: Since GRFT [5] searches the target's distance, velocity and acceleration to achieve the coherent integration, it has heavy computing burden due to the high-dimensional searching. Specifically, GRFT needs $N_{r2}N_{v2}N_{a2}M$ times complex multiplication and $N_{r2}N_{v2}N_{a2}(M-1)$ times complex addition. Its computational complexity can be expressed as:

$$I_{\text{GRFT}} = N_{r2}N_{v2}N_{a2}MI_{cm} + N_{r2}N_{v2}N_{a2}(M-1)I_{ca}, \tag{6.60}$$

where N_{r2}, N_{v2} and N_{a2} denote the number of searching range cell, searching velocity unit and searching acceleration unit within the GRFT operation, respectively.

HI: The HI [7] method performs the coherent integration for each sub-segment through MTD based on FFT, and its computational complexity is $N_{c3}N_{r3}I_{\text{FFT}}(N_{sub3})$, where N_{sub3}, N_{c3} denotes the pulse number in sub-segments and number of sub-segments for the HI method. Then, HI accomplishes the non-coherent integration within different sub-segments via searching of ambiguous integer and acceleration. Correspondingly, the computational complexity can be written as:

$$I_{\text{HI}} = N_{c3}N_{r3}I_{\text{FFT}}(N_{sub3}) + N_{r3}N_{sub3}N_{c3}I_{\text{abs}} \\ + N_{r3}N_{sub3}N_{a3}N_k(N_{c3}-1)I_{\text{ra}}, \tag{6.61}$$

where N_k and N_{a3} represent the number of searched ambiguous integer and acceleration and I_{abs} and I_{ra} denote the computational complexity of abs($\bullet$) and real addition, respectively.

SHI: SHI [8] divides the parameter space into subspaces via parameter space division (PSD) and subspace movement (SM) processing [8]; then, it achieves the integration by HI. The computational complexity of PSD and SM is $N_pI_{\text{FFT}}(N_{r4}) + N_{r4}N_pQI_{cm} + N_pQI_{\text{FFT}}(N_{r4})$, where Q is the number of subspaces. As a result, the computational complexity of SHI can be expressed as:

$$I_{\text{SHI}} = N_pI_{\text{FFT}}(N_{r4}) + N_{r4}N_pQI_{cm} + N_pQI_{\text{FFT}}(N_{r4}) \\ + QI'_{\text{HI}}, \tag{6.62}$$

TABLE 6.1   Computational Complexity

Method	Computational complexity
GRFT	$N_{r2}N_{v2}N_{a2}N_pI_{cm} + N_{r2}N_{v2}N_{a2}(N_p-1)I_{ca}$
Segmented processing	$(N_p + N_{v1}N_{a1})I_{\text{FFT}}(N_{r1}) + N_{r1}N_{c1}I_{\text{CZT}}(J)$
	$+ N_{r1}N_{v1}N_{a1}N_{c1}I_{cm} + N_{r1}N_{v1}N_{a1}(N_{c1}-1)I_{ca}$
HI	$N_{c3}N_{r3}I_{\text{FFT}}(N_{sub3}) + N_{r3}N_{sub3}N_{c3}I_{\text{abs}}$
	$+ N_{r3}N_{sub3}N_{a3}N_k(N_{c3}-1)I_{ra}$
SHI	$N_pI_{\text{FFT}}(N_{r4}) + N_{r4}N_pQI_{cm} + N_pQI_{\text{FFT}}(N_{r4}) + QI'_{\text{HI}}$

where I'_{HI} can be expressed as:

$$I'_{\text{HI}} = N_{c4}N_{r4}I_{\text{FFT}}(N_{sub4}) + N_{r4}N_{sub4}N_{c4}I_{\text{abs}}$$
$$+ N_{r4}N_{sub4}N_{a4}N'_k(N_{c4}-1)I_{ra}, \tag{6.63}$$

where N_{c4}, N_{sub4}, N_{a4} and N'_k denote the number of sub-segments, pulse in sub-segments, acceleration units and ambiguous integer, respectively.

Table 6.1 lists the computational complexity of the four methods above. The simulation experiments for the computational cost comparison are also shown in Section 6.4.4, from which we could find that the computational burden of the segmented integration method is much lower than that of GRFT.

6.4   SIMULATED ANALYSIS

In this section, we use simulation experiments to verify the effectiveness of the segmented processing-based coherent integration method. Radar parameter settings are shown in Table 6.2.

6.4.1   Single Target Integration Result

In this subsection, we conduct a simulation experiment to evaluate the integration performance of the segmented processing-based coherent integration method under a different SNR background. Suppose that the target's initial distance(r_0), velocity(v_0) and acceleration ($a'_0 = 2a_0$) are, respectively, 200 km (corresponding to the 100-th range cell), 450 m/s and 200 m/s^2.

1) $SNR = 10$ dB after pulse compression. The pulse compression result is shown in Fig. 6.3(a), and Fig. 6.3(b) is a partial enlargement of Fig. 6.3(a). It can be seen

TABLE 6.2   Radar Parameters

Parameters	Values
Carrier frequency f_c	150 MHz
Signal bandwidth B	10 MHz
Pulse duration T_p	20 μs
Sampling frequency f_s	10 MHz
Pulse repetition time T_r	2 ms

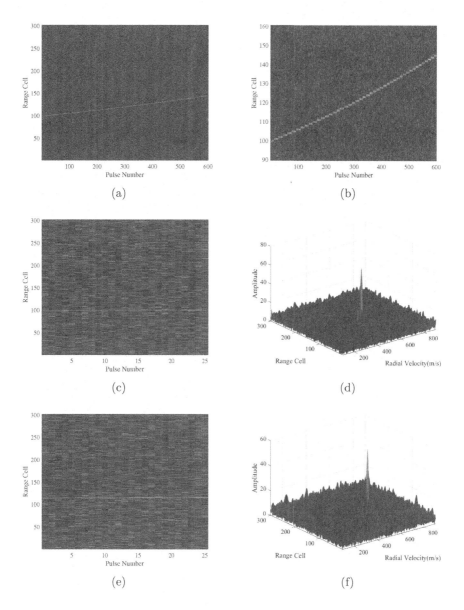

Figure 6.3 Single target integration results for SNR = 10 dB. (a) Compressed echoes. (b) Zoom in on compressed echoes. (c) Compressed echoes of the first sub-segment. (d) FBRFT result of the first sub-segment. (e) Compressed echoes of the 10th sub-segment. (f) FBRFT result of the 10th sub-segment.

that the target has crossed multiple range cells, and serious RM occurred because of the target's high speed and acceleration. After segmenting the echo using the criterion of (6.17), the pulse compression signals of the first sub-segment and the 10th sub-segment are as shown in Fig. 6.3(c) and Fig. 6.3(e), respectively, where the first-order RM caused by the target's acceleration in the sub-segment can be ignored.

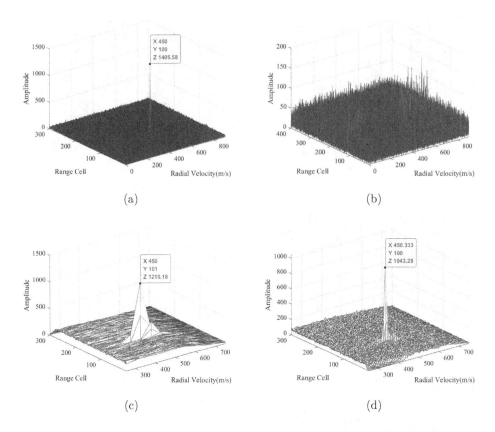

Figure 6.4 Single target integration results for SNR = 10 dB. (a) Segmented processing-based method. (b) RFT integration result. (c) HI integration result. (d) SHI integration result.

The FBRFT outputs of the first sub-segment and the 10th sub-segment are, respectively, shown in Fig. 6.3(d) and Fig. 6.3(f), while the target energy distributed in the sub-segment is accumulated. After that, the inter-subsegment coherent integration result of the multiple FBRFT output is shown in Fig. 6.4(a), from which we could see that the integration peak is higher. In addition, the integration results of RFT, HI and SHI are also, respectively, shown in Fig. 6.4(b)–Fig. 6.4(d). The RFT becomes invalid due to the RM and DS caused by the acceleration. Meanwhile, the integration peak value of the proposed algorithm is higher than that of SHI and HI, which indicates that the segmented processing-based coherent integration method could obtain higher integration gain than SHI and HI.

2) $SNR = -10$ dB after pulse compression. The result of pulse compression is shown in Fig. 6.5(a). Because the SNR after pulse compression is too low, the target is submerged in noise. Fig. 6.5(b) shows the target trajectory (under no noise condition). Fig. 6.5(c)–Fig. 6.5(e) show, respectively, the integration results of RFT, HI and SHI, from which we could see that the target is still submerged in noise after HI or SIII/RFT processing, which makes it difficult to be detected. Fig. 6.5(f) shows the processing result of the segmented processing-based coherent integration method. It

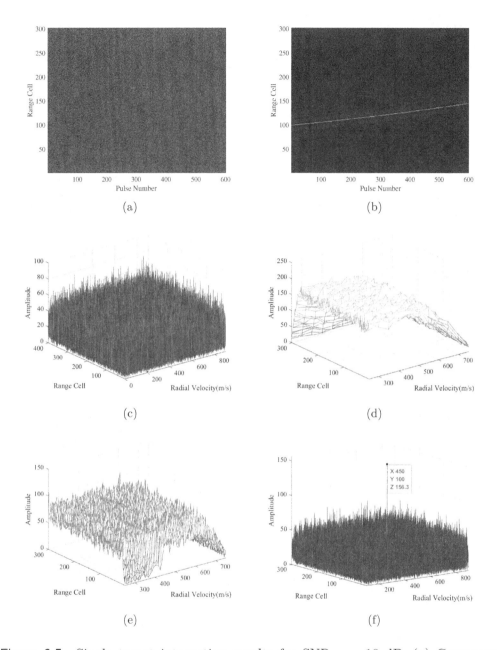

Figure 6.5 Single target integration results for SNR = −10 dB. (a) Compressed echoes. (b) Compressed echoes without noise. (c) RFT. (d) HI. (e) SHI. (f) Segmented processing-based method.

can be seen that the target energy is effectively accumulated and that the integration peak position corresponds to the target parameter, which is conducive to target detection. Hence, the segmented processing-based coherent integration method could be applied to target signal integration and detection under lower SNR environment, in comparison with RFT, HI and SHI methods.

6.4.2   Multiple Target Integration Result

We also conducted simulation experiments on the multi-target integration performance of the segmented processing-based coherent integration method. This subsection considers the following four scenarios.

Scenario 1): targets of different distances, same velocity and acceleration. The initial distances (r_0) of targets A and B are 199.7 and 200 km, respectively, corresponding to the 80th and 100th range cell. The velocity (v_0) and acceleration $(a_0' = 2a_0)$ of targets A and B are 450 m/s and 100 m/s^2. Fig. 6.6(a) shows the pulse compression signals of targets A and B, and since targets A and B have different initial radial distances, there is a fixed distance gap between the two trajectories. The result after processing by the segmented processing-based coherent integration method is shown in Fig. 6.6(b). It can be seen that the targets A and B are focused in the location of $(80, 450)$ and $(100, 450)$, which are consistent with the actual parameters of the targets. Fig. 6.6(c) shows the integration result of HI, and Fig. 6.6(d)

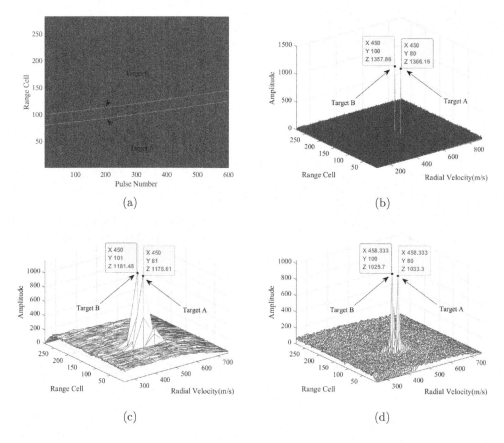

(a)

(b)

(c)

(d)

Figure 6.6   Simulation results for multiple target (scenario 1). (a) Compressed echoes. (b) Integration result of segmented processing-based method. (c) Integration result of HI. (d) Integration result of SHI.

shows the integration result of SHI, from which we could see that the two targets'
peak values of HI or SHI are smaller than those of the segmented processing-based
coherent integration method.

Scenario 2): targets of different velocities, same distance and acceleration: The
initial velocities (v_0) of targets C and D are 150 and 450 m/s^2, respectively. The
distance (r_0) and acceleration ($a'_0 = 2a_0$) of targets C and D are 200 km (100th range
cell) and 100 m/s^2. Fig. 6.7(a) shows the result of pulse compression. It can be seen
from Fig. 6.7(a) that the targets have the same starting range cell, and the distance
difference between the two targets increases with time due to the different target
velocity. Fig. 6.7(b) shows the processing result of the segmented processing-based
coherent integration method, and it can be seen that the targets C and D are focused
into $(100, 150)$ and $(100, 450)$, which are consistent with the actual parameters of the
targets. Fig. 6.7(c) shows the integration result of HI, and the results of SHI integration of the two targets are shown in Fig. 6.7(d) and Fig. 6.7(e), respectively. From
Fig. 6.7, we could notice that the integration peak values of the segmented processing-
based coherent integration method are higher, which indicates that the segmented
processing-based coherent integration method could achieve higher integration gain.

Scenario 3): targets of different distances and velocities, and same acceleration:
The initial distance (r_0) of target E is 199.85 km, corresponding to the 90th range
unit, and the initial velocity (v_0) is 450 m/s. The initial distance of target F is
200 km, corresponding to the 100th distance unit, and the initial velocity (v_0) is
150 m/s. The pulse compression signal of the targets and the integration result of
segmented processing-based coherent integration method are shown in Fig. 6.8(a)
and Fig. 6.8(b), respectively. The integrated peaks of targets E and F appear at
$(90, 450)$ and $(100, 150)$, respectively. Fig. 6.8(c) shows the integration result of HI.
The integration results of SHI for targets E and F are shown in Fig. 6.8(d) and
Fig. 6.8(e), respectively. The integration peak values of two targets of the segmented
processing-based coherent integration method are higher than those of HI or SHI.

Scenario 4): targets of different velocities and accelerations, and same distance:
The initial distance (r_0) of targets G and H is 200 km, corresponding to the 100th
range unit, and the initial velocities (v_0) of targets G and H are 150 and 450 m/s,
respectively. The initial accelerations ($a'_0 = 2a_0$) of targets G and H are 100 and -100
m/s^2, respectively. The pulse compression signal of the targets and the integration
result of the segmented processing-based coherent integration method are, respectively, shown in Fig. 6.9(a) and Fig. 6.9(b). The integrated peaks of targets G and H
appear at $(150, 100)$ and $(450, -100)$, respectively. Fig. 6.9(c) shows the integration
result of HI, and the results of SHI integration of the two targets (i.e., targets E and
F) are shown in Fig. 6.9(d) and Fig. 6.9(e), respectively.

From the multi-target simulation experiments results of the four scenarios shown
in Fig. 6.6–Fig. 6.9, we can see that the segmented processing-based coherent integration method can achieve better integration performance for multiple targets than
HI and SHI methods.

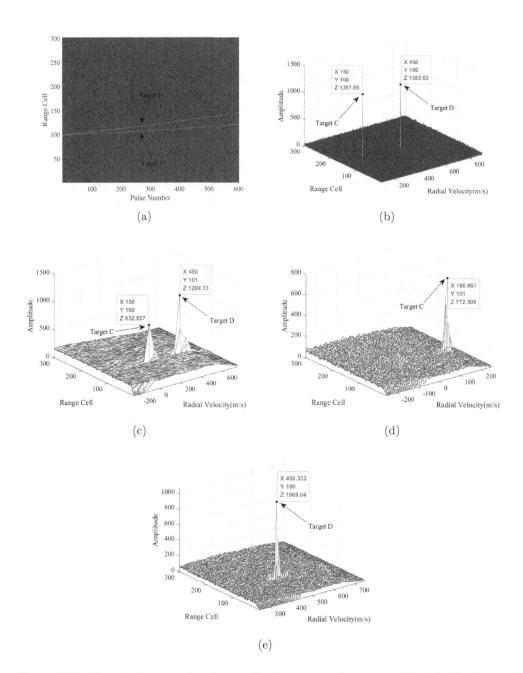

Figure 6.7 Simulation results for multiple target (scenario 2). (a) Compressed echoes. (b) Integration result of segmented processing-based method. (c) Integration result of HI. (d) Integration result of SHI for target C. (f) Integration result of SHI for target D.

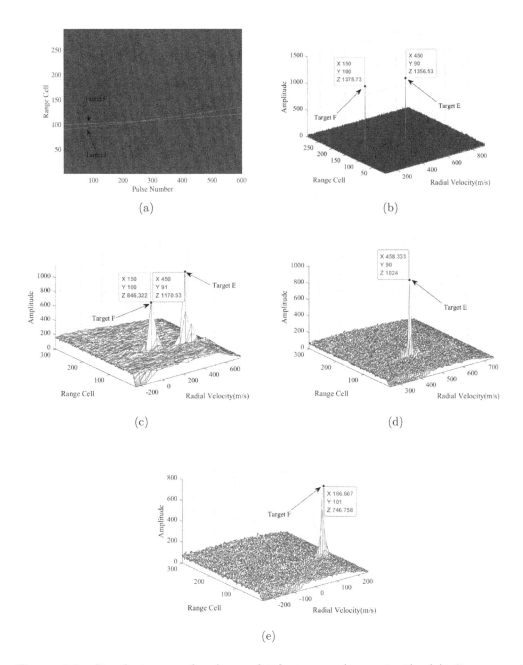

Figure 6.8 Simulation results for multiple target (scenario 3). (a) Compressed echoes. (b) Integration result of segmented processing-based method. (c) Integration result of HI. (d) Integration result of SHI for target E. (f)Integration result of SHI for target F.

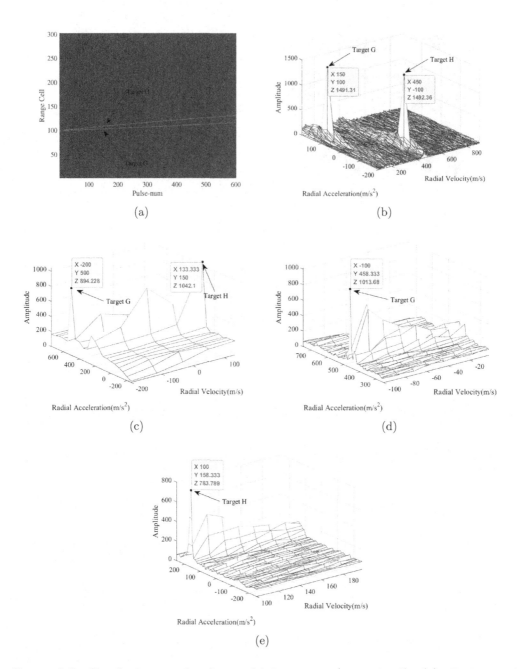

Figure 6.9 Simulation results for multiple target (scenario 4). (a) Compressed echoes. (b) Integration result of segmented processing-based method. (c) Integration result of HI. (d) Integration result of SHI for target G. (f)Integration result of SHI for target H.

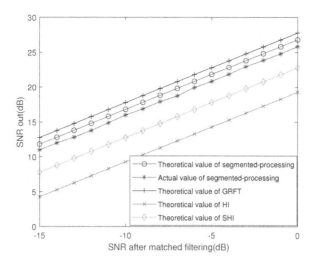

Figure 6.10 SNR gain.

6.4.3 SNR Gain and Detection Performance

This subsection analyzes the SNR gain and detection performance of the segmented processing-based coherent integration method through simulation experiments. The target parameters are consistent with those in Section 6.4.1. Fig. 6.10 shows the theoretical output SNR and actual output SNR of the segmented processing-based coherent integration method, as well as the theoretical output SNR curves of GRFT, HI and SHI. It can be seen that the segmented processing-based coherent integration method has a higher SNR gain than that of SHI and HI and is slightly inferior to GRFT. Fig. 6.11 shows the detection performance comparison of the segmented processing-based coherent integration method, GRFT, SHI, HI and RFT obtained from Monte Carlo experiments (1000 times) with a false alarm rate $P_{fa} = 10^{-3}$. The detection performance of the segmented processing-based coherent integration method is significantly better than SHI, HI and RFT, and the loss is about 1.5 dB compared with GRFT.

6.4.4 Computational Cost

We also evaluate the computational burden of the segmented processing-based coherent integration method and SHI, HI and GRFT. Fig. 6.12 shows the computational complexity curves of the four algorithms under different pulse numbers. It can be seen that the computational cost of GRFT is much higher than the segmented processing-based coherent integration method, SHI and HI. The calculation cost of the segmented processing-based coherent integration method is higher than SHI and HI. For example, when the number of pulses is 600, the computational complexity of GRFT is about 750 times the segmented processing-based coherent integration method. Compared with HI and SHI, the calculation complexity of the segmented

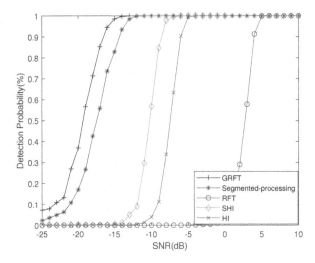

Figure 6.11   Detection probability.

processing-based coherent integration method is increased by about 110 and 5 times, respectively. Table 6.3 shows the running time of different algorithms (the processor is AMD R7 4800H, RAM is 16G, and simulation scenario is the same as Section 6.4.1), which indicates that the time cost of GRFT is much larger than that of the segmented processing-based coherent integration method.

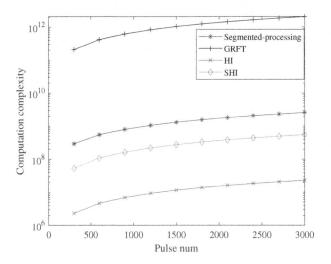

Figure 6.12   Computation complexity.

TABLE 6.3   Computational Time Cost

Algorithm	GRFT	Segment processing	HI	SHI
Time (s)	395.8	3.2123	1.5153	0.3923

6.5   SUMMARY

For the detection of high-speed moving targets with acceleration motion, this chapter introduces a computational efficient integration method based on segment processing. The segmented processing-based coherent integration method contained two steps: the first step is the integration within sub-segment and second step is the integration across different sub-segments. In particular, the method applies the FBRFT to achieve the signal integration output of each sub-segment. Then, based on the FBRFT output characteristics, the integrations across sub-segments are obtained via phase compensation and velocity location alignment. Numerical experimental results substantiate the efficacy of the proposed segment integration method for high-speed moving targets detection.

Bibliography

[1] P. H. Huang, G. S. Liao, Z. W. Yang, X. Xia, J. T. Ma, and J. T. Ma, "Long-time coherent integration for weak maneuvering target detection and high-order motion parameter estimation based on keystone transform," *IEEE Transactions on Signal Processing*, vol. 64, no. 15, pp. 4013–4026, August, 2016.

[2] X. L. Chen, J. Guan, Y. Huang, N. B. Liu, and Y. He, "Radon-linear canonical ambiguity function-based detection and estimation method for marine target with micromotion," *IEEE Transactions on Geoscience and Remote Sensing*, vol. 53, no. 4, pp. 2225–2240, April 2015.

[3] X. L. Chen, J. Guan, N. B. Liu, and Y. He, "Maneuvering target detection via Radon-Fractional Fourier transform-based long-time coherent integration," *IEEE Transactions on Signal Processing*, vol. 62, no. 4, pp. 939–953, February 2014.

[4] R. P. Perry, R. C. DiPietro, and R. L. Fante, "Sar imaging of moving targets," *IEEE Transactions on Aerospace and Electronic Systems*, vol. 35, no. 1, pp. 188–200, 1999.

[5] J. Yu, J. Xu, Y. N. Peng, and X. G. Xia, "Radon-fourier transform for radar target detection (iii): optimality and fast implementations," *IEEE Transactions on Aerospace and Electronic Systems*, vol. 48, no. 2, pp. 991–1004, 2012.

[6] J. Xu, X. G. Xia, S. B. Peng, J. Yu, Y. N. Peng, and L. C. Qian, "Radar maneuvering target motion estimation based on generalized radon-Fourier transform," *IEEE Transactions on Signal Processing*, vol. 60, no. 12, pp. 6190–6201, 2012.

[7] J. Xu, X. Zhou, L. C. Qian, X. G. Xia, and T. Long, "Hybrid integration for highly maneuvering radar target detection based on generalized radon-Fourier transform," *IEEE Transactions on Aerospace and Electronic Systems*, vol. 52, no. 5, pp. 2554–2561, 2016. .

[8] Z. G. Ding, P. J. You, L. C. Qian, X. Zhou, S. Y. Liu, and T. Long, "A subspace hybrid integration method for high-speed and maneuvering target detection," *IEEE Transactions on Aerospace and Electronic Systems*, vol. 56, no. 1, pp. 630–644, 2020.

[9] M. Guida, M. Longo, and M. Lops, "Biparametric CFAR procedures for log-normal clutter," in *IEEE Transactions on Aerospace and Electronic Systems*, vol. 29, no. 3, pp. 789–809, 1994.

[10] M. Richards, "Fundamentals of radar signal processing," *International Waveform Diversity and Design Conference*, New York, NY, USA: McGraw-Hill, 2005.

CHAPTER 7

SAF-SFT-Based Coherent Integration Processing

The previous three chapters discussed the coherent integration for radar high-speed target detection via multi-dimension searching in the time or frequency domain, which make the Radon-Lv's distribution (RLVD) method, keystone transform match filtering process (KTMFP) method and segmented processing method are of large computational complexity when the searching scope is huge. In this chapter, a computationally efficient coherent integration method via symmetric autocorrelation function (SAF) and scaled Fourier transform (i.e., SAF-SFT) is discussed. Due to the correlation operation, the cross term of SAF-SFT is also analyzed, and its characteristic indicates the applicability in the scenario of multi-targets. Detailed analysis of SAF-SFT with respect to computational cost, detection probability and parameter estimation ability is also shown, which indicates that the SAF-SFT could strike a balance between computational cost and detection probability as well as the estimation performance.

7.1 SIGNAL MODEL

Assume that the radar transmits the linear-frequency modulated signal [1–4], i.e.,

$$s_{trans}(t) = \text{rect}\left(\frac{t}{T_p}\right) \exp\left(j\pi\mu t^2\right) \exp(j2\pi f_c t), \qquad (7.1)$$

where

$$\text{rect}(x) = \begin{cases} 1, & |x| \leq \frac{1}{2} \\ 0, & |x| > \frac{1}{2} \end{cases},$$

where μ, T_p, t and f_c denote the frequency modulated rate, pulse duration, fast time and carrier frequency, respectively.

Without loss of generality, the signal model of the kth ($k = 1, 2, \cdots, K$) target is established for simplicity, where the instantaneous slant range $R_k(t_m)$ of the kth target satisfies [4]

$$R_k(t_m) = R_{0k} + v_k t_m + a_k t_m^2, \qquad (7.2)$$

DOI: 10.1201/9781003529101-7

where $t_m = mT \ (m = 0, \cdots, N)$ is the slow time; N and T are, respectively, the pulse number and pulse repetition time. a_i, v_i and R_{0i} are ith target's radial acceleration, velocity and initial slant range, respectively.

The received signal of the kth target after pulse compression (PC) could be formulated as [4]

$$s(t, t_m) = A_{1k}\text{sinc}\left[B\left(t - \frac{2(R_{0k} + v_k t_m + a_k t_m^2)}{c}\right)\right]$$
$$\times \exp\left(-j4\pi\frac{R_{0k} + v_k t_m + a_k t_m^2}{\lambda}\right), \tag{7.3}$$

where λ represents the wavelength, i.e.,

$$\lambda = c/f_c, \tag{7.4}$$

c, B and A_{1k} are, respectively, the speed of light, bandwidth and amplitude after PC.

Because of the target's high-speed and radar's low pulse repetition frequency (PRF), Doppler ambiguity would occur. As a result, the target's velocity could be expressed as

$$v_k = F_k v_a + v_{0k}, \tag{7.5}$$

where v_a represents the blind velocity, i.e.,

$$v_a = \lambda f_p/2, \tag{7.6}$$

v_{0k} is defined as the unambiguous velocity, and it satisfies $v_{0k} \in \left[-\frac{\lambda f_p}{4}, \frac{\lambda f_p}{4}\right]$. F_k denotes the fold factor (also called as the ambiguous number) of the kth target, and f_p is the PRF.

Instituting (7.5) into (7.3) and considering that $2\pi f_p F_k t_m$ is an integral multiple of 2π, we have

$$s(t, t_m) = A_{1k}\text{sinc}\left[B\left(t - \frac{2(R_{0k} + v_k t_m + a_k t_m^2)}{c}\right)\right]$$
$$\times \exp\left(-j4\pi\frac{R_{0k} + v_{0k} t_m + a_k t_m^2}{\lambda}\right). \tag{7.7}$$

Performing the Fourier transform (FT) on (7.7) along the fast time axis, we have

$$S(f, t_m) = A_{2k}\text{rect}\left(\frac{f}{B}\right)\exp\left(-j4\pi f\frac{F_k v_a t_m}{c}\right)$$
$$\times \exp\left[-j4\pi(f_c + f)\frac{(R_{0k} + v_{0k} t_m + a_k t_m^2)}{c}\right], \tag{7.8}$$

where A_{2i} denotes the signal's amplitude after FT, i.e.,

$$A_{2k} = A_{1k}/B. \tag{7.9}$$

Equation (7.8) shows that the target's acceleration and velocity are all coupled with f, which will result in Doppler spread (DS) and range cell migration (RCM) effect within the coherent integration (CI) time.

7.2 SAF-SFT METHOD

7.2.1 Range and Velocity Estimation

First, the generalized keystone transform (GKT) is used to correct the range migration caused by target's acceleration, which performs scaling as follows:

$$t_m = [f_c/(f + f_c)]^{1/2} u_m. \tag{7.10}$$

Apply the GKT on (7.8) yields

$$
\begin{aligned}
S(f, u_m) =& A_{2k} \text{rect}\left(\frac{f}{B}\right) \exp\left[-j4\pi(f_c + f)\frac{R_{0k}}{c}\right] \\
& \times \exp\left[-j4\pi f \frac{F_k v_a u_m}{c}\left(\frac{f_c}{f_c + f}\right)^{1/2}\right] \\
& \times \exp\left[-j4\pi\left(1 + \frac{f}{f_c}\right)^{1/2} f_c \frac{v_{0k} u_m}{c}\right] \\
& \times \exp\left(-j4\pi \frac{a_k u_m^2}{\lambda}\right).
\end{aligned} \tag{7.11}
$$

Taking the first-order Taylor series expansion on $f[f_c/(f_c+f)]^{1/2}$ and $(1+f/f_c)^{1/2}$, we have

$$f[f_c/(f_c + f)]^{1/2} \approx f, \tag{7.12}$$

$$(1 + f/f_c)^{1/2} \approx 1 + f/(2f_c). \tag{7.13}$$

Substituting (7.12) and (7.13) into (7.11), we have

$$
\begin{aligned}
S(f, u_m) \approx& A_{2k} \text{rect}\left(\frac{f}{B}\right) \exp\left(-j4\pi f \frac{R_{0k} + V_k u_m}{c}\right) \\
& \times \exp\left(-j4\pi \frac{R_{0k} + v_{0k} u_m + a_k u_m^2}{\lambda}\right),
\end{aligned} \tag{7.14}
$$

where

$$V_k = F_k v_a + 0.5 v_{0k}. \tag{7.15}$$

From (7.14), we can see that the coupling between the target's acceleration and range frequency has been eliminated. As a result, the range migration caused by acceleration has been eliminated. However, the range migration induced by the target's velocity still exists.

In order to remove the range migration caused by velocity, a SAF of (7.14) corresponding to f is defined as [5]

$$R(f, f_n, u_m) = S(f + f_n, u_m) S^*(f - f_n, u_m), \tag{7.16}$$

where $*$ denotes the complex conjugation operation and f_n represents the shift frequency variable with respect to f.

Substituting (7.14) into (7.16) yields

$$R(f, f_n, u_m) = A_{2k}^2 \exp\left(-j8\pi f_n \frac{R_{0k}}{c}\right) \exp\left(-j8\pi f_n \frac{V_k u_m}{c}\right) \exp(-j2\pi 0 f). \quad (7.17)$$

By (7.17), it can be seen that the energy along f dimension can be integrated via the addition operation. After the addition operation on (7.17) along f dimension, we have

$$
\begin{aligned}
R(f_n, u_m) &= add_f[R(f, f_n, u_m)] \\
&= add_f\left[A_{2k}^2 \exp\left(-j8\pi f_n \frac{R_{0k}}{c}\right) \exp\left(-j8\pi f_n \frac{V_k u_m}{c}\right) \exp(-j2\pi 0 f)\right] \\
&= add_f\left[\exp(-j2\pi 0 f)\right] A_{2k}^2 \exp\left(-j8\pi f_n \frac{R_{0k}}{c}\right) \exp\left(-j8\pi f_n \frac{V_k u_m}{c}\right) \\
&= Q A_{2k}^2 \exp\left(-j8\pi f_n \frac{R_{0k}}{c}\right) \exp\left(-j8\pi f_n \frac{V_k u_m}{c}\right),
\end{aligned}
$$

$$(7.18)$$

where $add_f[\cdot]$ represents the addition operation with respect to f and Q denotes the integration gain.

In (7.18), f_n and u_m are coupled with each other, and the SFT with respect to u_m could be employed to eliminate the coupling, which is defined as [6]

$$
\begin{aligned}
P(f_n, f_{ds}) &= SFT_{u_m}[R(f_n, u_m)] \\
&= \int_{u_m} R(f_n, u_m) \exp(-j2\pi f_{ds}\xi_1 f_n u_m) du_m,
\end{aligned}
\quad (7.19)
$$

where f_{ds} is the scaled Doppler frequency corresponding to u_m, $SFT_{u_m}[\cdot]$ represents the SFT operation and ξ_1 is a constant.

Substituting (7.18) into (7.19), we have

$$
\begin{aligned}
P(f_n, f_{ds}) &= Q A_{2k}^2 \exp\left(-j8\pi f_n \frac{R_{0k}}{c}\right) \\
&\quad \times \int_{u_m} \exp\left[-j2\pi\left(\frac{4V_k u_m}{c} f_n + f_{ds}\xi_1 f_n u_m\right)\right] du_m \\
&= Q A_{2k}^2 \exp\left(-j8\pi f_n \frac{R_{0k}}{c}\right) \delta\left(f_{ds} + \frac{4V_k}{c\xi_1}\right).
\end{aligned}
\quad (7.20)
$$

Applying the inverse Fourier transform (IFT) on (7.20) with respect to f_n, we have

$$P(t_r, f_{ds}) = A_{3k}\delta\left(t_r - \frac{4R_{0k}}{c}\right)\delta\left(f_{ds} + \frac{4V_k}{\xi_1 c}\right), \quad (7.21)$$

where t_r is the pseudo fast time corresponding to the shift-frequency and A_{3k} represents the signal's amplitude after IFT operation, i.e.,

$$A_{3k} = 2\pi Q A_{2k}^2. \quad (7.22)$$

The estimations of the target's initial range and velocity (i.e., $\hat{R}_{0k}$ and $\hat{V}_k$) could be obtained via the largest peak location of (7.21).

7.2.2 Acceleration Estimation

A compensation function could be generated based on the estimated velocity $\hat{V}_{0,k}$ as follows:

$$H(f, u_m; \hat{V}_k) = \exp\left(\frac{j4\pi f \hat{V}_k u_m}{c}\right). \tag{7.23}$$

Multiplying (7.14) with (7.22), we have

$$S_c(f, u_m) = A_{2k}\text{rect}\left(\frac{f}{B}\right)\exp\left(-j4\pi f\frac{R_{0k}}{c}\right)$$
$$\times \exp\left(-j4\pi\frac{R_{0k} + v_{0k}u_m + a_k u_m^2}{\lambda}\right). \tag{7.24}$$

Applying IFT on (7.24) with respect to f, we have

$$s_c(\hat{t}, u_m) = A_{4k}\text{sinc}\left[B\left(\hat{t} - \frac{2R_{0k}}{c}\right)\right]\exp\left(-j4\pi\frac{R_{0k} + v_{0k}u_m + a_k u_m^2}{\lambda}\right), \tag{7.25}$$

where

$$A_{4k} = 2\pi A_{2k}. \tag{7.26}$$

From (7.25), we can see that the range migration has been eliminated. Target energy is distributed in the range cell corresponding to R_{0k}. Thus, we could extract the target signal along the range cell based on estimation of R_{0k}, i.e.,

$$s(u_m) = A_{4k}\exp\left(-j4\pi\frac{R_{0k} + v_{0k}u_m + a_k u_m^2}{\lambda}\right). \tag{7.27}$$

The SAF of (7.27) is defined as

$$X(u_m, \tau_n) = s(u_m + \tau_n)s^*(u_m - \tau_n)$$
$$= A_{4k}^2\exp\left[-j2\pi\left(\frac{4v_{0k}\tau_n}{\lambda} + \frac{8a_k u_m \tau_n}{\lambda}\right)\right], \tag{7.28}$$

where τ_n represents the time lag.

Performing the SFT to (7.28) with respect to u_m, we have

$$P_2(f_t, \tau_n) = SFT_{u_m}[X(u_m, \tau_n)]$$
$$= \int_{u_m} X(u_m, \tau_n)\exp(-j2\pi\xi_2 f_t \tau_n u_m)du_m$$
$$= A_{4k}^2\exp\left(-j2\pi\tau_n\frac{4v_{0k}}{\lambda}\right) \tag{7.29}$$
$$\times \int_{u_m}\exp\left[-j2\pi\left(\frac{8a_k u_m}{\lambda}\tau_n + \xi_2 f_t \tau_n u_m\right)\right]du_m$$
$$= A_{4k}^2\exp\left(-j2\pi\tau_n\frac{4v_{0k}}{\lambda}\right)\delta\left(f_t + \frac{8a_k}{\xi_2\lambda}\right),$$

where ξ_2 is a constant.

Applying the FT on (7.29) with respect to τ_n, we have

$$X_f(f_t, f_\tau) = A_{5k}\delta\left(f_\tau + \frac{4v_{0k}}{\lambda}\right)\delta\left(f_t + \frac{8a_k}{\xi_2\lambda}\right), \tag{7.30}$$

where

$$A_{5k} = 2\pi A_{4k}^2. \tag{7.31}$$

By (7.30), the energy spectrum of the kth target is obtained in the $f_t - f_\tau$ domain. The unambiguous velocity and acceleration (i.e., $\hat{v_{0k}}$ and $\hat{a}_k$) could be estimated via the peak location of (7.30). After that, the kth target's true radial velocity can calculated as follows:

$$\hat{v}_k = \hat{V}_k + 0.5\hat{v}_{0k}. \tag{7.32}$$

The flow chart of the SAF-SFT-based algorithm is given in Fig. 7.1.

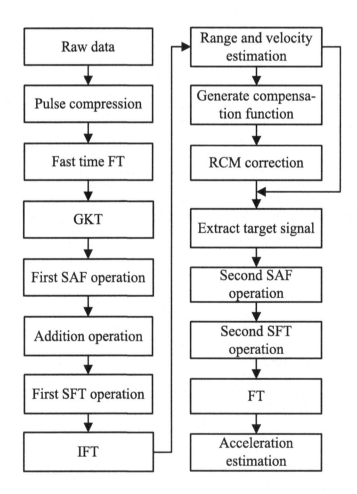

Figure 7.1   Flowchart of the SAF-SFT-based algorithm.

Remark on the selection of scaling factors ξ_1 and ξ_2: As shown in equations (7.21) and (7.30), the velocity and acceleration are obtained via the peak location of (7.21) and (7.30). To guarantee that the velocity and acceleration could be estimated without ambiguity-effect, we limit that maximum value of $\frac{4V_k}{\xi_1 c}$ and $\frac{8a_k}{\xi_2 \lambda}$ to be less than the ranges of the Doppler frequency and the chirp rate, i.e.,

$$\frac{4V_{k,\mathrm{max}}}{\xi_1 c} \leq \frac{f_p}{2}, \tag{7.33}$$

$$\frac{8a_{k,\mathrm{max}}}{\xi_2 \lambda} \leq \frac{f_p}{2}. \tag{7.34}$$

According to (7.33) and (7.34), ξ_1 and ξ_2 should satisfy

$$\xi_1 \geq \frac{8V_{k,\mathrm{max}}}{c f_p}, \tag{7.35}$$

$$\xi_2 \geq \frac{16a_{k,\mathrm{max}}}{\lambda f_p}. \tag{7.36}$$

When the ξ_1 and ξ_2 satisfy the condition shown in (7.35) and (7.36), respectively, smaller scaling factors can be chosen to increase estimation accuracy.

7.3 CROSS TERMS ANALYSIS

The above analysis is considered for the single target scene. For multi-targets, the cross terms will appear since the two SAF operations shown in (7.16) and (7.28). In the following, we will analyze the cross terms of the first SAF operation and second SAF operation.

7.3.1 Cross Terms of First SAF Operation

Similar to (7.8), the received signal of K targets after PC in the $f - t_m$ domain could be expressed as

$$\begin{aligned}
S(f, t_m) = \sum_{k=1}^{K} A_{2k} \mathrm{rect}\left(\frac{f}{B}\right) \exp\left(-j4\pi f \frac{F_k v_a t_m}{c}\right) \\
\times \exp\left[-j4\pi(f_c + f)\frac{(R_{0k} + v_{0k}t_m + a_k t_m^2)}{c}\right].
\end{aligned} \tag{7.37}$$

After the GKT operation, the signal will be changed into

$$\begin{aligned}
S(f, u_m) \approx \sum_{k=1}^{K} A_{2k} \mathrm{rect}\left(\frac{f}{B}\right) \exp\left(-j4\pi f \frac{R_{0k} + V_k u_m}{c}\right) \\
\times \exp\left(-j4\pi \frac{R_{0k} + v_{0k}u_m + a_k u_m^2}{\lambda}\right).
\end{aligned} \tag{7.38}$$

The SAF of (7.38) corresponding to f could be expressed as

$$R(f, f_n, u_m) = \sum_{k=1}^{K} A_{2k}^2 \exp\left(-j8\pi f_n \frac{R_{0k}}{c}\right) \exp\left(-j8\pi f_n \frac{V_k u_m}{c}\right) \tag{7.39}$$
$$+ R_{cross,1}(f, f_n, u_m) + R_{cross,2}(f, f_n, u_m),$$

where

$$R_{cross,1}(f, f_n, u_m) = \sum_{p=1}^{K} \sum_{q=1,q\neq p}^{K} A_{2p} A_{2q} \exp\left[-j4\pi f_n \frac{(R_{0p} + R_{0q}) + (V_p + V_q)u_m}{c}\right]$$
$$\times \exp\left[-j4\pi f \frac{(R_{0p} - R_{0q}) + (V_p - V_q)u_m}{c}\right]$$
$$\times \exp\left[-j4\pi \frac{(R_{0p} - R_{0q}) + (v_{0p} - v_{0q})u_m + 2(a_p - a_q)u_m^2}{\lambda}\right],$$
$$\tag{7.40}$$

$$R_{cross,2}(f, f_n, u_m) = \sum_{p=1}^{K} \sum_{q=1,q\neq p}^{K} A_{2p} A_{2q} \exp\left[-j4\pi f_n \frac{(R_{0p} + R_{0q}) + (V_p + V_q)u_m}{c}\right]$$
$$\times \exp\left[j4\pi f \frac{(R_{0p} - R_{0q}) + (V_p - V_q)u_m}{c}\right]$$
$$\times \exp\left[j4\pi \frac{(R_{0p} - R_{0q}) + (v_{0p} - v_{0q})u_m + 2(a_p - a_q)u_m^2}{\lambda}\right].$$
$$\tag{7.41}$$

According to the Euler's formula, i.e.,

$$\exp(j\varphi) + \exp(-j\varphi) = 2\cos\varphi, \tag{7.42}$$

we have

$$R_{cross,1}(f, f_n, u_m) + R_{cross,2}(f, f_n, u_m)$$
$$= 2 \sum_{p=1}^{K} \sum_{q=1,q\neq p}^{K} A_{2p} A_{2q}$$
$$\times \exp\left[-j4\pi f_n \frac{(R_{0p} + R_{0q}) + (V_p + V_q)u_m}{c}\right] \tag{7.43}$$
$$\times \cos\left[4\pi \left(f \frac{(R_{0p} - R_{0,q}) + (V_p - V_q)u_m}{c}\right.\right.$$
$$\left.\left. + \frac{(R_{0p} - R_{0q}) + (v_{0p} - v_{0q})u_m + 2(a_p - a_q)u_m^2}{\lambda}\right)\right].$$

From (7.43), it can be seen that the cosine function would disturb the integration of the cross term via addition operation along f distribution and SFT operation. Meanwhile, the self terms could be accumulated well via the addition operation along

f distribution and SFT operation. The consine function will be removed only when $R_{0p} = R_{0q}$, $V_p = V_q$, $v_{0p} = v_{0q}$ and $a_p = a_q$, which means that the pth target and the qth target are actually the same. In other words, the cross terms cannot be integrated as self terms. In the following, an example is given to show this more clearly.

Example 1: Two high-speed maneuvering targets (targets 1 and 2) are considered, and their motion parameters are set as: initial range cell number $n_{R01} = 200$, radial velocity $v_1 = 1500$ m/s, acceleration $a_1 = 15$ m/s^2 for target 1; initial range cell number $n_{R02} = 320$, radial velocity $v_2 = -1650$ m/s, acceleration $a_2 = -15$ m/s^2 for target 2. Radar parameters are carrier frequency $f_c = 1$ GHz, bandwith $B = 5$ MHz, sample frequency $f_s = 5$ MHz, pulse repetition frequency $f_p = 1000$ Hz, pulse duration $T_p = 20$ μs, pulse number $N = 2001$. $\xi_1 = 2.5 \times 10^{-7}$, $\xi_2 = 1$.

Fig. 7.2(a) shows the PC result. During the observation time, the targets' energy across several range cells. After the SAF operation shown in (7.16) and SFT operation shown in (7.20), the scaled Doppler frequency and shift-frequency distribution is given in Fig. 7.2(b). Then, the IFT operation is performed, and we can obtain the target energy spectrum in the range-velocity domain, as shown in Fig. 7.2(c). Two peaks are shown in Fig. 7.2(c), which implies that the self terms have been accumulated well.

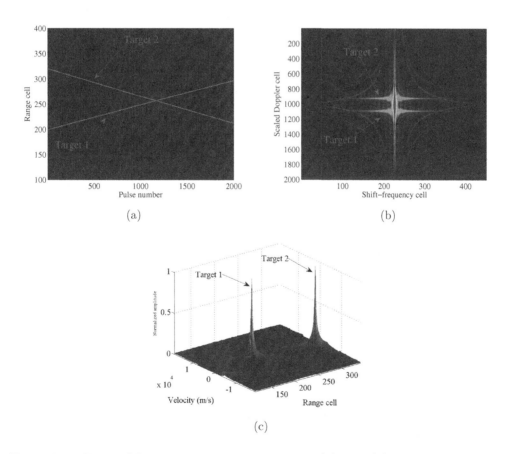

Figure 7.2 First SAF operation for multi-targets. (a) PC. (b) First SAF result. (c) Integration in range-velocity domain.

7.3.2 Cross Terms of Second SAF Operation

According to the analysis from (7.23)–(7.25), we can see that the extracted signal after range migration correction will be a sum of multi-linear frequency modulated (LFM) signals if the targets are within the same radial velocity and range cell, which can be expressed as follows:

$$s(u_m) = \sum_{k=1}^{K} A_{4k} \exp\left(-j4\pi \frac{R_{0k} + v_{0k}u_m + a_k u_m^2}{\lambda}\right). \tag{7.44}$$

The SAF of (7.44) is

$$
\begin{aligned}
X(u_m, \tau_n) &= s(u_m + \tau_n)s^*(u_m - \tau_n) \\
&= \sum_{i=1}^{K} A_{4i}^2 \exp\left[-j2\pi\left(\frac{4v_{0i}\tau_n}{\lambda} + \frac{8a_i u_m \tau_n}{\lambda}\right)\right] \\
&\quad + X_{cross}(u_m, \tau_n),
\end{aligned}
\tag{7.45}
$$

where

$$
\begin{aligned}
X_{cross}(u_m, \tau_n) &= \sum_{p=1}^{K}\sum_{q=1,q\neq p}^{K} A_{4p}A_{4q} \exp\left[-j\frac{4\pi}{\lambda}(R_{0p} - R_{0q})\right] \\
&\quad \times \exp\left[-j\frac{4\pi}{\lambda}(v_{0p} - v_{0q})u_m\right] \\
&\quad \times \exp\left[-j\frac{4\pi}{\lambda}(v_{0p} + v_{0q})\tau_n\right] \\
&\quad \times \exp\left[-j\frac{4\pi}{\lambda}(a_p - a_q)u_m^2\right] \\
&\quad \times \exp\left[-j\frac{4\pi}{\lambda}(a_p - a_q)\tau_n^2\right] \\
&\quad \times \exp\left[-j\frac{4\pi}{\lambda}2(a_p + a_q)u_m\tau_n\right].
\end{aligned}
\tag{7.46}
$$

Performing the SFT and FT operations on (7.45) with respect to u_m and τ_n, respectively, we can obtain

$$
\begin{aligned}
X_f(f_\tau, f_t) &= \sum_{k=1}^{K} A_{5k}\delta\left(f_\tau + \frac{4v_{0k}}{\lambda}\right)\delta\left(f_t + \frac{8a_k}{\xi_2\lambda}\right) \\
&\quad + \mathrm{FT}_{\tau_n}\left[\int_{u_m} X_{cross}(u_m, \tau_n)\exp(-j2\pi\xi_2 f_t \tau_n u_m)du_m\right].
\end{aligned}
\tag{7.47}
$$

From (7.45) and (7.46), it is observed that the self terms are all first-order functions of u_m and τ_n, whereas the cross terms are all second-order functions of u_m and τ_n. As a result, the self term could be integrated well via the SFT and FT operations; meanwhile, the cross terms could not integrate via the SFT and FT operations, which is helpful to the suppression of the cross terms. In the following, a simulation example is given to show that more clearly.

Example 2: Two high-speed maneuvering targets (targets 3 and 4) are considered, and their motion parameters are set as: initial range cell number $n_{R03} = 200$, radial velocity $v_3 = 1500$ m/s, acceleration $a_3 = -15$ m/s^2 for target 3; initial range cell number $n_{R04} = 200$, radial velocity $v_4 = 1500$ m/s, acceleration $a_4 = 15$ m/s^2 for target 4. Radar parameters are carrier frequency $f_c = 1$ GHz, bandwith $B = 5$ MHz, sample frequency $f_s = 5$ MHz, pulse repetition frequency $f_p = 1000$ Hz, pulse duration $T_p = 20$ μs, pulse number $N = 2001$. $\xi_1 = 2.5 \times 10^{-7}$, $\xi_2 = 1$.

The result of PC is given in Fig. 7.3(a). Note that the motion parameters of targets 3 and 4 are very similar (with the same initial range cell, same velocity, but only different on acceleration). As a result, the moving trajectories of targets 3 and 4 are very close. The result after first SAF operation is given in Fig. 7.3(b), and the energy spectrum in range-velocity domain after IFT operation is given in Fig. 7.3(c), which indicates that targets 3 and 4 are distributed in the same location because of their same range cell and velocity. With the estimation of velocity, the range migration could be corrected, as shown in Fig. 7.3(d), where the trajectory of targets 3 and 4 is distributed in the same range cell. Finally, after the second SAF and SFT operations, the energy spectrum in acceleration domain is shown in Fig. 7.3(e). It could be seen that the targets' energy is integrated as two peaks, which means that targets 3 and 4 could be distinguished clearly in the acceleration domain because of their different accelerations.

7.3.3 Cross Terms between Single Target and Noise

The kth target's signal with noise component after pulse compression could be stated as

$$
\begin{aligned}
s(t, t_m) =& A_{1k}\mathrm{sinc}\left[B\left(t - \frac{2(R_{0k} + v_{0k}t_m + a_k t_m^2)}{c}\right)\right] \\
& \times \exp\left(-j4\pi\frac{R_{0k} + v_{0k}t_m + a_i t_m^2}{\lambda}\right) + n(t, t_m),
\end{aligned}
\tag{7.48}
$$

where $n(\hat{t}, t_m)$ denotes the additive complex white Gaussian noise.

Performing FT on (7.48) along the fast time axis, we have

$$
\begin{aligned}
S(f, t_m) =& A_{2k}\mathrm{rect}\left(\frac{f}{B}\right)\exp\left(-j4\pi f\frac{F_k v_a t_m}{c}\right) \\
& \times \exp\left[-j4\pi(f_c + f)\frac{(R_{0k} + v_{0k}t_m + a_k t_m^2)}{c}\right] \\
& + N(f, t_m),
\end{aligned}
\tag{7.49}
$$

where $N(f, t_m)$ denotes the noise after the FT.

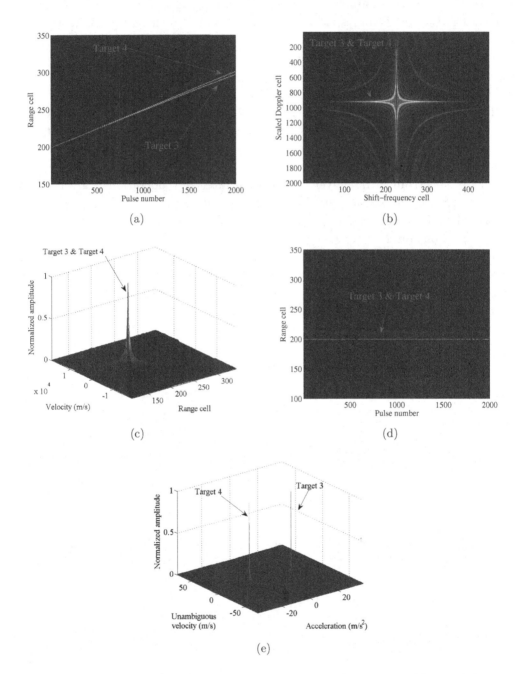

Figure 7.3 Second SAF-SFT operation under multi-target scene. (a) PC. (b) Scaled Doppler frequency and shift-frequency distribution. (c) The energy spectrum in range-velocity domain. (d) Range migration correction. (e) The energy spectrum in acceleration domain.

Applying the GKT operation on (7.49) yields

$$S(f, u_m) \approx A_{2k} \text{rect}\left(\frac{f}{B}\right) \exp\left(-j4\pi f \frac{R_{0k} + V_k u_m}{c}\right)$$

$$\times \exp\left(-j4\pi \frac{R_{0k} + v_{0k} u_m + a_k u_m^2}{\lambda}\right) + N(f, u_m) \qquad (7.50)$$

$$= S_k(f, u_m) + N(f, u_m),$$

where

$$S_k(f, u_m) = A_{2k} \text{rect}\left(\frac{f}{B}\right) \exp\left(-j4\pi f \frac{R_{0k} + V_k u_m}{c}\right)$$

$$\times \exp\left(-j4\pi \frac{R_{0k} + v_{0k} u_m + a_k u_m^2}{\lambda}\right). \qquad (7.51)$$

The SAF of (7.50) is

$$R(f, f_n, u_m) = S_k(f + f_n, u_m) S_k^*(f - f_n, u_m)$$

$$= A_{2k}^2 \exp\left(-j8\pi f_n \frac{R_{0k}}{c}\right) \exp\left(-j8\pi f_n \frac{V_k u_m}{c}\right)$$

$$+ S_k(f + f_n, u_m) N^*(f - f_n, u_m) \qquad (7.52)$$

$$+ S_k^*(f - f_n, u_m) N(f + f_n, u_m)$$

$$+ N(f + f_n, u_m) N^*(f - f_n, u_m).$$

Because of the randomness of noise component (e.g., amplitude and the phase), the theoretical derivation and analysis of the correlation between target's signal and noise (e.g., $S_k(f + f_n, u_m) N^*(f - f_n, u_m)$ and $S_k^*(f - f_n, u_m) N(f + f_n, u_m)$) may be difficult. Hence, it may be hard to obtain the analytical expressions of (7.52). Fortunately, according to the analysis for multi-target signal component in Sections 7.3.1 and 7.3.2, we could achieve some similar conclusions about the noise's effect: due to the noise component's random phase, it may be impossible that the target's signal and noise have the same phase modulation (note that the target's signal phase is a second-order function in terms of target's initial slant range, velocity and acceleration) for normal case. Thus, after the SAF operation, the cross term induced by the target's signal and noise would not affect the coherent integration performance and detection ability of the SAF-SFT-based method.

Detailed simulation experiments are also given to show the effectiveness of the SAF-SFT-based method in noise background (Sections 7.5.1 and 7.5.2).

7.4   DETAILED PROCEDURE AND COMPUTATIONAL COST

Denote the numbers of searching angle, velocity, acceleration, fold factor and range cell by N_θ, N_v, N_a, N_k and N_r, respectively. The main procedures of RFRFT/RLVD include the three-dimensional searching in the range-velocity-acceleration domain and the FRFT/LVD operation [4, 11]. Hence, the computational cost of RFRFT/RLVD is $O\left(N_r N_v N_a N^2 \log_2^N\right)$ [4, 11].

TABLE 7.1   Computational Complexity

Method	Computational complexity
RFT	$O\left(N_r N_v N\right)$
AR-MTD	$O\left(N_r N_\theta N\right)$
RFRFT/RLVD	$O\left(N_r N_v N_a N^2 \log_2^N\right)$
KTMFP	$O\left(N_k N_a N N_r \log_2^N\right)$
KTME	$O\left(N_r N_k N_a N\right)$
GKT-RFT	$O\left(N_r N_a N \log_2^N\right)$
SAF-SFT	$O\left(N_r^2 N\right)$

For the SAF-SFT-based coherent integration method, the main procedures include the GKT, SAF and SFT operations, where the computational burden of GKT is $O\left(N_r N \log_2 N\right)$ and the computational cost of SAF and SFT is $O\left(N_r^2 N\right)$ and $O\left(N_r N \log_2 N_r\right)$, respectively. As a consequence, the computational complexity of SAF-SFT-based algorithm is $O\left(N_r^2 N\right)$.

Besides, the computational burden of GKT-RFT is $O\left(N_r N_a N \log_2 N\right)$ [9], while the computational burden of KTMFP and KTME is $O\left(N_k N_a \left(N N_r \log_2^N + 2N N_r\right)\right)$ [10] and $O\left(N_r N_k N_a N\right)$ [8], respectively. As to the RFT and AR-MTD methods [3, 7], which can only deal with the RCM of target with a constant velocity, the computational complexities are $O\left(N_r N_v N\right)$ and $O\left(N_r N_\theta N\right)$, respectively.

Table 7.1 gives the computational burden of RFRFT, KTMFP, GKT-RFT, KTME, RFT, AR-MTD and the SAF-SFT-based method. Under the assumption that the simulation parameters are the same as those of Fig. 7.2 while the pulse number varies from 32 to 2048, the computational complexity curves of the SAF-SFT-based algorithm, RFRFT, KTMFP, GKT-RFT and KTME (which considers the acceleration motion) are shown in Fig. 7.4, which shows that the computational cost of the SAF-SFT algorithm is much less than that of other four algorithms.

7.5   SIMULATED ANALYSIS

In this subsection, simulation results for maneuvering target are performed to demonstrate the effectiveness of the SAF-SFT-based method, where the radar parameters are the same as those of example 1.

7.5.1   Coherent Integration in Noise Background

In Fig. 7.5, the coherent integration ability of the SAF-SFT algorithm is evaluated. The target's motion parameters are $n_{R0} = 200$, radial velocity $v = 1500$ m/s, acceleration $a = 15$ m/s^2. The SNR of the target's echo after pulse compression is 6 dB, as shown in Fig. 7.5(a). After the first SAF and SFT operations, the energy spectrum in the range-velocity domain is shown in Fig. 7.5(b), where the peak indicates the target's range and velocity. Make use of the velocity estimation, the range migration

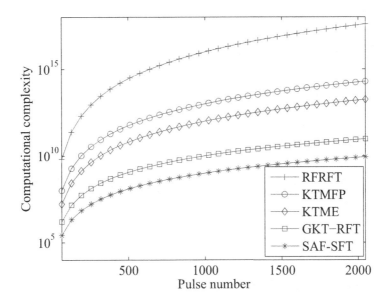

Figure 7.4   Computational complexity curves.

could be removed, as shown in Fig. 7.5(c). With the second SAF and SFT operations, the target energy spectrum in the acceleration domain could be achieved, as shown in Fig. 7.5(d).

7.5.2   Coherent Integration for Multiple Targets

In this section, the result of CI for multiple targets via the SAF-SFT-based method is given, where multiple targets are separated in different experiment scenes. Here, we add six targets (i.e., targets A, B, C, D, E and F) into the scenes, and their motion parameters are given in Table 7.2, where the SNR after PC in each scene is 6 dB. Fig. 7.6 shows the results of each scene. Particularly, Fig. 7.6(a) gives the PC result of the first scene, i.e., targets A and B (which have the same initial range cell and acceleration but with different velocities). Fig. 7.6(b) gives the CI result of targets A and B. Besides, Fig. 7.6(c) gives the PC result of the second scene, i.e., targets C and D (which have the same initial range cell and radial velocity but with different accelerations), where Fig. 7.6(d) gives the CI result of targets C and D. Finally, we consider the scene that targets have the same radial velocity and acceleration but with different range cells (i.e., targets E and F). The PC result of targets E and F is given in Fig. 7.6(e), and the corresponding CI result is given in Fig. 7.6(f). According to the simulation results above, we can see that the SAF-SFT-based method could realize the CI of multiple targets in different scenes.

7.5.3   Target Detection Ability

The detection probabilities of the SAF-SFT-based method and other representative algorithms (i.e., RFRFT, KTMFP, GKT-RFT, KTME, RFT, AR-MTD) under false

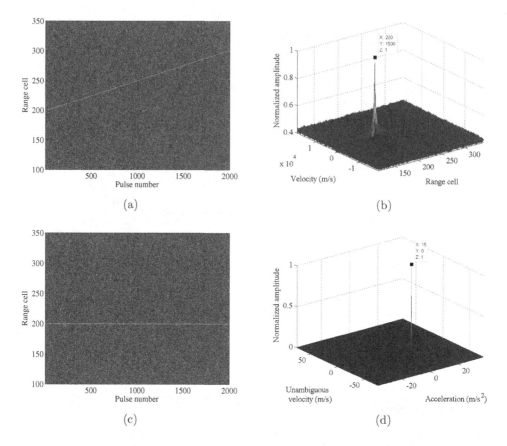

Figure 7.5 Coherent integration in noise background. (a) Pulse compression result. (b) The energy spectrum in range-velocity domain. (c) Range migration correction. (d) The energy spectrum in acceleration domain.

TABLE 7.2 Motion Parameters of Multiple Targets

Targets	Initial range cell	Radial velocity (m/s)	Radial acceleration (m/s^2)
A	320	900	15
B	320	1500	15
C	200	1500	16
D	200	1500	−16
E	320	1500	14
F	200	1500	14

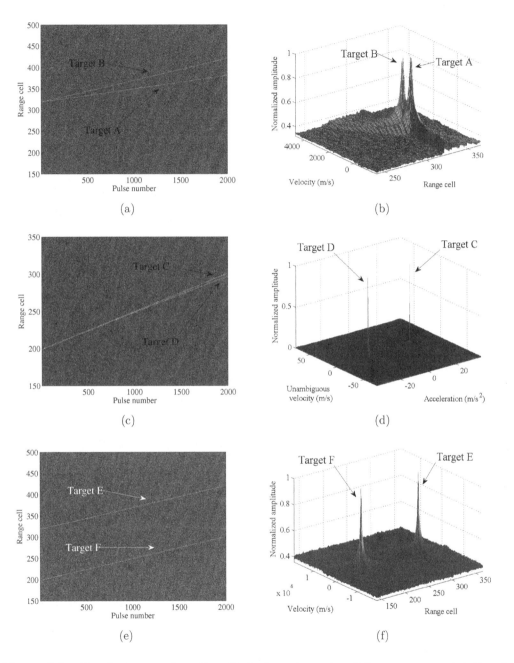

Figure 7.6 Simulation results of multiple targets under different scenes. (a) Result after PC with targets A and B. (b) CI result of targets A and B. (c) Result after PC with targets C and D. (d) CI result of targets C and D. (e) Result after PC with targets E and F. (f) CI result of targets E and F.

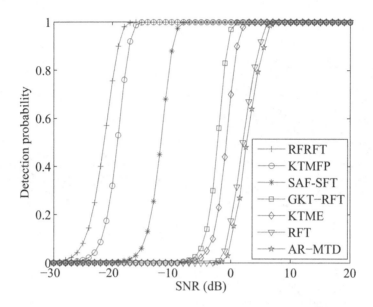

Figure 7.7 Detection performance.

alarm probability $P_{fa} = 10^{-6}$ are shown in Fig. 7.7. From this figure, we can conclude that (1) for the same detection probability Pd = 0.8, the required SNR of the SAF-SFT-based algorithm is 15 dB less than that of RFT, which indicates the integration gain for range migration correction and DS compensation of acceleration motion. (2) It could be seen from Fig. 7.7 that the required SNR of the SAF-SFT-based method is 11.5 dB less than that of GKT-RFT when detection probability Pd = 0.8, which represents the integration gain for DS compensation. (3) For the same detection probability Pd = 0.8, the required SNR of the SAF-SFT-based method is 7.4 dB larger than that of KTMFP, which is due to the correction operation of (7.16). In addition, it should be pointed out that the interpolation operation of KTMFP will lead to integration and detection performance loss, in comparison with the RFRFT method.(4) As to the same detection probability Pd = 0.8, the required SNR of SAF-SFT-based method is 8.9 dB larger than that of RFRFT, which is because the SAF-operation of (7.16) will bring about SNR loss. Nevertheless, the SAF-SFT-based approach removes the parameters-searching procedure, which reduces the computational burden significantly. Furthermore, as to the detection performance, the SAF-SFT-based method algorithm is superior to GKT-RFT, KTME, RFT and AR-MTD in low SNR background.

7.5.4 Parameter Estimation Performance

The motion parameter estimation performance of the SAF-SFT-based method, RFRFT, KT-MFP, GKT-RFT and KTME is compared in Fig. 7.8, where the time tag of the estimated parameters is $t_m = 0$. The input SNR after PC varies from -30 to 10 dB, and 1000 trials are performed for each SNR. The root-mean-square errors

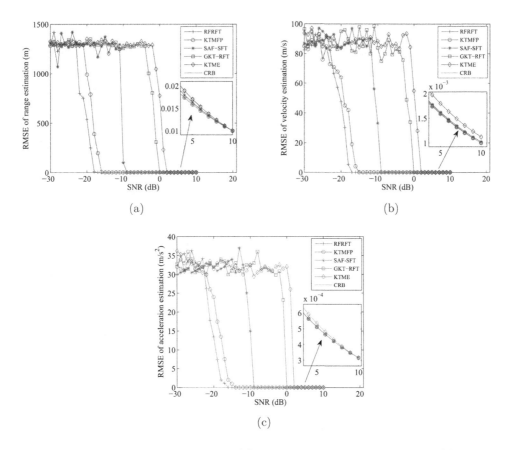

Figure 7.8 Parameters estimation. (a) RMSE of range estimation. (b) RMSE of velocity estimation. (c) RMSE of acceleration estimation.

(RMSEs) of target's motion parameters (i.e., range, velocity and acceleration) are given in Fig. 7.8(a), Fig. 7.8(b) and Fig. 7.8(c), respectively. The Cramer-Rao bound (CRB) is also given in the figure. We can see that RFRFT has the best estimation performance, where the RMSEs of the SAF-SFT-based method are gradually close to those of RFRFT and KTMFP when the input SNR is larger than -7 dB. It should be pointed out that the SAF-SFT-based method suffers 8.9 and 7.4 dB SNR loss, respectively, in comparison with RFRFT and KTMFP, which is because of the SAF operation. (Note that the interpolation operation of KTMFP will bring about integration and estimation performance loss, in comparison with RFRFT.) However, the estimation ability of SAF-SFT is better than that of GKT-RFT and KTME. Furthermore, the computational cost of the SAF-SFT-based algorithm is much less than that of RFRFT, KTMFP, GKT-RFT and KTME. Hence, the SAF-SFT-based approach is computationally efficient and strikes a balance among computation burden, target detection ability and parameter estimation performance.

7.6 SUMMARY

This chapter discussed a computationally efficient coherent integration method (i.e., SAF-SFT) for radar high-speed target detection and parameter estimation. The characteristics of the SAF-SFT method include the following: 1) after the generalized keystone transform, the first SAF and SFT operations are applied to achieve the range and velocity estimations. 2) With the estimations, the remaining range migration induced by target's velocity could be removed, and then the target signal could be extracted along the range cell. 3) Then, the second SAF and SFT operations are performed on the extracted target signal, where the target energy could be coherent accumulated and the acceleration estimation can be obtained. Analyses of the cross term, computational cost, detection probability and estimation RMSE demonstrate the high suitability of the SAF-SFT algorithm. Detailed simulation experiments verify the effectiveness of the SAF-SFT method.

Bibliography

[1] P. H. Huang, G. S. Liao, Z. W. Yang, X. Xia, J. T. Ma, and J. T. Ma, "Long-time coherent integration for weak maneuvering target detection and high-order motion parameter estimation based on keystone transform," *IEEE Transactions on Signal Processing*, vol. 64, no. 15, pp. 4013–4026, August, 2016.

[2] X. L. Chen, J. Guan, Y. Huang, N. B. Liu, and Y. He, "Radon-linear canonical ambiguity function-based detection and estimation method for marine target with micromotion," *IEEE Transactions on Geoscience and Remote Sensing*, vol. 53, no. 4, pp. 2225–2240, April 2015.

[3] X. Rao, H. H. Tao, J. Su, X. L. Guo, and J. Z. Zhang, "Axis rotation MTD algorithm for weak target detection," *Digital Signal Processing*, vol. 26, pp. 81–86, March 2014.

[4] X. L. Chen, J. Guan, N. B. Liu, and Y. He, "Maneuvering target detection via Radon-Fractional Fourier transform-based long-time coherent integration," *IEEE Transactions on Signal Processing*, vol. 62, no. 4, pp. 939–953, February 2014.

[5] J. B. Zheng, T. Su, H. W. Liu, G. S. Liao, Z. Liu, and Q. H. Liu,"Radar high-speed target detection based on the frequency-domain deramp-keystone transform," *IEEE Journal of Selected Topics in Applied Earth Observations and Remote Sensing*, vol. 9, no. 1, pp. 285–294, January 2016.

[6] D. Y. Zhu, Y. Li, and Z. D. Zhu, "A keystone transform without interpolation for SAR ground moving-target imaging," *IEEE Geoscience and Remote Sensing Letters*, vol. 4, no. 1, pp. 18–22, January 2007.

[7] L. C. Qian, J. Xu, X. G. Xia, W. F. Sun, T. Long, and Y. N. Peng, "Wideband-scaled Radon-Fourier transform for high-speed radar target detection," *IET Radar Sonar and Navigation*, vol. 8, no. 5, pp. 501–512, June 2014.

[8] M. D. Xing, J. H. Su, G. Y. Wang, and Z. Bao, "New parameter estimation and detection algorithm for high speed small target," *IEEE Transactions on Aerospace and Electronic Systems*, vol. 47, no. 1, pp. 214–224, January 2011.

[9] J. Tian, W. Cui, and S. Wu, "A novel method for parameter estimation of space moving targets," *IEEE Geoscience and Remote Sensing Letters*, vol. 11, no. 2, pp. 389–393, February 2014.

[10] Z. Sun, X. L. Li, W. Yi, G. L. Gui, and L. J. Kong, "Detection of weak maneuvering target based on keystone transform and matched filtering process," *Signal Processing*, vol. 140, pp. 127–138, November 2017.

[11] X. L. Li, G. L. Cui, W. Yi, and L. J. Kong, "Coherent integration for maneuvering target detection based on Radon-Lv's Distribution," *IEEE Signal Processing Letters*, vol. 22, no. 9, pp. 1467–1471, September 2015.

CHAPTER 8

Coherent Integration with Unknown Time Parameter

The previous chapters assume that the target's entry time and departure time are already known. However, a high-speed moving target may enter the radar coverage area unannounced and leave after an unspecified period, which implies that the target's entry time and departure time are unknown. In the absence of these time information, target detection and parameter estimation (DAPE) will be severely impacted.

In this chapter, we consider the coherent integration and detection problem for radar high-speed targets with unknown entry time and departure time (i.e., the time when the targets appears in/leaves the radar detection field is unknown). The window Radon fractional Fourier transform (WRFRFT)-based coherent integration method is discussed for the detection and estimation of the target's time parameters (i.e., entry time and departure time) and motion parameters (i.e., range, velocity and acceleration). Meanwhile, a computationally efficient coherent integration method based on extended generalized Radon Fourier transform (EGRFT) and window fractional fourier transform (WFRFT), i.e., EGRFT-WFRFT, is also discussed.

8.1 MATHEMATICAL MODEL OF RECEIVED SIGNAL

Assume that the linear frequency modulation is adopted as the radar's transmitted waveform, i.e., [1–3]

$$s_{trans}(t) = \text{rect}\left(\frac{t}{T_p}\right) \exp\left(j\pi\mu t^2\right) \exp(j2\pi f_c t), \tag{8.1}$$

where $\text{rect}(x) = \begin{cases} 1, & |x| \leq 0.5 \\ 0, & |x| > 0.5 \end{cases}$, t, μ, f_c and T_p denote, respectively, the fast time variable, chirp rate, carrier frequency and pulse duration.

Suppose that the total observation time of radar is from T_0 to T_1, where a moving target enters the radar detection area at time T_b and leaves the radar detection area at time T_e ($T_0 < T_b < T_e < T_1$). The instantaneous slant distance between the target

DOI: 10.1201/9781003529101-8

and the radar at time T_b is denoted as R_0, and the radial distance of target could be expressed as:

$$R(t_m) = R_0 + V(t_m - T_b) + A(t_m - T_b)^2, t_m \in [T_b, T_e], \tag{8.2}$$

where t_m denotes the slow time, while V and A are, respectively, the target's radial velocity and acceleration, T_b represents the beginning time (entry) of the target and T_e represents the ending time (departure) of the target, which are both unknown.

With the pulse compression (PC), the received signal within the observation time can be expressed as [4]

$$
\begin{aligned}
s(t, t_m) = w(t_m) A_0 \text{sinc} \left[B \left(t - \frac{2R(t_m)}{c} \right) \right] \\
\times \exp \left[-j4\pi \frac{R(t_m)}{\lambda} \right] + n_s(t, t_m),
\end{aligned}
\tag{8.3}
$$

where

$$w(t_m) = \text{rect} \left[\frac{t_m - 0.5(T_b + T_e)}{T_e - T_b} \right] = \begin{cases} 1, & T_b \leq t_m \leq T_e \\ 0, & \text{else} \end{cases}, \tag{8.4}$$

$n_s(t, t_m)$ represents noise, A_0 and λ are, respectively, the signal amplitude and the wavelength, while B and c represent the bandwidth and light speed, respectively.

From (8.3), it could be noticed that the radar echo contains target signal within the period $[T_b, T_e]$, while the radar echo only contains noise for the other time periods, i.e., the beginning time of target signal is T_b and the ending time of target signal is T_e. The sketch map of the radar echo in the t-t_m plane is given in Fig. 8.1.

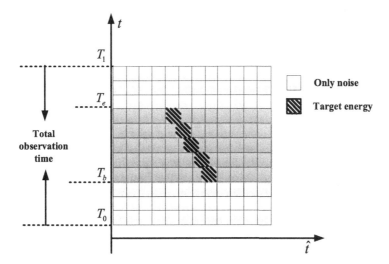

Figure 8.1   Sketch map of the radar's echo in the t-t_m plane.

8.2 WRFRFT-BASED METHOD

The definition of WRFRFT is given as

$$
\begin{aligned}
WR_{g(t_m)}(\alpha, u) &= F_\alpha[g(t_m)s\,(2r(t_m)/c, t_m)](u) \\
&= \int_{-\infty}^{\infty} g(t_m)s\,(2r(t_m)/c, t_m)\,K_\alpha(t_m, u)dt_m,
\end{aligned}
\tag{8.5}
$$

where $g(t_m)$ and $r(t_m)$ are, respectively, the window function and searching motion trajectory:

$$
r(t_m) = r_0 + v_0(t_m - \eta_0) + a_0(t_m - \eta_0)^2,
\tag{8.6}
$$

$$
g(t_m) = \text{rect}\left[\frac{t_m - 0.5(\eta_1 + \eta_0)}{\eta_1 - \eta_0}\right] = \begin{cases} 1, & \eta_0 \leq t_m \leq \eta_1 \\ 0, & \text{else} \end{cases}
\tag{8.7}
$$

$$
T_0 \leq \eta_0 \leq \eta_1 \leq T_1,
$$

η_0 is the beginning time of $g(t_m)$, η_1 is ending time of $g(t_m)$ and (r_0, v_0, a_0) denotes the searching motion parameters pair (i.e., searching initial range, searching radial velocity and radial acceleration). $\alpha = P\pi/2$ is the rotation angle, P is searching transform order, F_α represents the FRFT operator and the transform kernel $K_a(t_m, u)$ is given by

$$
K_a(t_m, u) =
\begin{cases}
A_\alpha \exp[j(0.5t^2 \cot \alpha - ut \csc \alpha + 0.5u^2 \cot \alpha)] & \alpha \neq n\pi \\
\delta[u - (-1)^n t_m] & \alpha = n\pi
\end{cases}
\tag{8.8}
$$

where $A_\alpha = \sqrt{(1 - j \cot \alpha)/2\pi}$.

Inserting (8.7) into (8.5) yields

$$
WR_{g(t_m)}(\alpha, u) = \int_{\eta_0}^{\eta_1} g(t_m)s\,(2r(t_m)/c, t_m)\,K_a(t_m, u)dt_m.
\tag{8.9}
$$

From the definition of WRFRFT, it could be interpreted as the transform of the target's intercepted and extracted signal in the FRFT domain. In particular, the WRFRFT includes three main steps: the first is an interception of the two-dimensional compressed signal based on function $g(t_m)$ (i.e., $g(t_m)$ determines the interception-operation's beginning/ending time). The second is the signal extraction process applied to the intercepted signal based on the resultant searching motion parameters pair (r_0, v_0, a_0). The third is the FRFT-based integration process.

For comparison, the definition of RFRFT [5] is also given:

$$
\begin{aligned}
RFRFT(\alpha, u) &= F_\alpha[s\,(2r(t_m)/c, t_m)](u) \\
&= \int_{-\infty}^{\infty} s\,(2r(t_m)/c, t_m)\,K_a(t_m, u)dt_m \\
&= \int_{T_0}^{T_1} s\,(2r(t_m)/c, t_m)\,K_a(t_m, u)dt_m.
\end{aligned}
\tag{8.10}
$$

From (8.9) and (8.10), we can notice that both WRFRFT and RFRFT extract the signal along target's motion trajectory and integrate with FRFT. The main difference is that RFRFT performs the extraction and integration operations within the total observation time, while the beginning/ending time of WRFRFT is adjustable (determined by η_0 and η_1); thus, it is able to better match with the moving target with unknown entry/departure time. In particular, the RFRFT can be considered a special case of WRFRFT (i.e., when $\eta_0 = T_0, \eta_1 = T_1$).

8.2.1 Some Properties of WRFRFT

1) Rotational Additivity: Note that the kernel of WRFRFT has the rotational additivity property, i.e.,

$$\int_{-\infty}^{\infty} K_\alpha(t_m, u) K_\beta(u, z) du = K_{\alpha+\beta}(t_m, z). \tag{8.11}$$

Thus, it could be easy for us to obtain the rotational additivity of WRFRFT, i.e.,

$$
\begin{aligned}
WR_{g(t_m)}&(\alpha + \beta, z) \\
&= F_\beta[WR_{g(t_m)}(\alpha, u)](z) \\
&= \int_{-\infty}^{\infty} K_\beta(u, z) \int_{-\infty}^{\infty} g(t_m) s\left(2r(t_m)/c, t_m\right) K_\alpha(t_m, u) dt du \\
&= \int_{-\infty}^{\infty} g(t_m) s\left(2r(t_m)/c, t_m\right) \int_{-\infty}^{\infty} K_\alpha(t_m, u) K_\beta(u, z) du dt \\
&= \int_{-\infty}^{\infty} g(t_m) s\left(2r(t_m)/c, t_m\right) K_{\alpha+\beta}(t_m, z) dt_m \\
&= F_{\alpha+\beta}[g(t_m) s\left(2r(t_m)/c, t_m\right)](z).
\end{aligned}
\tag{8.12}
$$

The rotational additivity of WRFRFT provides us the solution of the transform between WRFRFTs with different transform angles. That is to say, we only need to calculate WRFRFT with the total transform order for one time, which is a major advantage in terms of computational efficiency.

2) Inverse WRFRFT (IWRFRFT): According to the rotational additivity property above, it could also be concluded that the WRFRFT of angle $-\alpha$ is the inverse of the WRFRFT with angle α, since that $F_{-\alpha}(F_\alpha) = F_{\alpha-\alpha} = F_0 = I$. The IWFRFT is

$$g(t_m) s\left(2r(t_m)/c, t_m\right) = \int_{-\infty}^{\infty} WR_{g(t_m)}(\alpha, u) K_{-\alpha}(t_m, u) du. \tag{8.13}$$

3) Linear Additivity: It's easy to find that WRFRFT is linear. In particular, let ε_1 and ε_2 denote two constant coefficients, and we have

$$F_\alpha[\varepsilon_1 x_1 + \varepsilon_2 x_2](u) = \varepsilon_1 F_\alpha[x_1](u) + \varepsilon_2 F_\alpha[x_2](u). \tag{8.14}$$

This property shows that the WRFRFT satisfies the superposition principle, which is helpful in the analysis of multi-component signals.

4) Index Commutativity: Apply (8.5) for two orders, we have

$$
\begin{aligned}
&F_\beta[F_\alpha[g(t_m)s\left(2r(t_m)/c,t_m\right)](z)] \\
&= \int_{-\infty}^{\infty} K_\beta(u,z)\int_{-\infty}^{\infty} g(t_m)s\left(2r(t_m)/c,t_m\right)K_\alpha(t_m,u)dtdu \\
&= \int_{-\infty}^{\infty} K_\alpha(t_m,u)\left[\int_{-\infty}^{\infty} g(t_m)s\left(2r(t_m)/c,t_m\right)K_\beta(u,z)du\right]dt_m \\
&= F_\alpha[F_\beta[g(t_m)s\left(2r(t_m)/c,t_m\right)](z)].
\end{aligned}
\tag{8.15}
$$

Hence, the WRFRFT adheres to the index commutativity property.

5) Parseval Relation: The WRFRFT also holds the classical Parseval relation:

$$
\begin{aligned}
&\int_{-\infty}^{\infty} g(t_m)x\left(2r_x(t_m)/c,t_m\right)y\left(2r_y(t_m)/c,t_m\right)dt_m \\
&= \int_{-\infty}^{\infty} WR_x\left(\alpha,u\right)WR_y^*\left(\alpha,u\right)du,
\end{aligned}
\tag{8.16}
$$

where $WR_x\left(\alpha,u\right)=F_\alpha[g(t_m)x(t_m,2r_x(t_m)/c)](u)$, $WR_y\left(\alpha,u\right)=F_\alpha[g(t_m)y(t_m,2r_y(t_m)/c)](u)$. In particular, (8.16) will turn into the energy conservation property when $x=y$, i.e.,

$$
\int_{-\infty}^{\infty} g(t_m)|x\left(2r_x(t_m)/c,t_m\right)|^2dt=\int_{-\infty}^{\infty} |WR_x\left(\alpha,u\right)|^2du.
\tag{8.17}
$$

The squared magnitude of the WFRFT ($|WR_x\left(\alpha,u\right)|^2$) thus represents the signal energy spectrum with angle α and window function $g(t_m)$.

8.2.2 WRFRFT for Moving Target Detection and Estimation

Substituting (8.3) and (8.4) into (8.5) yields

$$
\begin{aligned}
WR_{g(t_m)}(\alpha,u) &= F_\alpha[g(t_m)s\left(2r(t_m)/c,t_m\right)](u) \\
&= \int_{-\infty}^{\infty} g(t_m)w(t_m)A_0\mathrm{sinc}\left[B\left(\frac{2r(t_m)}{c}-\frac{2R(t_m)}{c}\right)\right] \\
&\quad \times \exp\left[-j4\pi\frac{R(t_m)}{\lambda}\right]K_\alpha(t_m,u)dt_m \\
&\quad + \int_{-\infty}^{\infty} g(t_m)n_s(2r(t_m)/c,t_m)K_\alpha(t_m,u)dt_m,
\end{aligned}
\tag{8.18}
$$

where the first and second integral terms of (8.18) represent, respectively, the WRFRFT of target signal and the WRFRFT of noise.

Let C and D are respectively:

$$
C=\{t_m|g(t_m)=1\}, D=\{t_m|w(t_m)=1\}.
\tag{8.19}
$$

In other words, C denotes the function $g(t_m)$'s non-zero area, while D denotes the function $w(t_m)$'s non-zero area.

Case 1: When $C \cap D = \varnothing$, we have

$$g(t_m)w(t_m) = 0. \tag{8.20}$$

In this case, (8.18) could be expressed as

$$
\begin{aligned}
WR_{g(t_m)}(\alpha, u) &= 0 + \int_{-\infty}^{\infty} g(t_m)n_s(2r(t_m)/c, t_m)K_\alpha(t_m, u)dt_m \\
&= \int_{\eta_0}^{\eta_1} n_s(2r(t_m)/c, t_m)K_\alpha(t_m, u)dt_m.
\end{aligned} \tag{8.21}
$$

Hence, in this case, only noise is extracted and accumulated, but none of the target's signal is extracted and accumulated.

Case 2: When $C \cap D \neq \varnothing$, we have

$$
g(t_m)w(t_m) = \begin{cases} 1, & T' \leq t_m \leq T'' \\ 0, & \text{else} \end{cases}, \tag{8.22}
$$

where

$$T' = \max[T_b, \eta_0], T'' = \min[T_e, \eta_1]. \tag{8.23}$$

In this case, the WRFRFT of (8.18) can be recast as

$$
\begin{aligned}
WR_{g(t_m)}(\alpha, u) &= \int_{T'}^{T''} A_0 \mathrm{sinc}\left[B\left(\frac{2r(t_m)}{c} - \frac{2R(t_m)}{c}\right)\right] \\
&\quad \times \exp\left[-j4\pi \frac{R(t_m)}{\lambda}\right] K_\alpha(t_m, u)dt_m \\
&\quad + \int_{\eta_0}^{\eta_1} n_s(2r(t_m)/c, t_m)K_\alpha(t_m, u)dt_m.
\end{aligned} \tag{8.24}
$$

From (8.23) and (8.24), we notice that the values of η_0 and η_1 determine the extracted range of the target signal and noise. In order to ensure SNR improvement, on the one hand, we need to extract and integrate all the target signals, and at the same time, we need to extract as little noise as possible.

In particular, to guarantee that all the target signals are extracted and accumulated, the following equation should be satisfied:

$$T' \leq T_b, T'' \geq T_e. \tag{8.25}$$

Combining with (8.23), we have

$$\eta_0 \leq T_b, \eta_1 \geq T_e. \tag{8.26}$$

In addition to satisfying the inequality (8.26), we also need to ensure that as little noise as possible is extracted. Thus, we have

$$\eta_0 = T_b, \eta_1 = T_e. \tag{8.27}$$

Then, (8.24) could be rewritten as:

$$
\begin{aligned}
WR_{g(t_m)}(\alpha, u) &= \int_{T_b}^{T_e} A_0 \mathrm{sinc}\left[B\left(\frac{2r(t_m)}{c} - \frac{2R(t_m)}{c}\right)\right] \\
&\quad \times \exp\left[-j4\pi\frac{R(t_m)}{\lambda}\right] K_\alpha(t_m, u)dt_m \\
&\quad + \int_{T_b}^{T_e} n_s(2r(t_m)/c, t_m)K_\alpha(t_m, u)dt_m \\
&= \int_{T_b}^{T_e} A_0 \exp\left[-j4\pi\frac{R(t_m)}{\lambda}\right] K_\alpha(t_m, u)dt_m \\
&\quad + \int_{T_b}^{T_e} n_s(2r(t_m)/c, t_m)K_\alpha(t_m, u)dt_m
\end{aligned}
\tag{8.28}
$$

$$\text{when } r_0 = R_0, v_0 = V, a_0 = A.$$

Equation (8.28) shows that the entire target signal is extracted and could be coherently integrated in the FRFT domain when the motion parameters of the search match with the motion parameters of the target.

Based on the analysis above, it could be noticed that only when $\eta_0 = T_b, \eta_1 = T_e$, the target signal is totally extracted and accumulated in the FRFT domain, and it could ensure that as little noise as possible is extracted simultaneously, resulting in a peak value in the WRFRFT output (where the peak location corresponds to the time parameters and motion parameters of the target).

With different searching parameters (i.e., beginning/ending time, range, velocity and acceleration), different integration outputs of WRFRFT would be obtained, and the target signal will be focused as a peak when the searching time/motion parameters match with the target's time/motion parameters. Thus, the target's time parameters and motion parameters could be estimated by

$$
(\hat{T}_b, \hat{T}_e, \hat{R}_0, \hat{V}, \hat{A}) = \underset{(\eta_0, \eta_1, r_0, v_0, a_0)}{\arg\max} |WR_{g(t_m)}(\alpha, u)|.
\tag{8.29}
$$

A simulation experiment (Table 8.1 shows the radar parameters, and Table 8.2 gives the target parameters) is given to show how the WRFRFT performs with varying η_0 and η_1. The result of the radar echo after pulse compression is shown in Fig. 8.2(a). The WRFRFT result when $\eta_0 = 0.755$ s and $\eta_1 = 3$ s (i.e., the WRFRFT's window function matches the beginning/ending time of the target signal) is given in Fig. 8.2(b) (slice of velocity and acceleration). It is observed that the target signal is coherently accumulated as a peak, which is corresponding to the target's radial velocity and acceleration. Fig. 8.2(c) and Fig. 8.2(d) give, respectively, the WRFRFT results for the cases that $\eta_0 = 0.15$ s, $\eta_1 = 0.5$ s and $\eta_0 = 3.05$ s, $\eta_1 = 3.5$ s. In these two cases (Fig. 8.2(c) and Fig. 8.2(d)), only noise is extracted and accumulated, resulting in unfocused results for WRFRFT. Fig. 8.2(e) gives the WRFRFT output when $\eta_0 = 0.505$ s and $\eta_1 = 2.9$ s, where only part of the target signal is extracted in this case, and thus the peak value of Fig. 8.2(e) is smaller than that of Fig. 8.2(b). Fig. 8.2(f) shows the WRFRFT output when $\eta_0 = 0.755$ s, $\eta_1 = 3.4$ s, where the entire target signal is extracted and coherently accumulated in this case. However, it

TABLE 8.1   Radar Parameters

Parameters	Value
Carrier frequency	6 GHz
Bandwidth	10 MHz
Sample frequency	50 MHz
Pulse repetition frequency	200 Hz
Pulse duration	10 μs
Beginning time of target	0.755 s
SNR after PC	4 dB

is worth pointing out that compared with Fig. 8.2(b), much more noise is extracted and accumulated in Fig. 8.2(f). As a result, the peak value of Fig. 8.2(f) is smaller than the peak value of Fig. 8.2(b).

In Fig. 8.2(g), we examine the WRFRFT outputs for a fixed η_1 ($\eta_1 = 3$) but varying η_0. In other words, with the ending time of WRFRFT' window function $g(t_m)$ matches the target signal's ending time, the peak value reaches its maximum value when $\eta_0 = 0.755$ s (i.e., the beginning time of $g(t_m)$ equals to the target signal's beginning time). Similarly, Fig. 8.2(h) shows the integrated peak value of WRFRFT for a fixed η_0 ($\eta_0 = 0.755$ s) but varying η_1. That is to say, the beginning time of $g(t_m)$ matches the target signal's beginning time. We could see that the peak value reaches its maximum value when $\eta_1 = 3$ s (i.e., the ending time of $g(t_m)$ equals to the target signal's ending time). Table 8.3 shows the amplitude values of the peak in Fig. 8.2(b)−Fig. 8.2(f).

Based on the results of Fig. 8.2, it can be assured that only when the beginning/ending time of WRFRFT's window function equals to the target signal's beginning/ending time, the integrated output of WRFRFT will reach its maximum value.

8.2.3   Procedure of the WRFRFT-Based Method

The main steps of the WRFRFT-based approach can be summarized as follows:

Step 1: Based on the relative prior information of targets to be expected (such as moving status and varieties) and the radar parameters, the searching scope of initial range, velocity, acceleration and beginning/ending time can be obtained (denoted respectively as $[r_{\min}, r_{\max}]$, $[v_{\min}, v_{\max}]$, $[a_{\min}, a_{\max}]$, $[\eta_{0\min}, \eta_{0\max}]$ and $[\eta_{1\min}, \eta_{1\max}]$). In addition, the searching interval of the beginning/ending time, initial range, velocity

TABLE 8.2   Moving Target's Time Parameters and Motion Parameters

Parameters	Value
Initial range cell	287
Radial velocity (m/s)	90
Radial acceleration (m/s^2)	26
Beginning time (s)	0.755
Ending time (s)	3

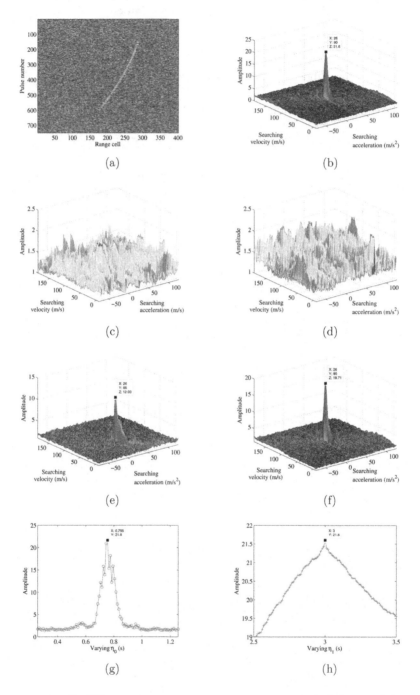

Figure 8.2 WRFRFT response with respect to η_0 and η_1. (a) PC. (b) $\eta_0 = 0.755$ s, $\eta_1 = 3$ s. (c) $\eta_0 = 0.15$ s, $\eta_1 = 0.5$ s. (d) $\eta_0 = 3.05$ s, $\eta_1 = 3.5$ s. (e) $\eta_0 = 0.505$ s, $\eta_1 = 2.9$ s. (f) $\eta_0 = 0.755$ s, $\eta_1 = 3.4$ s. (g) Integrated peak value curve for fixed η_1 but varying η_0. (h) Integrated peak value curve for fixed η_0 but varying η_1.

TABLE 8.3 Peak Values for Fig. 8.2(b)–Fig. 8.2(f)

Figures	Fig. 8.2(b)	Fig. 8.2(c)	Fig. 8.2(d)	Fig. 8.2(e)	Fig. 8.2(f)
Peak value	21.6	2.042	2.207	12.03	19.71

and acceleration can be set as [5]:

$$\rho_\eta = PRT, \tag{8.30}$$

$$\rho_r = c/2B, \tag{8.31}$$

$$\rho_v = \lambda/2(T_1 - T_0), \tag{8.32}$$

$$\rho_a = \lambda/2(T_1 - T_0)^2, \tag{8.33}$$

where PRT is radar pulse repetition time.

Step 2: With the searching parameters $(r_s, v_s, a_s, \eta_{0s}, \eta_{1s})$, the moving trajectory to be searched and the window function $g(t_m)$ could be respectively expressed as:

$$r_s(t_m) = r_{0s} + v_s(t_m - \eta_{0s}) + a_s(t_m - \eta_{0s})^2, t_m \in [\eta_{0s}, \eta_{1s}], \tag{8.34}$$

$$g(t_m) = \text{rect}\left[\frac{t_m - 0.5(\eta_{1s} + \eta_{0s})}{\eta_{1s} - \eta_{0s}}\right], \tag{8.35}$$

where $r_{0s} = r_{\min} : \rho_r : r_{\max}$, $v_s = v_{\min} : \rho_v : v_{\max}$, $a_s = a_{\min} : \rho_a : a_{\max}$, $\eta_{0s} = \eta_{0\min} : \rho_\eta : \eta_{0\max}$ and $\eta_{1s} = \eta_{1\min} : \rho_\eta : \eta_{1\max}$.

Step 3: Intercept and extract the target signal from the compressed signal based on the window function and searching motion trajectory, i.e.,

$$s_e(t_m) = g(t_m)s\left(2r_s(t_m)/c, t_m\right). \tag{8.36}$$

Step 4: Apply the WRFRFT operation on the extracted signal.

Step 5: Go through all the searching parameters and obtain the corresponding WRFRFT output $WR_{g(t_m)}(\alpha, u)$.

Step 6: Take the amplitude of WRFRFT output in step 5 as test statistic and compare with the adaptive threshold for a given false alarm probability

$$\left|WR_{g(t_m)}(\alpha, u)\right| \underset{H_0}{\overset{H_1}{\gtrless}} \gamma, \tag{8.37}$$

where γ dentoes the detection threshold [5], which could be obtained via the reference unit after WRFRFT. If $|WR_{g(t_m)}(\alpha, u)|$ is larger than γ, target is confirmed. In addition, the target's time parameters and motion parameters can be estimated via the peak location of $WR_{g(t_m)}(\alpha, u)$.

Fig. 8.3 gives the flow chart of the WRFRFT-based method.

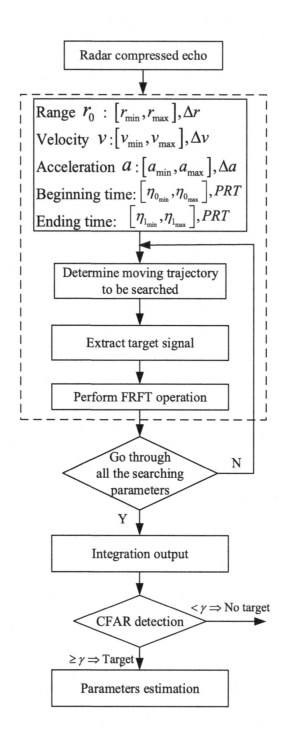

Figure 8.3   Flowchart of the WRFRFT method.

8.2.4 Some Important Discussions

8.2.4.1 *Discussion of the Beginning/Ending Time's Influence*

During the coherent integration process of WRFRFT, two time parameters (i.e., beginning time η_0 and ending time η_1) have important influence on the integration result of WRFRFT. In particular, the beginning time determines that whether the target's across range cell (ARC) could be removed or not and the ending time determines the length of the extracted signal. Relatively speaking, the beginning time η_0 is more important than the ending time η_1 for the coherent integration of WRFRFT. The reason is analyzed in the following.

Note that the coherent integration of the target's extracted signal using WRFRFT (as shown in (8.18)) is obtained based on the fact that ARC elimination is accomplished when the searching motion parameters equal to the target's motion parameters [4, 5], i.e.,

$$r(t_m) - R(t_m) \equiv 0, \text{ when } r_0 = R_0, v_0 = V, a_0 = A. \tag{8.38}$$

That is to say, $r(t_m) - R(t_m)$ is equal to zero at every instantaneous time t_m.

Combing (8.2) and (8.6), we have

$$\begin{aligned} r(t_m) - R(t_m) =& (r_0 - R_0) + (v - V)t_m + (a_0 - A)t_m^2 \\ &+ 2(AT_b - a_0\eta_0)t_m + (a_0\eta_0^2 - AT_b^2) \\ &+ (VT_b - v\eta_0). \end{aligned} \tag{8.39}$$

When $r_0 = R_0$, $v_0 = V$ and $a_0 = A$, the expression of $r(t_m) - R(t_m)$ could be further written as

$$\begin{aligned} r(t_m) - R(t_m) &= 2A(T_b - \eta_0)t_m + A(\eta_0^2 - T_b^2) + V(T_b - \eta_0) \\ &= A(\eta_0 - T_b)(A\eta_0 + AT_b - 2At - V). \end{aligned} \tag{8.40}$$

From (8.40), we could notice that in order to make $r(t_m) - R(t_m)$ equal to zero for every instantaneous time t_m, the following condition should be satisfied:

$$\eta_0 - T_b = 0. \tag{8.41}$$

Thus, if the value of η_0 matches with the target signal's beginning time, then the ARC of the extracted target signal could be removed and the coherent integration could be obtained via WRFRFT. However, if the value of η_0 does not match with the target signal' beginning time, the ARC could not be totally removed and the coherent integration of the extracted target signal could not be obtained. That is to say, the value of the beginning time η_0 directly determines whether the ARC elimination and coherent integration of the extracted target signal could be achieved or not.

As to the ending time (η_1), it has no direct relation with the ARC elimination and the realization of coherent integration for the extracted target signal. Actually, the ending time (η_1) generally determines the length of the extracted signal. Hence, the beginning time (η_0) is more important (relatively speaking) than the ending time (η_1) for the coherent integration of WRFRFT.

8.2.4.2 *Discussion of Multiple Targets Scene*

Consider K targets in the scene, where the entry time of the ith ($i = 1, 2 \cdots, K$) target is $T_{b,i}$ and the corresponding departure time is $T_{e,i}$, while the instantaneous slant range of the ith target could be expressed as

$$R_i(t_m) = R_{0,i} + v_{0,i}(t_m - T_{b,i}) + a_{0,i}(t_m - T_{b,i})^2, \tag{8.42}$$

where $R_{0,i}$, $v_{0,i}$ and $a_{0,i}$ are, respectively, the ith target's initial range, radial velocity and acceleration.

After PC, the radar's echo of the K targets could be recast as

$$
\begin{aligned}
s(t, t_m) &= \sum_{i=1}^{K} w_i(t_m)\sigma_{0,i}\text{sinc}\left[B\left(t - \frac{2R_i(t_m)}{c}\right)\right] \\
&\quad \times \exp\left[-j4\pi\frac{R_i(t_m)}{\lambda}\right] + n_s(t, t_m) \\
&= \sum_{i=1}^{K} s_i(t, t_m) + n_s(t, t_m),
\end{aligned}
\tag{8.43}
$$

where

$$w_i(t_m) = \text{rect}\left[\frac{t_m - 0.5(T_{b,i} + T_{e,i})}{T_{e,i} - T_{b,i}}\right] = \begin{cases} 1, & T_{b,i} \le t_m \le T_{e,i} \\ 0, & \text{else} \end{cases} \tag{8.44}$$

$\sigma_{0,i}$ represents the signal amplitude of the ith target and $s_i(t, t_m)$ denotes the signal component of the ith target, i.e.,

$$
\begin{aligned}
s_i(t, t_m) &= w_i(t_m)\sigma_{0,i}\text{sinc}\left[B\left(t - \frac{2R_i(t_m)}{c}\right)\right] \\
&\quad \times \exp\left[-j4\pi\frac{R_i(t_m)}{\lambda}\right].
\end{aligned}
\tag{8.45}
$$

Then, the WRFRFT of the multiple targets' compressed signal is

$$
\begin{aligned}
WR_{g(t_m)}(\alpha, u) &= F_\alpha[g(t_m)s\left(2r(t_m)/c, t_m\right)](u) \\
&= \int_{\eta_0}^{\eta_1} g(t_m)s\left(2r(t_m)/c, t_m\right) K_\alpha(t_m, u)dt_m \\
&= \int_{\eta_0}^{\eta_1} g(t_m)s_1\left(2r(t_m)/c, t_m\right) K_\alpha(t_m, u)dt_m \\
&\quad + \int_{\eta_0}^{\eta_1} g(t_m)s_2\left(2r(t_m)/c, t_m\right) K_\alpha(t_m, u)dt_m \\
&\quad + \cdots + \int_{\eta_0}^{\eta_1} g(t_m)s_i\left(2r(t_m)/c, t_m\right) K_\alpha(t_m, u)dt_m \\
&\quad + \cdots + \int_{\eta_0}^{\eta_1} g(t_m)s_K\left(2r(t_m)/c, t_m\right) K_\alpha(t_m, u)dt_m \\
&\quad + \int_{\eta_0}^{\eta_1} g(t_m)n_s\left(2r(t_m)/c, t_m\right) K_\alpha(t_m, u)dt_m,
\end{aligned}
\tag{8.46}
$$

where $\int_{\eta_0}^{\eta_1} g(t_m) s_i \left(2r(t_m)/c, t_m\right) K_\alpha(t_m, u) dt_m$ and $\int_{\eta_0}^{\eta_1} g(t_m) n_s \left(2r(t_m)/c, t_m\right) K_\alpha$ $(t_m, u) dt_m$ denote, respectively, the WRFRFT of the ith target's signal component and the WRFRFT of noise.

With different searching parameters, different integration outputs will be obtained via (8.46). Only when the searching parameters match with the ith target's motion parameters, the signal energy of the ith target will be accumulated as a peak via WRFRFT, while the target's parameters then could be estimated via the peak location. Note that the integrated peak intensity of the ith target is proportional to the product of its initial amplitude (i.e., $\sigma_{0,i}$) and the effective dwell time. Thus, there are normally two cases for the multi-target scene:

1) If the products of the targets' initial amplitude and the effective dwell time are close, then the integrated peak values of different targets are close. Thus, we could accomplish the multi-target detection and parameter estimation based on the WRFRFT output, where the processing strategy is similar to that shown in Section 8.2.3.

2) If the products of the targets' initial amplitude and the effective dwell time differ significantly, then the integrated peak of the weak target may be shaded by the integrated output of the strong target and may make it difficult to achieve the detection and parameter estimation of the weak targets. In this case, we should use the WRFRFT algorithm coming with the CLEAN technique [6]. By this way, the coherent detection and parameter estimation of the strong targets and weak ones could be accomplished iteratively.

8.3 EXPERIMENTS AND ANALYSIS FOR WRFRFT-BASED METHOD

The experiments with simulated data (Sections 8.3.1–8.3.4) and real data (Section 8.3.5) are given to demonstrate the effectiveness of the WRFRFT method, where the radar parameters of Sections 8.3.1–8.3.4 are set as the same as those in Table 8.1. In addition, the target's time parameters and motion parameters of Sections 8.3.1–8.3.2 are given in Table 8.2. Several typical coherent detection algorithms (RFRFT, GRFT, RFT) are used for comparison.

8.3.1 WRFRFT of Weak Target

In Fig. 8.4, the WRFRFT's response for a weak target is given, where the signal-to-noise ratio (SNR) after PC is 0 dB (as shown in Fig. 8.4(a)). Fig. 8.4(b)−Fig. 8.4(f) give respectively different projections of the WRFRFT output. More specifically, the projection in range cell-velocity domain is shown in Fig. 8.4(b), and the projection in range cell-acceleration domain is given in Fig. 8.4(c). Meanwhile, the projection in velocity-acceleration space is shown in Fig. 8.4(d), and the projection in acceleration-beginning time domain is given in Fig. 8.4(e). It can be noticed that the peak locations of different projections (e.g., Fig. 8.4(b), Fig. 8.4(c), Fig. 8.4(d)) indicate the corresponding parameters (e.g., range, velocity, acceleration, beginning time) of the target. Also, the projection in beginning time-ending time space is shown in Fig. 8.4(f), where Fig. 8.4(g) and Fig. 8.4(h) give, respectively, the beginning-time

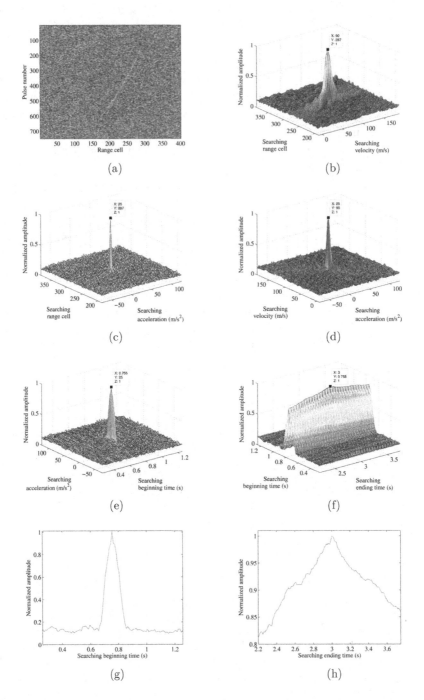

Figure 8.4 WRFRFT for a weak target. (a) PC. (b) Projection in range cell-velocity domain. (c) Projection in range cell-acceleration domain. (d) Projection in velocity-acceleration domain. (e) Projection in acceleration-beginning time domain. (f) Projection in beginning time-ending time domain. (g) Beginning time response slice. (h) Ending time response slice.

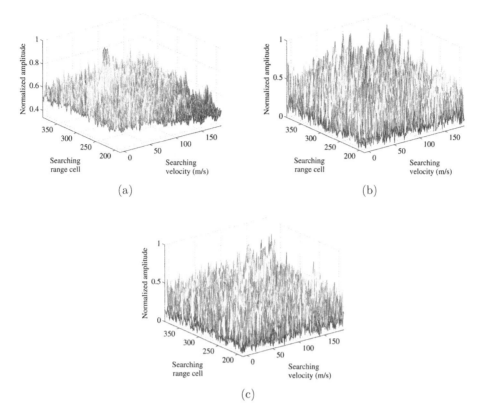

Figure 8.5 Integration results of (a) RFRFT, (b) GRFT and (c) RFT.

response slice and the ending-time response slice of Fig. 8.4(f), from which we could obtain the estimations of the beginning/ending time of the target signal.

In order to make comparison, the processing results of RFRFT, GRFT and RFT are shown in Fig. 8.5(a)−Fig. 8.5(c). On the whole, the outputs of these three algorithms are all defocused, due to the mismatch among the beginning/ending time of the target signal and RFRFT, GRFT as well as the RFT. Compared to the integration results of GRFT and RFT (which are seriously defocused), the integration result of RFRFT seems slightly better, but there is still no obvious peak in Fig. 8.5(a). More importantly, the relatively high peak position of Fig. 8.5(a) does not appear at the position corresponding to the target's motion parameters, which means that the target's motion parameters cannot be accurately estimated.

8.3.2 Parameters Estimation Performance

The estimation performances (i.e., root-mean-square error [RMSE]) of the WRFRFT algorithm for target's motion parameters (range, velocity and acceleration) and time parameters (beginning/ending time) under different SNR levels are evaluated by the Monte Carlo experiment. The target's parameters are initial range $R_0 = 35.5$ km, radial velocity $V = 91.33$ m/s and radial acceleration $A = 26.12$ m/s^2. The searching

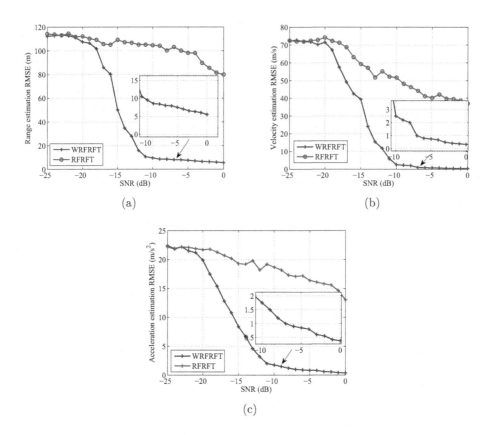

Figure 8.6 RMSE of the estimated motion parameters. (a) Range. (b) Velocity. (c) Acceleration.

area of range is [30 km : 15 m : 39 km], and the searching area of velocity is [80 m/s : 0.5 m/s : 100 m/s], while the searching area of acceleration is [0 m/s^2 : 0.5 m/s^2 : 40 m/s^2]. Thus, the target parameters would not be on the search grids. Fig. 8.6 gives the estimation performance curves of the target's motion parameters, where 200 times Monte Carlo trails are performed for each SNR. From Fig. 8.6, we can notice that the WRFRFT has a better estimation ability than RFRFT, and it could obtain good estimation performance when the SNR is larger than −10 dB. Additionally, Fig. 8.7 shows the RMSE curves of the target's time parameters where similar behaviors are observed.

8.3.3 Detection Ability

The detection performances of the WRFRFT method, RFRFT, GRFT, MTD and RFT under different SNR are shown in Fig. 8.8, where the false alarm rate is set as $P_f = 10^{-5}$. From Fig. 8.8, it could be noticed that the WRFRFT method could obtain better detection probability than RFRFT, GRFT and RFT thanks to its ability of matching with the target's beginning time and ending time as well as the

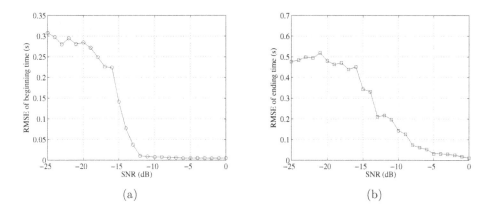

Figure 8.7 RMSE of the estimated time parameters. (a) Beginning time (entry). (b) Ending time (departure).

correction of ARC and DS. For example, WRFRFT is able to detect a target with 0.8 probability at SNR 10/11.5/15.6 dB lower than RFRFT/GRFT/RFT, respectively.

Moreover, for the RFRFT/GRFT/RFT methods, their integral intervals equal to the observation time. However, longer observation time leads to larger noise power, while the target power does not change. Thus, for a target with fixed dwell time, the performance of RFRFT/GRFT/RFT methods would be affected by the observation time. In particular, the longer the observation time is, the larger noise power will be presented during the integration output of RFRFT/GRFT/RFT, which results in worse detection performance. To show the observation-time's influence on the detection of RFRFT/GRFT/RFT more clear, a simulation experiment is carried on, during which two different values of the observation time are addressed (i.e., 3.75 and

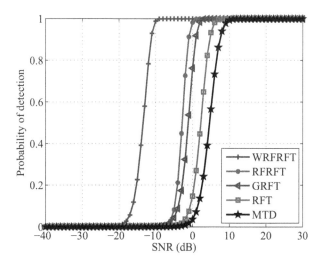

Figure 8.8 Detection ability curves.

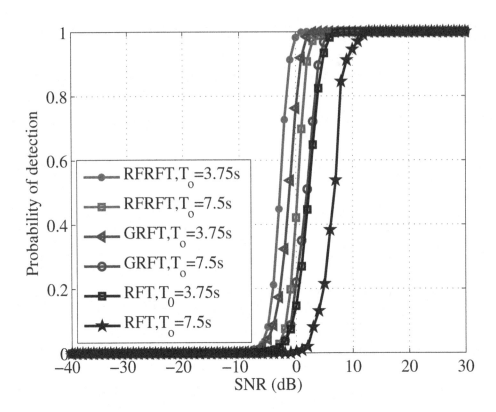

Figure 8.9 The observation time's influence on the detection ability of RFRFT/GRFT/RFT.

7.5 s), and the other simulation parameters are the same to that of Fig. 8.8. The detection performance curves of RFRFT/GRFT/RFT under different observation times are plotted in Fig. 8.9. It could noticed that with the increase of the observation time (e.g., from 3.75 to 7.5 s), the detection performance of RFRFT/GRFT/RFT decreased.

8.3.4 WRFRFT for Multiple Targets

Fig. 8.10 shows the simulation results of WRFRFT for multiple targets. Four moving targets (denoted as T_1, T_2, T_3 and T_4) are considered, and the parameters are listed in Table 8.4. The targets' motion trajectories after range compression are shown in Fig. 8.10(a), where the four curved trajectories are observed. Fig. 8.10(b)−Fig. 8.10(d) give different slices of WRFRFT output.

More specifically, Fig. 8.10(b) shows the WRFRFT result with $a_s = 25$ m/s^2, $\eta_{0s} = 0.755$ s and $\eta_{1s} = 3$ s. Note that the searching values of acceleration, beginning/ending time match with the corresponding parameters of T_1 and T_2. Hence, the signal energy of T_1 and T_2 is coherently accumulated, and the targets are well focused as seen from the two peaks in this slice (Fig. 8.10(b)). However, because of

TABLE 8.4   Parameters of the Four Moving Targets

Parameter	T_1	T_2	T_3	T_4
Initial range cell	287	323	269	305
Velocity (m/s)	90	70	75	95
Acceleration (m/s^2)	25	25	17	13
Beginning time (s)	0.705	0.705	0.905	1.005
Ending time (s)	3	3	3.4	3.2
SNR after PC (dB)	6	6	6	6

the searching values of acceleration, beginning time and ending time in this slice do not matched with T_3 and T_4, and thus the signal energy of T_3 and T_4 cannot be coherently integrated in this slice.

Fig. 8.10(c) shows the WRFRFT result with $a_s = 17$ m/s^2, $\eta_{0s} = 0.905$ s, $\eta_{1s} = 3.4$ s, which matches with the acceleration and beginning/ending time of T_3. Correspondingly, the target signal of T_3 is coherently focused as a peak in this slice

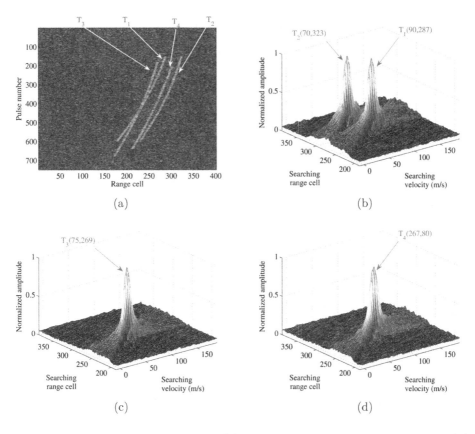

Figure 8.10   Multiple targets scene. (a) Results after pulse compression. (b) The focusing result of T_1 and T_2 with $a_s = 25$ m/s^2, $\eta_{0s} = 0.755$ s, $\eta_{1s} = 3$ s. (c) The focusing result of T_3 with $a_s = 17$ m/s^2, $\eta_{0s} = 0.905$ s, $\eta_{1s} = 3.4$ s. (d) The focusing result of T_4 with $a_s = 13$ m/s^2, $\eta_{0s} = 1.005$ s, $\eta_{1s} = 3.2$ s.

TABLE 8.5   Radar Parameters for Real Data

Parameter	Value
Wave band	C
Bandwidth	20 MHz
Sample frequency	60 MHz
Pulse repetition frequency	500 Hz
Pulse duration	18 μs

(Fig. 8.10(c)). Moreover, Fig. 8.10(d) shows the WRFRFT result with $a_s = 13$ m/s^2, $\eta_{0s} = 1.005$ s, $\eta_{1s} = 3.2$ s. In this slice (Fig. 8.10(d)), the searching values of acceleration and beginning/ending time match with the parameters of T_4 and thus we could notice that there is a peak formed, which is corresponding to T_4.

8.3.5   Real Data Results

To verify the effectiveness of the WRFRFT method, an evaluation is made using real collected data. The data set is the target signal of an unmanned aerial vehicle (UAV). Its maximum speed is 60 m/s, and its maximum acceleration is 10 m/s^2. The detection system was a linear frequency modulated (LFM) pulse radar, and the detailed parameters are listed in Table 8.5. During the experiment, the target did not appear in the radar detection beam for the early stage of radar startup, so there was only noise and clutter, but no target echo signal. Then, the UAV flew into the radar detection area, and the radar received the echo signal of the UAV. The UAV flew out of the radar detection area a few seconds later, but the radar was still transmitting and receiving signals after the UAV's departure.

The selected data to be processed consists of 2000 pulses (i.e., coherent processing interval is 4 s), as show in Fig. 8.11(a), from which we could see that the target motion trajectory crosses several range cells. In particular, we could choose "two points" (as shown in Fig. 8.11(a)) from the target's moving trajectory to obtain a rough estimation of target's velocity, i.e., $v_0 = (79 - 72) \times 2.5$ m/(1.428 s $-$ 0.974 s) $= 27.533$ m/s. After WRFRFT, the focusing result of target is given in Fig. 8.11(b). We could notice that the target signal is focused as a peak and its position represents the target's velocity and acceleration (i.e., 29 m/s and 4 m/s^2). For comparison, the processing results of RFRFT and GRFT are also given in Fig. 8.11(c) and Fig. 8.11(d), respectively. For the focusing result of RFRFT (Fig. 8.11(c)), it could be noticed that there is no significant peak in the RFRFT output and that the "pseudo or false peak" location is corresponding to -9 m/s and 0 m/s^2. In addition, the "pseudo or false peak" location of GRFT (Fig. 8.11(d)) is corresponding to -25 m/s and 0 m/s^2. Therefore, both RFRFT and GRFT could not focus on the target signal correctly, resulting in estimation error and even false alarm. Furthermore, the beginning-time response slice and ending-time response slice of WRFRFT are given in Fig. 8.11(e) and Fig. 8.11(f), respectively. From Fig. 8.11(e) and Fig. 8.11(f), we could obtain the estimations of target signal's beginning/ending time, i.e., 0.602 and 3.406 s.

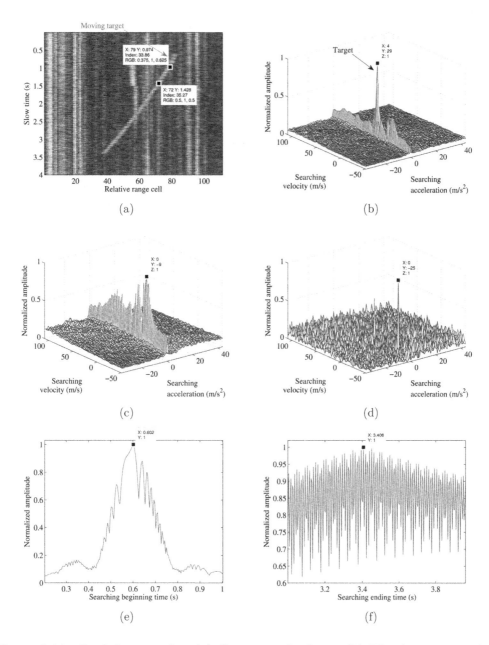

Figure 8.11   Real data results. (a) Compressed echoes. (b) The focusing result of WRFRFT. (c) Integration result of RFRFT. (d) Integration result of GRFT. (e) Beginning time response slice. (f) Ending time response slice.

8.4 EGRFT-BASED COHERENT INTEGRATION METHOD

The EGRFT in our chapter is defined as follows:

$$
\begin{aligned}
EGRFT_{h(t_m)}(r_0, v_0, a_0) = &\int_{\eta_0}^{T_1} s\left(2r(t_m)/c, t_m\right) h(t_m) \\
&\times \exp\left(j4\pi \frac{v_0(t_m - \eta_0) + a_0(t_m - \eta_0)^2}{\lambda}\right) dt_m,
\end{aligned}
\tag{8.47}
$$

where

$$
r(t_m) = r_0 + v_0(t_m - \eta_0) + a_0(t_m - \eta_0)^2
\tag{8.48}
$$

$$
h(t_m) = \mathrm{rect}\left[\frac{t_m - 0.5(T_1 + \eta_0)}{T_1 - \eta_0}\right] = \begin{cases} 1, & \eta_0 \le t_m \le T_1 \\ 0, & \text{else} \end{cases}
\tag{8.49}
$$

$$
T_0 \le \eta_0 \le T_1,
$$

r_0 represents the initial range to be searched, v_0 represents the radial velocity to be searched, a_0 represents the radial acceleration to be searched and η_0 denotes the beginning time of window function $h(t_m)$.

Using equation (8.3) inside equation (8.47), we have

$$
\begin{aligned}
EGRFT_{h(t_m)}(r_0, v_0, a_0) = &\int_{\eta_0}^{T_1} \sigma_0 \mathrm{sinc}\left[B\left(\frac{2(r(t_m) - R(t_m))}{c}\right)\right] \\
&\times w(t_m) h(t_m) \exp\left[j4\pi \frac{r(t_m) - R(t_m)}{\lambda}\right] dt_m \\
&+ EGRFT_n(r_0, v_0, a_0),
\end{aligned}
\tag{8.50}
$$

where $\sigma_0 = \sigma \exp(-j4\pi r_0/\lambda)$, $EGRFT_n(r_0, v_0, a_0)$ denotes the EGRFT of noise, i.e.,

$$
\begin{aligned}
EGRFT_n(r_0, v_0, a_0) = &\int_{\eta_0}^{T_1} n_s(2r(t_m)/c, t_m) \exp(-j4\pi r_0/\lambda) \\
&\times \exp\left(j4\pi \frac{r(t_m)}{\lambda}\right) dt_m.
\end{aligned}
\tag{8.51}
$$

In order to compare, the expressions of GRFT [4] and STGRFT [7] are also given:

$$
\begin{aligned}
GRFT(r_0, v_0, a_0) = &\int_{T_0}^{T_1} s\left(2r(t_m)/c, t_m\right) \\
&\times \exp\left(j4\pi \frac{v_0(t_m - T_0) + a_0(t_m - T_0)^2}{\lambda}\right) dt_m,
\end{aligned}
\tag{8.52}
$$

$$
\begin{aligned}
STGRFT(r_0, v_0, a_0) = &\int_{\eta_0}^{\eta_1} s\left(2r(t_m)/c, t_m\right) \\
&\times \exp\left(j4\pi \frac{r_0 + v_0(t_m - \eta_0) + a_0(t_m - \eta_0)^2}{\lambda}\right) dt_m,
\end{aligned}
\tag{8.53}
$$

where η_1 denotes the ending time of function $g(t_m)$.

Based on the definitions of EGRFT, GRFT and STGRFT shown in (8.47), (8.52) and (8.53), respectively, we could observe that the differences between EGRFT and GRFT/STGRFT are as follows: (i) compared with the original GRFT (where the beginning time of GRFT operation is fixed, i.e., T_0), the beginning time of EGRFT operation is adjustable, which means that the EGRFT has more freedom on the beginning-time dimension. (ii) Different from STGRFT (which needs five-dimensional searching), the EGRFT only requires four-dimensional searching and could avoid the compensation operation for the phase term induced by the initial distance of the target. As a result, EGRFT is more computational efficient than STGRFT, and the computational complexity of EGRFT is much lower than that of STGRFT.

8.4.1 Integration Response with Respect to η_0

Similar to GRFT, the prerequisite for the coherent integration of the entire target signal via EGRFT is that the range cell migration (RCM) correction and Doppler frequency migration (DFM) elimination must be completed when $r_0 = R_0, v_0 = V$ and $a_0 = A$ [1–4], i.e.,

$$
\begin{aligned}
r(t_m) - R(t_m) &= (r_0 - R_0) + (v_0 - V)t_m + (a_0 - A)t_m^2 \\
&\quad + 2(AT_b - a\eta_0)t_m + (a\eta_0^2 - AT_b^2) \\
&\quad + (VT_b - v\eta_0) \equiv 0, \quad t_m \in [T_b, T_e] \\
&\text{when } r_0 = R_0, v_0 = V, a_0 = A.
\end{aligned} \tag{8.54}
$$

Note that the symbol $\equiv$ means constantly equal.

According to the relationship between η_0 and T_b, the refocusing response of EGRFT under four different cases is discussed as follows:

Case 1: $\eta_0 = T_b$, then equation (8.50) could be recast as

$$
\begin{aligned}
EGRFT_{h(t_m)}(r_0, v_0, a_0) &= \int_{T_b}^{T_e} \sigma_0 \text{sinc}\left[B\left(\frac{2(r(t_m) - R(t_m))}{c}\right)\right] \\
&\quad \times \exp\left[j4\pi \frac{r(t_m) - R(t_m)}{\lambda}\right] dt_m \\
&\quad + EGRFT_n(r_0, v_0, a_0).
\end{aligned} \tag{8.55}
$$

When $r_0 = R_0, v_0 = V$ and $a_0 = A$, the RCM correction and DFM compensation of target signal will be accomplished (i.e., (8.54) is satisfied). Correspondingly, (8.55) could be expressed as

$$
\begin{aligned}
EGRFT_{h(t_m)}(r_0, v_0, a_0) &= \int_{T_b}^{T_e} \sigma_0 \exp\left[j4\pi 0 t_m\right] dt_m \\
&\quad + EGRFT_n(r_0, v_0, a_0) \\
&= \sigma_0(T_e - T_b) + EGRFT_n(r_0, v_0, a_0).
\end{aligned} \tag{8.56}
$$

Thus, in case 1, the coherent refocusing of the target signal is achieved, where the peak located in $r_0 = R_0, v_0 = V$ and $a_0 = A$.

Case 2: $\eta_0 < T_b$, then (8.50) could be recast as

$$EGRFT_{h(t_m)}(r_0, v_0, a_0) = \int_{T_b}^{T_e} \sigma_0 \text{sinc}\left[B\left(\frac{2(r(t_m) - R(t_m))}{c}\right)\right]$$
$$\times \exp\left[j4\pi\frac{r(t_m) - R(t_m)}{\lambda}\right] dt_m \qquad (8.57)$$
$$+ EGRFT_n(r_0, v_0, a_0).$$

When $r_0 = R_0, v_0 = V$ and $a_0 = A$, the $r(t_m) - R(t_m)$ shown in (8.54) could be expressed as

$$r(t_m) - R(t_m) = 2A(T_b - \eta_0)t_m + A(\eta_0^2 - T_b^2)$$
$$+ V(T_b - \eta_0) \not\equiv 0. \qquad (8.58)$$

Hence, the coherent refocusing of the moving target signal cannot be obtained in case 2.

Case 3: $T_e \geq \eta_0 > T_b$, then the EGRFT of (8.50) can be recast as

$$EGRFT_{h(t_m)}(r_0, v_0, a_0) = \int_{\eta_0}^{T_e} \sigma_0 \text{sinc}\left[B\left(\frac{2(r(t_m) - R(t_m))}{c}\right)\right]$$
$$\times \exp\left[j4\pi\frac{r(t_m) - R(t_m)}{\lambda}\right] dt_m \qquad (8.59)$$
$$+ EGRFT_n(r_0, v_0, a_0).$$

Noting that when $r_0 = R_0, v_0 = V$ and $a_0 = A$, the $r(t_m) - R(t_m)$ shown in (8.54) is

$$r(t_m) - R(t_m) = 2A(T_b - \eta_0)t_m + A(\eta_0^2 - T_b^2)$$
$$+ V(T_b - \eta_0) \not\equiv 0. \qquad (8.60)$$

Thus, the coherent refocusing of target signal cannot be achieved via EGRFT in case 3.

Case 4: $T_1 \geq \eta_0 > T_e$, then equation (8.50) can be rewritten as

$$EGRFT_{h(t_m)}(r_0, v_0, a_0) = 0 + EGRFT_n(r_0, v_0, a_0). \qquad (8.61)$$

From (8.61), we can notice that in case 4 only noise is integrated.

Table 8.6 summarizes the EGRFT's response of the four cases. Based on Table 8.6 and the discussions of the four cases above, we could deduce that only when $\eta_0 = T_b$, the coherent refocusing of the entire target signal can be obtained via EGRFT, where the focused peak location is corresponding to $r_0 = R_0, v_0 = V$ and $a_0 = A$. Hence, T_b, R_0, V and A could be estimated by

$$(\hat{T}_b, \hat{R}_0, \hat{V}, \hat{A}) = \underset{(\eta_0, r_0, v_0, a_0)}{\arg\max} |EGRFT_{h(t_m)}(r_0, v_0, a_0)|. \qquad (8.62)$$

A simulation example is given to show the refocusing response of the EGRFT-based method under the four cases mentioned above. Tables 8.7–8.9 give, respectively,

TABLE 8.6   EGRFT Response under Four Cases

Different cases	Relationship of η_0 and T_b	Coherent refoucsing
Case 1	$\eta_0 = T_b$	Yes
Case 2	$\eta_0 < T_b$	No
Case 3	$T_e \geq \eta_0 > T_b$	No
Case 4	$T_b \leq T_e < \eta_0 \leq T_1$	No

TABLE 8.7   Window Functions of EGRFT under Four Cases

Four cases	Window function
Case 1	$h(t_m) = \begin{cases} 1, & t_m \in [0.805\text{ s}, 3.78\text{ s}] \\ 0, & \text{otherwise} \end{cases}$
Case 2	$h(t_m) = \begin{cases} 1, & t_m \in [0.75\text{ s}, 3.78\text{ s}] \\ 0, & \text{otherwise} \end{cases}$
Case 3	$h(t_m) = \begin{cases} 1, & t_m \in [1.055\text{ s}, 3.78\text{ s}] \\ 0, & \text{otherwise} \end{cases}$
Case 4	$h(t_m) = \begin{cases} 1, & t_m \in [3.53\text{ s}, 3.78\text{ s}] \\ 0, & \text{otherwise} \end{cases}$

TABLE 8.8   Radar Parameters

Parameter	Value
Carrier frequency	0.6 GHz
Bandwidth	50 MHz
Sample frequency	100 MHz
Pulse repetition frequency	200 Hz
Pulse duration	4 μs
Observation time	0~3.78 s

TABLE 8.9   Motion Parameters of Target

Parameter	Value
Initial range cell	268
Initial radial velocity	75 m/s
Radial acceleration	21 m/s^2
Entry time of target	0.805 s
Departure time of target	3.28 s

the window functions, the radar parameters and the motion parameters of the target. Fig. 8.12 gives the result. In particular, the compressed result of radar's received echo (SNR = 5 dB) is given in Fig. 8.12(a), from which we can see that the target's residence time is only part of the entire observation time. Fig. 8.12(b), Fig. 8.12(c), Fig. 8.12(d) and Fig. 8.12(e) give, respectively, the EGRFT's responses of case 1, case 2, case 3 and case 4. It is observed from Fig. 8.12(b)–Fig. 8.12(e) that only in case 1 the target signal is focused as a peak (Fig. 8.12(b)) in the EGRFT's output, while the target signal is unfocused for the other three cases (Fig. 8.12(c)–Fig. 8.12(e)).

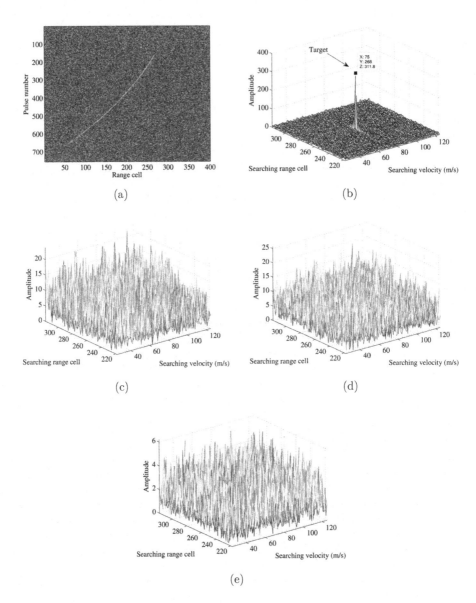

Figure 8.12 EGRFT response for the four cases. (a) PC. (b) Case 1. (c) Case 2. (d) Case 3. (e) Case 4.

8.4.2 Ending Time Estimation via WFRFT

With the estimations $(\hat{T}_b, \hat{R}_0, \hat{V}, \hat{A})$, we can extract the radar echoes (corresponding to the "entry time") containing target signal from the compressed signal (8.3), i.e.,

$$s_e(t_m) = s(2\hat{r}(t_m)/c, t_m), t_m \geq \hat{T}_b, \tag{8.63}$$

where

$$\hat{r}(t_m) = \hat{R}_0 + \hat{V}(t_m - \hat{T}_b) + \hat{A}(t_m - \hat{T}_b)^2. \tag{8.64}$$

Substituting (8.3) into (8.63) yields

$$s_e(t_m) = w(t_m)\sigma_0 \exp\left[-j4\pi\frac{(R_0 + Vt_m + At_m^2)}{\lambda}\right] + w_1(t_m)n_e(t_m), \tag{8.65}$$

$$t_m \in [T_b, T_1],$$

where

$$w_1(t_m) = \mathrm{rect}\left[\frac{t_m - 0.5(T_b + T_1)}{T_1 - T_b}\right] = \begin{cases} 1, & T_b \leq t_m \leq T_1 \\ 0, & \text{else} \end{cases}, \tag{8.66}$$

$n_e(t_m)$ denotes the extracted noise.

From (8.65), we notice that the extracted echoes contain target signal (which is an LFM signal) and noise. However, the target signal only exists within the period $[T_b, T_e]$, whereas there is only noise present for the other time. In other words, the target signal ends at time T_e.

The FRFT, which has excellent performance in analyzing LFM and sinusoidal signals, has been widely used in the radar signal processing field [8–10]. Inspired by the FRFT, we propose a method named WFRFT to achieve the estimation of the ending (departure) time T_e. First, to eliminate the signal's spectrum spans problem induced by target's acceleration [11–13], the quadratic phase term in (8.65) is compensated by the de-chirp technique. More specifically, a reference function is constructed based on the estimated acceleration, i.e.,

$$H(t_m) = \exp\left(j4\pi\hat{A}t_m^2/\lambda\right). \tag{8.67}$$

Multiplying (8.65) with (8.67) yields

$$s_e(t_m) = w(t_m)\sigma_0 \exp\left[-j4\pi\frac{(R_0 + Vt_m)}{\lambda}\right] + w_1(t_m)n_s(t_m), \tag{8.68}$$

where $n_s(t_m)$ denotes the noise after multiplication operation.

The WFRFT of $s_e(t_m)$ is defined as follows:

$$\begin{aligned} X_\alpha(u, \eta_1) &= F_\alpha[w_2(t_m)s_e(t_m)] \\ &= \int_{-\infty}^{\infty} w_2(t_m)s_e(t_m)K_\alpha(t_m, u)dt_m, \end{aligned} \tag{8.69}$$

where $w_2(t_m)$ is the window function of the WFRFT, i.e.,

$$w_2(t_m) = \text{rect}\left[\frac{t_m - 0.5(T_b + \eta_1)}{\eta_1 - T_b}\right] = \begin{cases} 1, & T_b \leq t_m \leq \eta_1 \\ 0, & \text{else} \end{cases}, \quad (8.70)$$

$\alpha = P\pi/2$ is the rotation angle of FRFT, where $P \in [0, 2]$ is the fractional order. That is to say, parameter $\alpha \in [0, \pi]$, and it is determined by the fractional order P [5]. Normally, the fractional order P is unknown, and thus we need to search within the scope $[0, 2]$. The searching interval could be set as: $\Delta_P = -\frac{2}{\pi}\text{arccot}(\frac{1}{f_s(T_1 - T_0)})$ [5], where f_s denotes the sampling frequency, F_α denotes the FRFT operator, η_1 is the ending time of function $w_2(t_m)$ and $K_\alpha(t_m, u)$ is as follows:

$$K_\alpha(t_m, u) = \begin{cases} A_\alpha \exp[j(0.5t_m^2 \cot\alpha - ut_m \csc\alpha + 0.5u^2 \cot\alpha)] & \alpha \neq n\pi \\ \delta[u - (-1)^n t_m] & \alpha = n\pi, \end{cases} \quad (8.71)$$

where $A_\alpha = \sqrt{(1 - j\cot\alpha)/2\pi}$.

From (8.69), we could see that the WFRFT of $s_e(t_m)$ includes two operations: the first is the signal interception process via the window function $w_2(t_m)$, whose length is determined by η_1. The second is the FRFT operation, which is performed on the intercepted signal to achieve the accumulation of target signal in the fractional-order domain. With respect to different values of η_1, signals of different lengths will be intercepted and accumulated. Only when $\eta_1 = T_e$, the target signal is entirely extracted and refocused, resulting in a maximum value in the FRFT domain. Thus, the ending time (i.e., departure time) of the target signal could be estimated by

$$\hat{T}_e = \arg\max_{\eta_1} |X_\alpha(u, \eta_1)|. \quad (8.72)$$

8.4.3 Detailed Procedure of EGRFT-WFRFT

Fig. 8.13 shows the flowchart of the EGRFT-WFRFT method, where the main steps are as follows:

1) Apply PC on the radar received echo, and we could achieve the compressed echo, as shown in (8.3).

2) According to the prior information such as types and moving status of targets to be expected, the searching scope of the range, velocity and acceleration are, respectively, set as $[r_{\min}, r_{\max}]$, $[v_{\min}, v_{\max}]$ and $[a_{\min}, a_{\max}]$, where the searching intervals of range, velocity and acceleration are respectively [5]:

$$\rho_r = c/2B, \quad (8.73)$$

$$\rho_v = \lambda/2(T_1 - T_0), \quad (8.74)$$

$$\rho_a = \lambda/2(T_1 - T_0)^2. \quad (8.75)$$

3) Based on the total observation time and the radar's pulse repetition time (i.e., PRT), the search scope of the beginning time could be determined, i.e., $\eta_0 \in [T_0, T_1]$. Besides, the searching step of the beginning time is $\Delta\eta = PRT$.

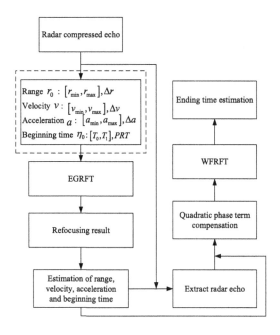

Figure 8.13 Flowchart of the EGRFT-WFRFT method.

4) With the searching parameters determined in steps 2 and 3, perform the corresponding EGRFT on the compressed radar echo.

5) Based on the peak location of EGRFT outputs, the estimations of T_b, R_0, V and A could be obtained.

6) Extract the radar echoes based on the estimated parameters of step 5 and then perform quadratic phase term compensation.

7) Apply WFRFT to achieve the estimation of the ending time.

8.4.4 Comparison of Computational Cost

The computational complexity of the major steps of the EGRFT-based method, STGRFT, WRFRFT and GRFT is analyzed in terms of the total number of multiplication operations and addition operations (i.e., the times of complex multiplications and additions). Let N_{η_0} denote the searching number of beginning time and N_r represent the number of searching range cell. Assume that N_v and N_a denote, respectively, the number of searching velocity and the number of searching acceleration.

The main steps of the EGRFT-based method include the EGRFT operation (four-dimensional searching process and the computational complexity are in the order of $N_{\eta_0} N_r N_v N_a M$) and the departure time estimation using WFRFT (with the computational burden in order of $N_{\eta_1} M^2 \log_2 M$), where M is the pulse number within the observation time. Thus, the computational burden of EGRFT-WFRFT is $O(N_{\eta_0} N_r N_v N_a M + N_{\eta_1} M^2 \log_2 M)$.

For the STGRFT-based method, which mainly depends on the five-dimensional searching process, the computational cost is in the order of $O(N_{\eta_1} N_{\eta_0} N_r N_v N_a M)$,

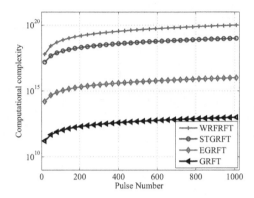

Figure 8.14   Computational complexity curves.

TABLE 8.10   Comparison of the Three Methods

Methods	Computational complexity
EGRFT	$O(N_{\eta_0} N_r N_v N_a M + N_{\eta_1} M^2 \log_2 M)$
STGRFT	$O(N_{\eta_1} N_{\eta_0} N_r N_v N_a M)$
WRFRFT	$O(N_{\eta_0} N_{\eta_1} N_r N_v N_a M \log_2 M)$
GRFT	$O(N_r N_v N_a M)$

where N_{η_1} is the searching number of ending time. As to the WRFRFT approach, which involves the searching operation in the beginning time-end time-acceleration-velocity-range space, the computational burden is $O(N_{\eta_0} N_{\eta_1} N_r N_v N_a M \log_2 N)$. For GRFT, the computational burden is in the order of $O(N_r N_v N_a N)$. Table 8.10 shows the computational burden of EGRFT, STGRFT, WRFRFT and GRFT.

Suppose that the scope of target's radial acceleration and velocity is, respectively, $[-20, 20 \text{ m/s}^2]$ and $[-200, 200 \text{ m/s}]$. The range cell number is 400, while the radar parameters were the same as those in Fig. 8.12. Then, the computational complexity curves of the four methods with different pulse numbers are shown in Fig. 8.14. We can notice from Table 8.10 and Fig. 8.14 that the EGRFT-based method has a much lower computational complexity than WRFRFT and STGRFT. Besides, although GRFT has a lower computational cost than the WRFRFT method, it could not obtain the refocusing and parameter estimation of target with unknown time information, as shown in Section 8.5.

8.5   EXPERIMENTS AND DISCUSSIONS FOR EGRFT-BASED METHOD

The simulation experiments and real data test in this section are carried out to validate the EGRFT-WFRFT method. Table 8.8 lists the radar parameters during simulation experiments.

8.5.1 Refocusing under Low SNR Background

The coherent focusing ability of the EGRFT-based method for a moving target (Table 8.9 lists the parameters of the target) under a low SNR environment is evaluated in Fig. 8.15. Fig. 8.15(a) shows the compressed echo, where the SNR is set as -5 dB. Because of the low SNR, the target signal is overwhelmed by noise. To clearly show the target's moving trajectory, the PC result without noise is also given in Fig. 8.15(b).

After coherent integration via EGRFT, the focused results are shown in Fig. 8.15(c)–Fig. 8.15(e) (corresponding to range cell-velocity slice, velocity-acceleration slice and acceleration-beginning time slice, respectively). It could be seen that the target energy is well focused as a peak in the output, from which we could obtain the estimations of target's entry time, initial range cell, radial velocity and acceleration. With the estimations of target's motion parameters and the beginning time obtained by Fig. 8.15(c)–Fig. 8.15(e), we could extract the radar echoes containing target's entire signal and noise (in time period beyond the ending time), as shown in Fig. 8.15(f). After quadratic phase compensation, WFRFT is performed on the extracted echoes, and the result is given in Fig. 8.15(g), from which the estimation of target departure time could be obtained.

In order to make comparison, Fig. 8.15(h), Fig. 8.15(i) and Fig. 8.15(j) give, respectively, the integration results of WRFRFT, GRFT and STGRFT. It is observed from Fig. 8.15(i)–Fig. 8.15(j) that the target's energy is unfocused after GRFT/STGRFT operation.

8.5.2 Refocusing of Multiple Targets

The refocusing ability of the EGRFT-based method for multiple targets is evaluated in this subsection, where Table 8.11 shows the motion parameters of the targets. We consider four situations, i.e., targets with different ranges, targets with different velocities, targets with different accelerations and targets with different entry times.

Situation 1: We consider targets 1 and 2 in this situation, which are different in initial range cell. Fig. 8.16(a) shows the compressed echo, and Fig. 8.16(b) shows the focused result in range cell-velocity domain.

Situation 2: Targets 1 and 3 are considered in this situation, where these two targets are different in radial velocity. Fig. 8.16(c) shows the compressed echo, and Fig. 8.16(d) shows the focused result in range cell-velocity domain.

Situation 3: Targets 1 and 4 are considered in this situation. Fig. 8.16(e) shows the compressed echo. We could see that the moving trajectories of targets 1 and 4 are very close. This is because that targets 1 and 4 have very similar motion parameters (i.e., only slight difference in radial acceleration). Fig. 8.16(f) shows the focused result in the velocity-acceleration domain. We could notice that these two targets could be separated very well.

Situation 4: Targets 1 and 5 are considered in this situation, which are of difference with respect to the entry time. Fig. 8.16(g) gives the PC result, and Fig. 8.16(h) shows the focused result in acceleration-beginning time domain.

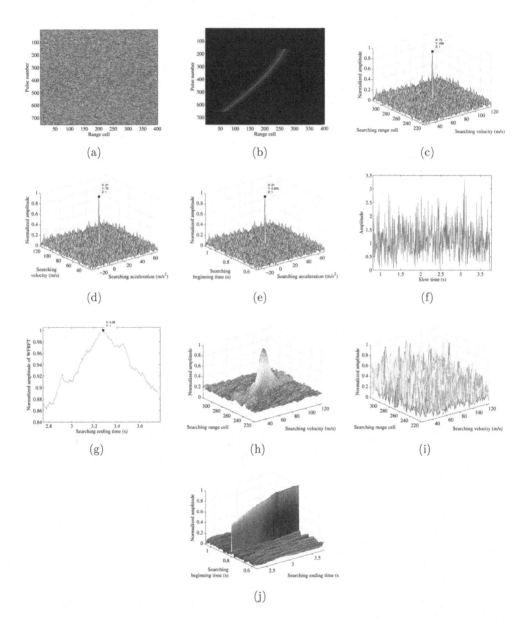

Figure 8.15 Refocusing under low SNR. (a) PC. (b) Target's moving trajectory. (c) Slice of searching initial range cell and velocity. (d) Slice of searching velocity and acceleration. (e) Slice of searching acceleration and beginning time. (f) Extracted radar echo. (g) Ending time estimation via WFRFT. (h) WRFRFT. (i) GRFT. (j) STGRFT.

TABLE 8.11 Parameters of the Multiple Targets

Targets	Target 1	Target 2	Target 3	Target 4	Target 5
Initial range cell	268	240	268	268	268
Radial velocity (m/s)	75	75	50	75	75
Radial acceleration (m/s^2)	21	21	21	19	21
Entry time (s)	0.805	0.805	0.805	0.805	1.305
Departure time (s)	3.28	3.28	3.28	3.28	3.28
SNR after PC (dB)	6	6	6	6	6

It could be noticed from Fig. 8.16 that the EGRFT-based approach could realize the focusing of multiple targets signal effectively.

Furthermore, the refocusing performance of the EGRFT-based method for targets with different parameters has been shown in Fig. 8.17, while the targets' motion parameters are given in Table 8.12. Fig. 8.17(a) shows the pulse compression result. Fig. 8.17(b) shows the refocusing result (slice) of target 6 with $a_0 = 20$ m/s^2 and $\eta_0 = 0.805$ s. Since the searching acceleration and searching beginning time are, respectively, equal to the acceleration and beginning time of target 6, the signal energy of target 6 is well focused in this slice. In addition, Fig. 8.17(c) shows the refocusing result (slice) of target 7 with $a_0 = -20$ m/s^2 and $\eta_0 = 1.055$ s, where the searching beginning time and searching acceleration are, respectively, equal to the beginning time and acceleration of target 7. Hence, target 7's signal energy is well focused in this slice. From Fig. 8.17, we could see that targets 6 and 7 are well focused in the corresponding slices which matched the targets' motion parameters.

8.5.3 Motion Parameters Estimation Performance

We also evaluate the motion parameters estimation performance of the EGRFT-WFRFT method via Monte Carlo trials. The motion parameters of the moving target are the same as those in Section 8.5.1. Fig. 8.18(a)–Fig. 8.18(c) show the motion parameters (i.e., target's radial acceleration, radial velocity and initial range) estimation performance (i.e., estimation RMSE), where the RMSE curves versus different SNRs are given. The estimation results of STGRFT, WRFRT and GRFT methods are shown as benchmarks. It could be noticed from the figures that the EGRFT-based method could achieve good motion estimation performance when the input SNR is bigger than −11 dB. It is further noted that the performance of the EGRFT-based

TABLE 8.12 Parameters of the Target 6 and Target 7

Targets	Target 6	Target 7
Initial range cell	302	302
Radial velocity (m/s)	72	52
Radial acceleration (m/s^2)	20	−10
Entry time (s)	0.805	1.055
Departure time (s)	3.28	3.53
SNR after PC (dB)	6	6

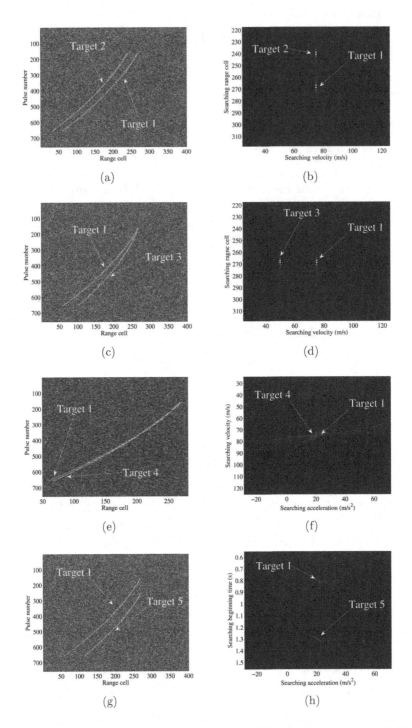

Figure 8.16 Focused result of multiple targets in different situations. (a) Compressed echo in situation 1. (b) Focused result in situation 1. (c) Compressed echo in situation 2. (d) Focused result in situation 2. (e) Compressed echo in situation 3. (f) Focused result in situation 3. (g) Compressed echo in situation 4. (h) Focused result in situation 4.

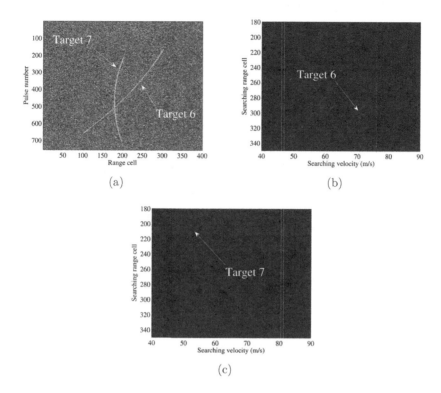

Figure 8.17 Focused of multiple targets with different values of parameters. (a) Compressed echo. (b) Slice of focused result for target 6. (c) Slice of focused result for target 7.

method is comparable to that of STGRFT and WRFRFT. That is to say, in comparison with STGRFT and WRFRFT, the EGRFT-based method has comparable estimation performance and superior computational efficiency.

Remark 3: The moving target's signal begins at time T_b and ends at time T_e. For the WRFRFT method, it performs beginning time and ending time searching and matches them with the target signal's beginning time and ending time. As such, only the target signal and noise distributed in $[T_b, T_e]$ would be extracted and accumulated. On the other hand, the EGRFT-based method only performs beginning time searching and its ending time is set as T_1 (which is larger than T_e). Hence, the EGRFT-based method could only match the target signal's beginning time. Then, the target signal distributed in $[T_b, T_e]$ and the noise distributed in $[T_b, T_1]$ would be extracted and integrated during the EGRFT operation. That is to say, much more noise is accumulated for the EGRFT-based algorithm. Hence, the anti-noise performance of the WRFRFT is better than that of the EGRFT-based method.

8.5.4 Real Dataset Processing

The refocusing ability of the EGRFT-based method is also evaluated via real measured radar data. Table 8.13 lists the radar parameters. In the experiment, a UAV

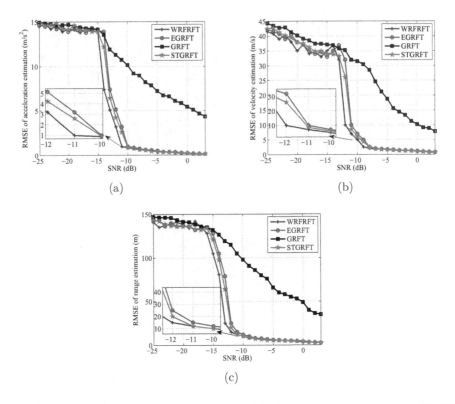

Figure 8.18 Motion parameter estimation. (a) Acceleration estimation RMSE. (b) Velocity estimation RMSE. (c) Range estimation RMSE.

was employed as the target for radar refocusing. After the radar is turned on, it did not immediately entry the radar coverage, so that the received echo only contained noise and clutter during this time. The selected data to be processed have a total of 2001 pulses and 71 range cells. Fig. 8.19(a) gives the PC echoes of radar in the range-azimuth domain. It is obvious from Fig. 8.19(a) that the target signal only exists in part of the whole radar observation time and that the radar echoes contain only noise and clutter at the beginning/ending time of the observation period. Fig. 8.19(b)−Fig. 8.19(d) show different slices of the focusing result of the EGRFT-based method, from which we can notice that the target signal is well focused.

TABLE 8.13 Radar Parameters for Real Measured Data

Parameter	Value
Wave band	C
Bandwidth	20 MHz
Sample frequency	60 MHz
Pulse repetition frequency	500 Hz
Pulse duration	18 μs
Pulse number of received echoes	2001

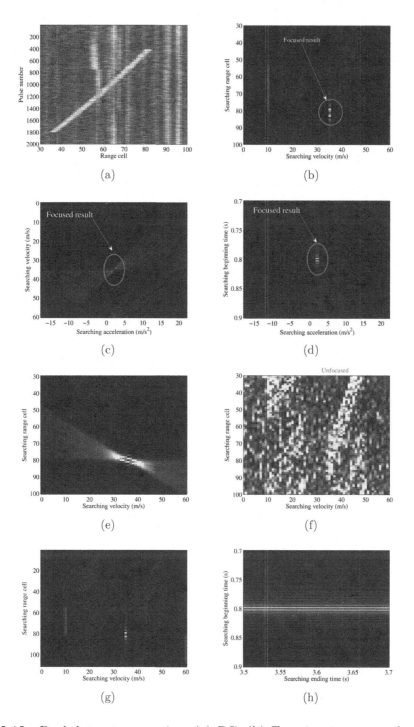

Figure 8.19 Real dataset processing. (a) PC. (b) Focusing in range-velocity slice. (c) Focusing in velocity-acceleration slice. (d) Focusing in acceleration-beginning time slice. (e) WRFRFT. (f) GRFT. (g) Range-velocity slice of STGRFT. (h) Beginning time-end time slice of STGRFT.

TABLE 8.14   The Estimated Values and Real Values of Kinematic Parameters

Parameter	Initial range cell	Velocity (m/s)	Acceleration (m/s^2)
Real value	80	36.20	3.10
EGRFT	80	36.30	3.13
STGRFT	80	36.10	3.13
WRFRFT	80	36.15	3.12
GRFT	70	51.30	1.15

The integration outputs of WRFRFT and GRFT are, respectively, shown in Fig. 8.19(e) and Fig. 8.19(f) for the purpose of comparison. It could be seen that WRFRFT could also realize the refocusing, but GRFT becomes invalid, and the corresponding output is unfocused. In addition, the integration results of STGRFT are shown in Fig. 8.19(g)−Fig. 8.19(h). It could be noticed that the result of STGRFT will be unfocused along the ending time dimension. Furthermore, the estimated values and the real kinematic parameters of the target have been given in Table 8.14. We could see from the table that the EGRFT-based method, STGRFT and WRFRFT can estimate the target's motion parameters accurately. In particular, the estimation errors of acceleration and velocity are, respectively, 0.03 m/s^2 and 0.1 m/s for the EGRFT-based method, which are very minor. However, the estimation errors of acceleration and velocity are, respectively, 1.95 m/s^2 and 15.1 m/s for the GRFT method.

8.6   SUMMARY

In this chapter, two coherent integration transform-based methods, i.e., WRFRFT and EGRFT, are introduced for radar detection of high-speed targets with unknown entry/departure time, including the transform definition, detailed coherent integration processing flow and computational complexity analysis. The performances are validated by detailed experiments.

Bibliography

[1] J. Xu, J. Yu, Y. N. Peng, and X. G. Xia, "Radon-Fourier transform (RFT) for radar target detection (I): generalized Doppler filter bank processing," *IEEE Transactions on Aerospace and Electronic Systems*, vol. 47, no. 2, pp. 1186–1202, April 2011.

[2] J. Xu, J. Yu, Y. N. Peng, and X. G. Xia, "Radon-fourier transform (RFT) for radar target detection (II): blind speed sidelobe suppression," *IEEE Transactions on Aerospace and Electronic Systems*, vol. 47, no. 4, pp. 2473–2489, October 2011.

[3] J. Yu, J. Xu, Y. N. Peng, and X. G. Xia, "Radon-fourier transform (RFT) for radar target detection (III): optimality and fast implementations," *IEEE Transactions on Aerospace and Electronic Systems*, vol. 48, no. 2, pp. 991–1004, April 2012.

[4] J. Xu, X. G. Xia, S. B. Peng, J. Yu, Y. N. Peng, and L. C. Qian, "Radar maneuvering target motion estimation based on generalized radon-Fourier transform," *IEEE Transactions on Signal Processing*, vol. 60, no. 12, pp. 6190–6201, December 2012.

[5] X. L. Chen, J. Guan, N. B. Liu, and Y. He, "Maneuvering target detection via Radon-Fractional Fourier transform-based long-time coherent integration," *IEEE Transactions on Signal Processing*, vol. 62, no. 4, pp. 939–953, February 2014.

[6] X. Li, L. Kong, G. Cui, and W. Yi, "CLEAN-based coherent integration method for high-speed multi-targets detection," *IET Radar Sonar and Navigation*, vol. 10, no. 9, pp. 1671–1682, December 2016.

[7] X. L. Li, Z. Sun, T. S. Yeo, T. X. Zhang, W. Yi, G. L. Cui, and L. J. Kong, "STGRFT for detection of maneuvering weak target with multiple motion models," *IEEE Transactions on Signal Processing*, vol. 67, no. 7, pp. 1902–1917, April 2019.

[8] S. C. Pei and J. Ding, "Fractional Fourier transform, wigner distribution, and filter design for stationary and nonstationary random processes," *IEEE Transactions on Signal Processing*, vol. 58, no. 8, pp. 4079–4092, August 2010.

[9] E. Sejdic, I. Djurovic, and L. Stankovic, "Fractional Fourier transform as a signal processing tool: an overview of recent developments," *Signal Processing*, vol. 91, no. 6, pp. 1351–1369, June 2011.

[10] L. Qi, R. Tao, S, Y. Zhou, and Y. Wang, "Detection and parameter estimation of multicomponent LFM signal based on the fractional Fourier transform," *Science in China Series F: Information Sciences*, vol. 47, no. 2, pp. 184–198, 2004.

[11] H. B. Sun, G. S. Liu, H. Gu, and W. M. Su, "Application of the Fractional Fourier transform to moving target detection in airborne SAR," *IEEE Transactions on Aerospace and Electronic Systems*, vol. 38, no. 4, pp. 1416–1424, October 2002.

[12] S. Q. Zhu, G. S. Liao, Y. Qu, Z. G. Zhou, and X. Y. Liu, "Ground moving targets imaging algorithm for synthetic aperture radar," *IEEE Transactions on Geoscience and Remote Sensing*, vol. 49, no. 1, pp. 463–477, January 2011.

[13] G. C. Sun, M. D. Xing, X. G. Xia, Y. R. Wu, and Z. Bao, "Robust ground moving-target imaging using deramp-keystone processing," *IEEE Transactions on Geoscience and Remote Sensing*, vol. 51, no. 2, pp. 966–982, February 2013.

CHAPTER 9

Fast Non-Searching Coherent Integration Processing

The previous chapters only consider the high-speed target with constant velocity or constant acceleration but do not consider the high-speed target with high-order motion (such as jerk). This chapter considers the coherent integration problem for high-speed maneuvering targets with jerk motion. A fast non-searching method based on adjacent cross correlation function (ACCF) and Lv's distribution (LVD) is discussed, where the adjacent correlation operation is first employed to remove the range migration (RM) and reduce the order of Doppler frequency migration (DFM). After that, LVD is applied to realize the coherent integration, target detection and parameters estimation. In addition, at the cost of some performance loss, another fast method via ACCF iteratively is also introduced to further reduce the computational complexity and obtain the motion parameters estimation.

9.1 SIGNAL MODEL

Suppose that the radar transmits a linear frequency modulated (LFM) signal, i.e.,

$$s_{trans}(t, t_m) = \text{rect}\left(\frac{t}{T_p}\right) \exp\left(j\pi\mu t^2\right) \exp(j2\pi f_c t_m), \qquad (9.1)$$

where

$$\text{rect}\,(x) = \begin{cases} 1 & |x| \leq \frac{1}{2}, \\ 0 & |x| > \frac{1}{2}, \end{cases}$$

T_p is the pulse duration, μ is the frequency modulated rate, f_c is the carrier frequency, $t_m = mT_r$ is the slow time, $m = 0, 1, \cdots, M - 1$; T_r is the pulse repetition time, M denotes the number of coherent integrated pulses and t is the fast time.

Assume that there are K targets in the scene. Neglecting the high-order components, the instantaneous slant range $r_k(t_m)$ of the kth target with complex motions

DOI: 10.1201/9781003529101-9

satisfies [1, 2]

$$r_k(t_m) = r_{0,k} + v_{0,k}t_m + a_{0,k}t_m^2 + a_{1,k}t_m^3, \tag{9.2}$$

where $r_{0,k}$ is the initial slant range from the radar to the kth target and $v_{0,k}$, $a_{0,k}$ and $v_{0,k}$ denote, respectively, the kth target's radial velocity, acceleration and jerk.

The received baseband signal of K targets can be stated as [1, 3]

$$
\begin{aligned}
s_r(t, t_m) = \sum_{k=1}^{K} A_{0,k}\mathrm{rect}\left(\frac{t - 2r_k(t_m)/c}{T_p}\right) \exp\left(-j\frac{4\pi r_k(t_m)}{\lambda}\right) \\
\times \exp\left[j\pi\mu\left(t - \frac{2r_k(t_m)}{c}\right)^2\right],
\end{aligned}
\tag{9.3}
$$

where $A_{0,k}$ is the kth target's reflectivity, c is the light speed and $\lambda = c/f_c$ denotes the wavelength.

After pulse compression, the compressed signal can be represented as

$$
\begin{aligned}
s(t, t_m) = \sum_{k=1}^{K} A_{1,k}\mathrm{sinc}\left[B\left(t - \frac{2r_k(t_m)}{c}\right)\right] \\
\times \exp\left(-j\frac{4\pi f_c r_k(t_m)}{c}\right),
\end{aligned}
\tag{9.4}
$$

where B denotes the bandwidth of the transmitted signal and $A_{1,k}$ is the amplitude after compression.

Equation (9.4) shows that the target envelope changes with the slow time after pulse compression. When the offset exceeds the range resolution, i.e., $\rho_r = c/2B$, the RM effect would occur. Additionally, the phase of (9.4) is a cubic phase function of slow time due to the jerk motion of target, which would lead to DFM and make the signal energy defocused. Both RM and DFM will create difficulties during coherent integration and parameters estimation for the maneuvering target.

9.2 COHERENT INTEGRATION AND PARAMETERS ESTIMATION VIA ACCF-LVD

In this section, ACCF-LVD is presented to achieve the coherent integration and obtain the parameters estimation for maneuvering targets. We first discuss the ACCF-LVD with mono-target, and then the analysis of ACCF-LVD with multi-targets is introduced.

9.2.1 ACCF-LVD with Mono-Target

9.2.1.1 *RM Correction via ACCF*

Consider the compressed signal of the kth target, i.e.,

$$
\begin{aligned}
s_k(t, t_m) = A_{1,k}\mathrm{sinc}\left[B\left(t - \frac{2r_k(t_m)}{c}\right)\right] \\
\times \exp\left[-j\frac{4\pi f_c r_k(t_m)}{c}\right].
\end{aligned}
\tag{9.5}
$$

The ACCF of signal (9.5) can be stated as [4]

$$R_{1,k}(\tau_1, t_m) = \int_0^T s_k(t, t_m)s_k^*(t - \tau_1, t_{m+1})dt, \qquad (9.6)$$

where $*$ denotes the complex conjugation and T is the coherent integration time.

Note that the correlation of time domain signals is equal to the inverse Fourier transform (IFT) of the conjugate multiplication of their corresponding frequency domain signals, i.e.,

$$R_{1,k}(\tau_1, t_m) = \underset{f_r}{\mathrm{IFT}}[S_k(f_r, t_m)S_k^*(f_r, t_{m+1})], \qquad (9.7)$$

where

$$
\begin{aligned}
S_k(f_r, t_m) &= \underset{t}{\mathrm{FT}}[s_k(t, t_m)] \\
&= A_{2,k}^{[1]}\mathrm{rect}\left(\frac{f_r}{B}\right)\exp\left[-j\frac{4\pi(f_r + f_c)r_k(t_m)}{c}\right],
\end{aligned} \qquad (9.8)
$$

$$
\begin{aligned}
S_k(f_r, t_{m+1}) &= \underset{t}{\mathrm{FT}}[s_k(t, t_{m+1})] \\
&= A_{2,k}^{[1]}\mathrm{rect}\left(\frac{f_r}{B}\right)\exp\left[-j\frac{4\pi(f_r + f_c)r_k(t_{m+1})}{c}\right],
\end{aligned} \qquad (9.9)
$$

$\underset{t}{\mathrm{FT}}(\cdot)$ and $\underset{f_r}{\mathrm{IFT}}(\cdot)$ denote, respectively, the FT over t and IFT over f_r and $A_{2,k}^{[1]}$ represents the amplitude after FT.

Substituting (9.8) and (9.9) into (9.7), we have

$$
\begin{aligned}
R_{1,k}(\tau_1, t_m) &= A_{2,k}\mathrm{sinc}\left[B\left(\tau_1 + \frac{2N_{0,k}}{c}\right)\right]\exp\left(j\frac{4\pi f_c N_{1,k}}{c}\right) \\
&\times \exp\left(j\frac{4\pi f_c N_{2,k}}{c}t_m\right)\exp\left(j\frac{4\pi f_c N_{3,k}}{c}t_m^2\right),
\end{aligned} \qquad (9.10)
$$

where $A_{2,k}$ denotes the amplitude of the ACCF, and

$$N_{0,k} = N_{1,k} + N_{2,k}t_m + N_{3,k}t_m^2, \qquad (9.11)$$

$$N_{1,k} = v_{0,k}T_r + a_{0,k}T_r^2 + a_{1,i}T_r^3, \qquad (9.12)$$

$$N_{2,k} = 2a_{0,k}T_r + 3a_{1,i}T_r^2, \qquad (9.13)$$

$$N_{3,k} = 3a_{1,i}T_r. \qquad (9.14)$$

From (9.10)−(9.14), we can see that the envelope of the sinc function in (9.10) is a second-order function of t_m and that the peak is located at

$$
\begin{aligned}
\tau_1 = &-\frac{2(v_{0,k}T_r + a_{0,k}T_r^2 + a_{1,k}T_r^3)}{c} \\
&-\frac{2(2a_{0,k}T_r + 3a_{1,k}T_r^2)t_m}{c} - \frac{6a_{1,k}T_r t_m^2}{c}.
\end{aligned} \qquad (9.15)
$$

There are three terms in (9.15). The first term is a constant, which is independent of the slow time. The second term is a linear function of t_m. The third term is a second-order function of t_m. Hence, we only need to consider the second term and the third term's effect on the envelope migration. Although the changes of $\frac{2(2a_{0,k}T_r+3a_{1,k}T_r^2)t_m}{c}$ and $\frac{6a_{1,k}T_r t_m^2}{c}$ are different with respect to the slow time t_m, the peak location of ACCF can be considered to be within the same range cell as long as the shift of the target between the adjacent time is less than a range cell, i.e., the following condition should be satisfied

$$|a_{0,k}| < \frac{cf_p^2}{4f_s}, \tag{9.16}$$

$$|a_{1,k}| < \frac{cf_p^3}{6f_s}, \tag{9.17}$$

where f_p is the radar pulse repetition frequency and f_s denotes the sampling rate.

Generally speaking the conditions in (9.16) and (9.17) can be satisfied [4]. Hence, the RM correction has been finished, which is helpful to the coherent integration and parameters estimation. If (9.16) and (9.17) are not satisfied, the keystone transform [5–7] can be employed to remove the residual RM.

9.2.1.2 *Acceleration and Jerk Estimation via LVD*

As shown in (9.10), the envelope of the sinc function is a constant with respect to the slow time t_m, i.e., the target energy concentrated in the same range cell after RM correction. Without loss of generality, only one range cell signal is considered for simplicity, and then the sinc term in (9.10) is a constant. Therefore, the signal in (9.10) can be rewritten as follows:

$$
\begin{aligned}
R_{1,k}(t_m) =& A_{3,k} \exp\left(j\frac{4\pi f_c N_{1,k}}{c}\right) \exp\left(j\frac{4\pi f_c N_{2,k}}{c}t_m\right) \\
& \times \exp\left(j\frac{4\pi f_c N_{3,k}}{c}t_m^2\right),
\end{aligned}
\tag{9.18}
$$

where $A_{3,k}$ denotes the amplitude of the signal.

From (9.18), it is observed that $R_{1,k}(t_m)$ is a chirp signal, where the centroid frequency is $\frac{2N_{2,k}f_c}{c}$ and the chirp rate is $\frac{4f_c N_{3,k}}{c}$. Hence, the LVD, which has excellent performance in analyzing chirp signal without using any searching operation [8–10], can be applied for estimating the centroid frequency and chirp rate. The detailed discussion would be shown in the following.

The parametric symmetric instantaneous autocorrelation function (PSIAF) of (9.18) is defined by

$$
\begin{aligned}
R_C(t_m, \tau) =& R_{1,k}\left(t_m + \frac{\tau+a}{2}\right) R_{1,k}^*\left(t_m - \frac{\tau+a}{2}\right) \\
=& A_{3,k}^2 \exp\left[j\frac{4\pi N_{2,k}f_c}{c}(\tau+a) + j\frac{8\pi f_c N_{3,k}}{c}(\tau+a)t_m\right],
\end{aligned}
\tag{9.19}
$$

where τ is a delay variable and a denotes a constant time-delay related to a scaling operator.

It can be seen from (9.19) that the time variable t_m and lag variable τ couple with each other in the exponential phase term. To remove the coupling, do a variable transform as follows:

$$t_m = \frac{t_n}{h(\tau + a)},\tag{9.20}$$

where h is a scaling factor.

Then, by inserting (9.20) into (9.19), we have

$$R_C(t_n, \tau) = A_{3,k}^2 \exp\left[j\frac{4\pi N_{2,k} f_c}{c}(\tau + a) + j\frac{8\pi f_c N_{3,k}}{hc} t_n\right].\tag{9.21}$$

Equation (9.21) shows that the coupling term between t_n and τ has been removed. According to [8–10], we generally use the parameters $a = 1$ and $h = 1$ for obtaining a desirable centroid frequency-chirp rate (CFCR) representation. Then, by performing two-dimensional (2-D) FT on (9.21) with respect to τ and t_n, we obtain the LVD of $R_{1,k}(t_m)$ as follows:

$$\begin{aligned}L_{R_{1,k}(t_m)}(f_{ce}, \gamma) =&\, A_{4,k} \exp(j2\pi f_{ce})\mathrm{sinc}\left(f_{ce} - \frac{2N_{2,k} f_c}{c}\right)\\ &\times \mathrm{sinc}\left(\gamma - \frac{4f_c N_{3,k}}{c}\right),\end{aligned}\tag{9.22}$$

where $A_{4,k}$ denotes the amplitude after 2-D FT. By (9.22), the parameters $(N_{2,k}, N_{3,k})$ can be estimated via the peak location of $L_{R_{1,k}(t_m)}(f_{ce}, \gamma)$. Then, the estimations of the acceleration and jerk, i.e., $\hat{a}_{0,k}$ and $\hat{a}_{1,k}$, can be obtained via (9.13) and (9.14).

9.2.1.3 Velocity Estimation

With the estimated $\hat{N}_{2,k}$ and $\hat{N}_{3,k}$, a reference function can be generated as follows:

$$S_{ref}(f_r, t_m) = \exp\left[-j\frac{4\pi}{c}(f_c + f_r)\left(\hat{N}_{2,k} t_m + \hat{N}_{3,k} t_m^2\right)\right].\tag{9.23}$$

Applying FT to the signal $R_{1,k}(\tau_1, t_m)$ shown in (9.10) with respect to the variable τ_1, we have

$$\begin{aligned}&R_{1,k}(f_r, t_m)\\ &= A_{5,k} \exp\left[j\frac{4\pi}{c}(f_c + f_r)\left(N_{1,k} + N_{2,k} t_m + N_{3,k} t_m^2\right)\right],\end{aligned}\tag{9.24}$$

where $A_{5,k}$ denotes the amplitude after FT.

Multiplying (9.24) with (9.23) yields

$$R_{1,k}(f_r, t_m) = A_{5,k} \exp\left[j\frac{4\pi}{c}(f_c + f_r)N_{1,k}\right]. \tag{9.25}$$

Taking IFT to (9.25) with respect to f_r and FT with respect to t_m, respectively, we have

$$R_{1,k}(\tau_1, f_{t_m}) = A_{6,k}\text{sinc}\left[B\left(\tau_1 + \frac{2N_{1,k}}{c}\right)\right]\exp\left(j\frac{4\pi f_c N_{1,k}}{c}\right) \\ \times \text{sinc}[\pi(N-1)T_r f_{t_m}], \tag{9.26}$$

where $A_{6,k}$ denotes the amplitude after IFT and FT.

It can be seen from (9.26) that the coherent accumulation of target energy is achieved and the target is detected if the ratio of the peak value to noise is larger than a given threshold. Besides, the estimation of $N_{1,k}$ can be obtained via the peak location of $R_{1,k}(\tau_1, f_{t_m})$. Then, based on the definition of $N_{1,k}$ in (9.12), the velocity, i.e., $v_{0,k}$, can be estimated. Below, an example is given to demonstrate how this ACCF-LVD method works to accomplish the RM correction and maneuvering target detection.

Example 1: A fast maneuvering target is used in this example. Radar parameters are set as follows: the carrier frequency $f_c = 1$ GHz, the bandwidth $B = 1$ MHz, the sample frequency $f_s = 5$ MHz, pulse repetition frequency $f_p = 200$ Hz and the number of integration pulses $M = 513$. The motion parameters of the target are initial slant range $r_0 = 200$ km, radial velocity $v_0 = 1600$ m/s, acceleration $a_0 = 50$ m/s^2 and jerk $a_1 = 10$ m/s^3. Simulation results of this example are given in Fig. 9.1.

Fig. 9.1(a) shows the result of pulse compression, which indicates that serious RM occurs because of target's high-speed. Fig. 9.1(b) shows the result of ACCF. From this figure, we can see that the RM has been removed and that the target energy is located in the same range cell, which is helpful to the coherent integration. The integration result of LVD is given in Fig. 9.1(c). It can be seen that the target energy is well accumulated as one peak in the output. Based on the LVD output, the maneuvering target detection and motion parameters estimation can be accomplished.

9.2.2 ACCF-LVD with Multi-Target

Consider the compressed signal for multi-target, i.e.,

$$s(t, t_m) = \sum_{k=1}^{K} A_{1,k}\text{sinc}\left[B\left(t - \frac{2r_k(t_m)}{c}\right)\right] \\ \times \exp\left[-j\frac{4\pi f_c r_k(t_m)}{c}\right]. \tag{9.27}$$

The ACCF of (9.27) can be expressed as

$$R_1(\tau_1, t_m) = R_{1,self}(\tau_1, t_m) + R_{1,other}(\tau_1, t_m), \tag{9.28}$$

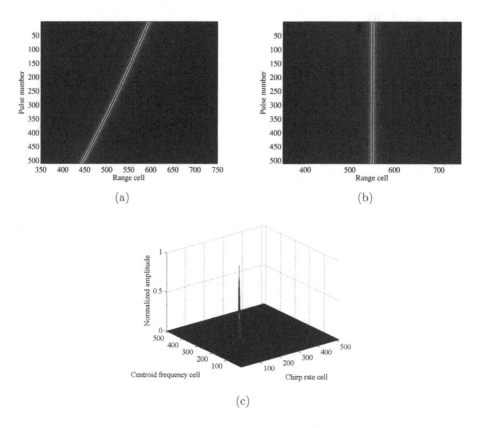

Figure 9.1 Simulation results of Example 1. (a) Result of pulse compression. (b) Result of ACCF. (c) Integration result of ACCF-LVD.

where

$$
\begin{aligned}
R_{1,self}(\tau_1, t_m) = \sum_{k=1}^{K} A_{2,k} \mathrm{sinc}\left[B\left(\tau_1 + \frac{2N_{0,k}}{c} \right) \right] \\
\times \exp\left(j\frac{4\pi f_c N_{1,k}}{c} \right) \exp\left(j\frac{4\pi f_c N_{2,k}}{c} t_m \right) \\
\times \exp\left(j\frac{4\pi f_c N_{3,k}}{c} t_m^2 \right),
\end{aligned}
\tag{9.29}
$$

$$
\begin{aligned}
R_{1,other}(\tau_1, t_m) = \sum_{l=1}^{K} \sum_{n=1,n\neq l}^{K} \sigma_{l,n} \mathrm{sinc}\left[B\left(\tau_1 + \frac{2C_{l,n}}{c} \right) \right] \\
\times \exp\left(j\frac{4\pi f_c C_{l,n}^{[0]}}{c} \right) \exp\left(j\frac{4\pi f_c C_{l,n}^{[1]}}{c} t_m \right) \\
\times \exp\left(j\frac{4\pi f_c C_{l,n}^{[2]}}{c} t_m^2 \right) \exp\left(j\frac{4\pi f_c C_{l,n}^{[3]}}{c} t_m^3 \right),
\end{aligned}
\tag{9.30}
$$

$$C_{l,n} = C_{l,n}^{[0]} + C_{l,n}^{[1]}t_m + C_{l,n}^{[2]}t_m^2 + C_{l,n}^{[3]}t_m^3, \tag{9.31a}$$

$$C_{l,n}^{[0]} = r_{0,l} - r_{0,n} + v_{0,l}T_r + a_{0,l}T_r^2 + a_{1,l}T_r^3, \tag{9.31b}$$

$$C_{l,n}^{[1]} = v_{0,l} - v_{0,n} + 2a_{0,l}T_r + 3a_{1,l}T_r^2, \tag{9.31c}$$

$$C_{l,n}^{[2]} = a_{0,l} - a_{0,n} + 3a_{1,l}T_r, \tag{9.31d}$$

$$C_{l,n}^{[3]} = a_{1,l} - a_{1,n}, \tag{9.31e}$$

$R_{1,self}(\tau_1, t_m)$ denotes the self terms, $R_{1,ohter}(\tau_1, t_m)$ denotes the cross terms and $\sigma_{l,n}$ is the amplitude of the cross term generated by the lth target and the nth target.

Similarly, only one range cell signal is considered for simplicity, and then the signal in (9.28) can be rewritten as follows:

$$
\begin{aligned}
R_1(t_m) =& R_{1,self}(t_m) + R_{1,ohter}(t_m) \\
=& \sum_{k=1}^{K} A_{3,k} \exp\left(j\frac{4\pi f_c N_{1,k}}{c}\right) \exp\left(j\frac{4\pi f_c N_{2,k}}{c}t_m\right) \\
& \times \exp\left(j\frac{4\pi f_c N_{3,k}}{c}t_m^2\right) \\
& + \sum_{l=1}^{K} \sum_{n=1,n\neq l}^{K} \sigma_{l,n}^{[1]} \exp\left(j\frac{4\pi f_c C_{l,n}^{[0]}}{c}\right) \exp\left(j\frac{4\pi f_c C_{l,n}^{[1]}}{c}t_m\right) \\
& \times \exp\left(j\frac{4\pi f_c C_{l,n}^{[2]}}{c}t_m^2\right) \exp\left(j\frac{4\pi f_c C_{l,n}^{[3]}}{c}t_m^3\right).
\end{aligned}
\tag{9.32}
$$

After the instantaneous autocorrelation operation (as shown in (9.19)) and the variable transform (as shown in (9.20)), the PSIAF of (9.32) can be expressed as

$$
\begin{aligned}
R_C(t_n, \tau) =& R_{C,self}(t_n, \tau) + R_{C,cross}^{[1]}(t_n, \tau) \\
& + R_{C,cross}^{[2]}(t_n, \tau) + R_{C,cross}^{[3]}(t_n, \tau) \\
& + R_{C,cross}^{[4]}(t_n, \tau) + R_{C,cross}^{[5]}(t_n, \tau),
\end{aligned}
\tag{9.33}
$$

where

$$
\begin{aligned}
& R_{C,self}(t_n, \tau) \\
& = \sum_{k=1}^{K} A_{3,k}^2 \exp\left[j\frac{4\pi N_{2,k} f_c(\tau + a)}{c} + j\frac{8\pi f_c N_{3,k} t_n}{hc}\right],
\end{aligned}
\tag{9.34}
$$

$$R_{C,cross}^{[1]}(t_n, \tau) = \sum_{l=1}^{K} \sum_{n=1, n \neq l}^{K} A_{3,l} A_{3,n}$$

$$\times \exp\left[j\frac{4\pi f_c(N_{1,l} - N_{1,n})}{c}\right] \exp\left\{j\frac{4\pi f_c}{c}\left[N_{2,n}(\tau + a)\right]\right\}$$

$$\times \exp\left\{j\frac{4\pi f_c}{c}\left[(N_{2,l} - N_{2,n})\left(\frac{t_n}{h\tau + ha} + \frac{\tau + a}{2}\right)\right]\right\}$$

$$\times \exp\left\{j\frac{4\pi f_c}{c}\left[\frac{2N_{3,n}t_n}{h}\right]\right\}$$

$$\times \exp\left\{j\frac{4\pi f_c}{c}\left[(N_{3,l} - N_{3,n})\left(\frac{t_n}{h\tau + ha} + \frac{\tau + a}{2}\right)^2\right]\right\}, \qquad (9.35)$$

$$R_{C,cross}^{[2]}(t_n, \tau) = \sum_{l=1}^{K} \sum_{n=1, n \neq l}^{K} \left(\sigma_{l,n}^{[1]}\right)^2 \exp\left[j\frac{4\pi f_c}{c}C_{l,n}^{[1]}(\tau + a)\right]$$

$$\times \exp\left(j\frac{8\pi f_c}{hc}C_{l,n}^{[2]}t_n\right) \qquad (9.36)$$

$$\times \exp\left\{j\frac{4\pi f_c}{c}C_{l,n}^{[3]}\left[\frac{3t_n^2}{h^2(\tau + a)} + \frac{(\tau + a)^3}{4}\right]\right\},$$

$$R_{C,cross}^{[3]}(t_n, \tau) = \sum_{l=1}^{K} \sum_{n=1, n \neq l}^{K} \sum_{d=1, d \neq l}^{K} \sum_{e=1, e \neq d, e \neq n}^{K} \sigma_{l,n}^{[1]} \sigma_{d,e}^{[1]}$$

$$\times \exp\left[j\frac{4\pi f_c}{c}\left(C_{l,n}^{[0]} - C_{d,e}^{[0]}\right)\right] \exp\left\{j\frac{4\pi f_c}{c}\left[C_{d,e}^{[1]}(\tau + a)\right]\right\}$$

$$\times \exp\left\{j\frac{4\pi f_c}{c}\left[\left(C_{l,n}^{[1]} - C_{d,e}^{[1]}\right)\left(\frac{t_n}{h\tau + ha} + \frac{\tau + a}{2}\right)\right]\right\}$$

$$\times \exp\left\{j\frac{4\pi f_c}{c}\left[\frac{2C_{d,e}^{[2]}t_n}{h}\right]\right\} \qquad (9.37)$$

$$\times \exp\left\{j\frac{4\pi f_c}{c}\left[\left(C_{l,n}^{[2]} - C_{d,e}^{[2]}\right)\left(\frac{t_n}{h\tau + ha} + \frac{\tau + a}{2}\right)^2\right]\right\}$$

$$\times \exp\left\{j\frac{4\pi f_c}{c}\left[\frac{3C_{d,e}^{[3]}t_n^2}{h^2(\tau + a)} + C_{d,e}^{[3]}(\tau + a)^3\right]\right\}$$

$$\times \exp\left\{j\frac{4\pi f_c}{c}\left[\left(C_{l,n}^{[3]} - C_{d,e}^{[3]}\right)\left(\frac{t_n}{h\tau + ha} + \frac{\tau + a}{2}\right)^3\right]\right\},$$

$$R_{C,cross}^{[4]}(t_n, \tau) = \sum_{k=1}^{K} \sum_{l=1}^{K} \sum_{n=1,n\neq l}^{K} A_{3,k} \sigma_{l,n}^{[1]}$$

$$\times \exp\left[j\frac{4\pi f_c}{c} \left(N_{1,k} - C_{l,n}^{[0]} \right) \right]$$

$$\times \exp\left\{ j\frac{4\pi f_c}{c} \left[C_{l,n}^{[1]}(\tau + a) \right] \right\}$$

$$\times \exp\left\{ j\frac{4\pi f_c}{c} \left[\left(N_{2,k} - C_{l,n}^{[1]} \right) \left(\frac{t_n}{h\tau + ha} + \frac{\tau + a}{2} \right) \right] \right\}$$

$$\times \exp\left\{ j\frac{4\pi f_c}{c} \left[\frac{2C_{l,n}^{[2]} t_n}{h} \right] \right\} \tag{9.38}$$

$$\times \exp\left\{ j\frac{4\pi f_c}{c} \left[\left(N_{3,k} - C_{l,n}^{[2]} \right) \left(\frac{t_n}{h\tau + ha} + \frac{\tau + a}{2} \right)^2 \right] \right\}$$

$$\times \exp\left\{ j\frac{4\pi f_c}{c} \left[\frac{3C_{l,n}^{[3]} t_n^2}{h^2(\tau + a)} + C_{l,n}^{[3]}(\tau + a)^3 \right] \right\}$$

$$\times \exp\left\{ j\frac{4\pi f_c}{c} \left[-C_{l,n}^{[3]} \left(\frac{t_n}{h\tau + ha} + \frac{\tau + a}{2} \right)^3 \right] \right\},$$

$$R_{C,cross}^{[5]}(t_n, \tau) = \sum_{k=1}^{K} \sum_{l=1}^{K} \sum_{n=1,n\neq l}^{K} A_{3,k} \sigma_{l,n}^{[1]}$$

$$\times \exp\left[j\frac{4\pi f_c}{c} \left(C_{l,n}^{[0]} - N_{1,k} \right) \right]$$

$$\times \exp\left\{ j\frac{4\pi f_c}{c} \left[N_{2,k}(\tau + a) \right] \right\}$$

$$\times \exp\left\{ j\frac{4\pi f_c}{c} \left[\left(C_{l,n}^{[1]} - N_{2,k} \right) \left(\frac{t_n}{h\tau + ha} + \frac{\tau + a}{2} \right) \right] \right\} \tag{9.39}$$

$$\times \exp\left\{ j\frac{4\pi f_c}{c} \left[\frac{2C_{l,n}^{[2]} t_n}{h} \right] \right\}$$

$$\times \exp\left\{ j\frac{4\pi f_c}{c} \left[\left(N_{3,k} - C_{l,n}^{[2]} \right) \left(\frac{t_n}{h\tau + ha} + \frac{\tau + a}{2} \right)^2 \right] \right\}$$

$$\times \exp\left\{ j\frac{4\pi f_c}{c} \left[C_{l,n}^{[3]} \left(\frac{t_n}{h\tau + ha} + \frac{\tau + a}{2} \right)^3 \right] \right\}.$$

Performing 2-D FT on (9.33) with respect to t_n and τ, we can obtain the LVD for multi-target, i.e.,

$$L_{R_1(t_m)}(f_{ce}, \gamma) = \sum_{k=1}^{K} A_{4,k} \exp(j2\pi f_{ce}) \mathrm{sinc}\left(f_{ce} - \frac{2N_{2,k}f_c}{c} \right)$$

$$\times \mathrm{sinc}\left(\gamma - \frac{4f_c N_{3,k}}{c} \right) + L_{R_{1,ohter}(t_m)}(f_{ce}, \gamma), \tag{9.40}$$

where

$$
\begin{aligned}
\mathrm{L}_{R_{1,ohter}(t_m)}(f_{ce},\gamma) = \underset{t_n}{\mathrm{FT}} \Big\{ \underset{\tau}{\mathrm{FT}} \Big[& R_{C,cross}^{[1]}(t_n,\tau) + R_{C,cross}^{[2]}(t_n,\tau) \\
& + R_{C,cross}^{[3]}(t_n,\tau) + R_{C,cross}^{[4]}(t_n,\tau) + R_{C,cross}^{[5]}(t_n,\tau) \Big] \Big\},
\end{aligned}
\tag{9.41}
$$

$\mathrm{FT}_{t_n}(\cdot)$ and $\mathrm{FT}_\tau(\cdot)$ denote, respectively, the FT over t_n dimension and the FT over τ dimension.

From (9.33)−(9.39), we can see that $R_C(t_n,\tau)$ contains both self terms (i.e., $R_{C,self}(t_n,\tau)$) and cross terms (i.e., $R_{C,cross}^{[1]}(t_n,\tau)$, $R_{C,cross}^{[2]}(t_n,\tau), \cdots, R_{C,cross}^{[5]}(t_n,\tau)$). Fortunately, note that the self terms are all first-order phase functions of t_n and τ, whereas the cross terms are either quadratic phase functions or cubic phase functions of t_m and τ. As a result, the energy of self terms can be integrated via 2D-FT; meanwhile, the energy of cross terms cannot be accumulated due to the mismatch with FT. Therefore, $\mathrm{L}_{R_{1,ohter}(t_m)}(f_{ce},\gamma)$ is very small and can be ignored in comparison with the self terms, i.e., (9.40) can be approximated as

$$
\begin{aligned}
\mathrm{L}_{R_1(t_m)}(f_{ce},\gamma) \approx \sum_{k=1}^{K} & A_{4,k}\exp(j2\pi f_{ce})\mathrm{sinc}\left(f_{ce} - \frac{2N_{2,k}f_c}{c}\right) \\
& \times \mathrm{sinc}\left(\gamma - \frac{4f_c N_{3,k}}{c}\right).
\end{aligned}
\tag{9.42}
$$

In (9.42), each target is completely accumulated and corresponding to a sole peak in the output of LVD. Based on the peak location of (9.42), $N_{2,k}$ and $N_{3,k}$ can be estimated. Below, a noise-free example is given to show how the ACCF-LVD algorithm accumulates the self term and suppresses the cross term under multi-targets.

Example 2: Two maneuvering targets A and B are used in this example. Radar parameters of this example are the same as those in Example 1. The motion parameters are set as: $A_{1,1} = 1$, $r_{0,1} = 200$ km, $v_{0,1} = 1600$ m/s, $a_{0,1} = 50$ m/s^2, $a_{1,1} = 10$ m/s^3 for target A; $A_{1,2} = 3$, $r_{0,2} = 203$ km, $v_{0,2} = 3600$ m/s, $a_{0,2} = -50$ m/s^2, $a_{1,2} = -10$ m/s^3 for target B. Fig. 9.2 gives the simulation results of this example.

Fig. 9.2(a) shows the result of pulse compression, which indicates that the strength of target B is stronger than that of target A. Fig. 9.2(b) shows the envelope of the ACCF. It can be seen that the self terms are located in the same range cell (i.e., RM has been corrected), while the cross terms are distributed in different range cells, which is helpful to the suppression of cross terms. After adjacent correlation operation, the LVD result is given in Fig. 9.2(c). We can see that the target's energy is coherently accumulated as two peaks in the output of LVD, which indicates that the cross term effect has been suppressed well.

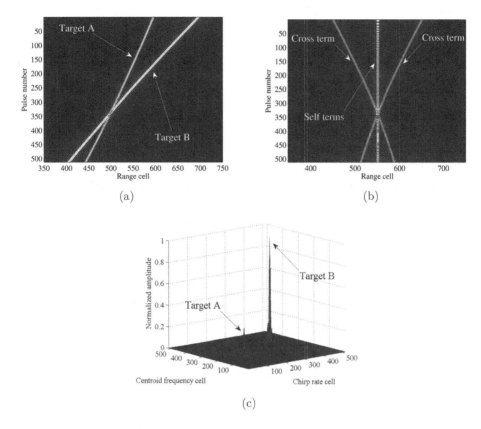

Figure 9.2 Simulation results of Example 2. (a) Result after pulse compression. (b) Result of ACCF. (c) Integration result of LVD.

9.3 PARAMETERS ESTIMATION VIA ACCF ITERATIVELY

In this section, another fast algorithm-based on ACCF iteratively is introduced to further reduce the computational complexity and obtain the motion parameters estimation. We first discuss the ACCF iteratively with mono-target, and then the analysis of ACCF iteratively with multi-target is given.

9.3.1 ACCF Iteratively for Mono-Target

Without loss of generality, we only consider the compressed signal of the kth target for simplicity, which can be rewritten as

$$
\begin{aligned}
s_k\left(t, t_m\right)= & A_{1,k}\mathrm{sinc}\left[B\left(t-\frac{2\left(r_{0,k}+v_{0,k}t_m+a_{0,k}t_m^2+a_{1,k}t_m^3\right)}{c}\right)\right] \\
& \times \exp\left(-j\frac{4\pi f_c r_{0,k}}{c}\right)\exp\left(-j\frac{4\pi f_c v_{0,k}}{c}t_m\right) \\
& \times \exp\left(-j\frac{4\pi f_c a_{0,k}}{c}t_m^2\right)\exp\left(-j\frac{4\pi f_c a_{1,k}}{c}t_m^3\right),
\end{aligned}
\tag{9.43}
$$

and the ACCF of $s_k(t, t_m)$ is

$$
\begin{aligned}
R_{1,k}(\tau_1, t_m) =& A_{2,k}\text{sinc}\left[B\left(\tau_1 + \frac{2(N_{1i} + N_{2,k}t_m + N_{3,k}t_m^2)}{c}\right)\right] \\
&\times \exp\left(j\frac{4\pi f_c N_{1,k}}{c}\right)\exp\left(j\frac{4\pi f_c N_{2,k}}{c}t_m\right) \\
&\times \exp\left(j\frac{4\pi f_c N_{3,k}}{c}t_m^2\right).
\end{aligned}
\tag{9.44}
$$

Comparing the original compressed signal (9.43) and ACCF (9.44), it can be seen that the adjacent correlation operation reduces the order of envelope migration and phase over slow time. Inspired by this, we employ another adjacent correlation operation on (9.44) and the corresponding ACCF can be expressed as

$$
\begin{aligned}
R_{2,k}(\tau_2, t_m) =& \int_0^T R_{1,k}(\tau_1, t_m)R_{1,k}^*(\tau_1 - \tau_2, t_{m+1})d\tau_1 \\
=& A_{7,k}\text{sinc}\left[B\left(\tau_2 + \frac{2M_{0,k}}{c}\right)\right]\exp\left(j\frac{4\pi f_c M_{1,k}}{c}\right) \\
&\times \exp\left(j\frac{4\pi f_c M_{2,k}}{c}t_m\right),
\end{aligned}
\tag{9.45}
$$

where $A_{7,k}$ denotes the amplitude after second ACCF operation, and

$$
M_{0,k} = M_{1,k} + M_{2,k}t_m,
\tag{9.46}
$$

$$
M_{1,k} = -N_{2,k}T_r - N_{3,k}T_r^2,
\tag{9.47}
$$

$$
M_{2,k} = -2N_{3,k}T_r.
\tag{9.48}
$$

From (9.45), we can see that the DFM has been removed, and $R_{2,k}(\tau_2, t_m)$ becomes a complex sinusoid signal over t_m. Applying FT on (9.45) with respect to t_m yields

$$
\begin{aligned}
R_{2,k}(\tau_2, f_{t_m}) =& A_{8,k}\text{sinc}\left[B\left(\tau_2 + \frac{2M_{0,k}}{c}\right)\right]\exp\left(j\frac{4\pi f_c M_{1,k}}{c}\right) \\
&\times \text{sinc}\left[(N-2)T_r\left(f_{t_m} - \frac{2f_c M_{2,k}}{c}\right)\right],
\end{aligned}
\tag{9.49}
$$

where $A_{8,k}$ denotes the amplitude after FT along the t_m dimension.

It is observed from (9.49) that the sinc-like point spread function (PSF) is reconstructed and that the coherent integration is achieved. Based on (9.49), the target is detected if the ratio of the peak value to noise is larger than a given threshold. In addition, parameter M_{2i} can be estimated via the peak location of $R_{2,k}(\tau_2, f_{t_m})$. Then, according to the definition of $M_{2,k}$ in (9.48), the estimation of $N_{3,k}$ can be obtained as follows:

$$
\hat{N}_{3,k} = -\frac{\hat{M}_{2,k}}{2T_r}.
\tag{9.50}
$$

With the estimated $\hat{N}_{3,k}$, a reference function is generated as follows:

$$
s_{ref}(t_m) = \exp\left(-j\frac{4\pi f_c \hat{N}_{3,k}}{c}t_m^2\right).
\tag{9.51}
$$

Multiplying (9.44) with (9.51) yields

$$
\begin{aligned}
R_{1,k}(\tau_1, t_m) =& A_{2,k}\operatorname{sinc}\left[B\left(\tau_1 + \frac{2N_{0,k}}{c}\right)\right]\exp\left(j\frac{4\pi f_c N_{1,k}}{c}\right) \\
&\times \exp\left(j\frac{4\pi f_c N_{2,k}}{c}t_m\right).
\end{aligned}
\tag{9.52}
$$

Applying FT on (9.52) with respect to t_m, we have

$$
\begin{aligned}
R_{1,k}(\tau_1, f_{t_m}) =& A_{9,k}\operatorname{sinc}\left[B\left(\tau_1 + \frac{2N_{0,k}}{c}\right)\right]\exp\left(j\frac{4\pi f_c N_{1,k}}{c}\right) \\
&\times \operatorname{sinc}\left[(N-1)T_r\left(f_{t_m} - \frac{2f_c N_{2,k}}{c}\right)\right].
\end{aligned}
\tag{9.53}
$$

By the peak location of (9.53), the estimation of $N_{2,k}$ can be achieved. Then, the estimations of acceleration and jerk (i.e., $\hat{a}_{0,k}$ and $\hat{a}_{1,k}$) can be obtained via (9.13) and (9.14). After that, the velocity estimation can be accomplished with the processing procedure shown in (9.23)−(9.26).

9.3.2 ACCF Iteratively for Multi-Target

After the first adjacent cross correlation operation, the ACCF of multiple-targets can be rewritten as follows:

$$
\begin{aligned}
R_1(\tau_1, t_m) =& \sum_{k=1}^{K} A_{2,k}\operatorname{sinc}\left[B\left(\tau_1 + \frac{2N_{0,k}}{c}\right)\right] \\
&\times \exp\left(j\frac{4\pi f_c N_{1,k}}{c}\right)\exp\left(j\frac{4\pi f_c N_{2,k}}{c}t_m\right) \\
&\times \exp\left(j\frac{4\pi f_c N_{3,k}}{c}t_m^2\right) + \\
&\sum_{l=1}^{K}\sum_{n=1,n\neq l}^{K} \sigma_{l,n}\operatorname{sinc}\left[B\left(\tau_1 + \frac{2C_{l,n}}{c}\right)\right] \\
&\times \exp\left(j\frac{4\pi f_c C_{l,n}^{[0]}}{c}\right)\exp\left(j\frac{4\pi f_c C_{l,n}^{[1]}}{c}t_m\right) \\
&\times \exp\left(j\frac{4\pi f_c C_{l,n}^{[2]}}{c}t_m^2\right)\exp\left(j\frac{4\pi f_c C_{l,n}^{[3]}}{c}t_m^3\right).
\end{aligned}
\tag{9.54}
$$

Perform the second adjacent cross correlation operation on (9.54), and the ACCF of (9.54) can expressed as

$$
\begin{aligned}
R_2(\tau_2, t_m) =& \int_0^T R_1(\tau_1, t_m)R_1^*(\tau_1 - \tau_2, t_{m+1})d\tau_1 \\
=& R_{2,self}(\tau_2, t_m) + R_{2,cross}^{[1]}(\tau_2, t_m) \\
&+ R_{2,cross}^{[2]}(\tau_2, t_m) + R_{2,cross}^{[3]}(\tau_2, t_m) \\
&+ R_{2,cross}^{[4]}(\tau_2, t_m) + R_2^{[5]}(\tau_2, t_m),
\end{aligned}
\tag{9.55}
$$

where

$$R_{2,self}(\tau_2, t_m) = \sum_{k=1}^{K} A_{7,k} \mathrm{sinc}\left[B\left(\tau_2 + \frac{2M_{0,k}}{c}\right)\right]$$
$$\times \exp\left(j\frac{4\pi f_c M_{1,k}}{c}\right) \exp\left(j\frac{4\pi f_c M_{2,k}}{c}t_m\right), \quad (9.56)$$

$$R_{2,cross}^{[1]}(\tau_2, t_m) = \sum_{l=1}^{K} \sum_{n=1,n\neq l}^{K} \delta_{l,n}^{[1]} \mathrm{sinc}\left[B\left(\tau_2 + \frac{2D_{l,n}}{c}\right)\right]$$
$$\times \exp\left(j\frac{4\pi f_c D_{l,n}^{[0]}}{c}\right) \exp\left(j\frac{4\pi f_c D_{l,n}^{[1]}}{c}t_m\right) \quad (9.57)$$
$$\times \exp\left(j\frac{4\pi f_c D_{l,n}^{[2]}}{c}t_m^2\right),$$

$$R_{2,cross}^{[2]}(\tau_2, t_m) = \sum_{l=1}^{K} \sum_{n=1,n\neq l}^{K} \delta_{l,n}^{[2]} \mathrm{sinc}\left[B\left(\tau_2 + \frac{2E_{l,n}}{c}\right)\right]$$
$$\times \exp\left(j\frac{4\pi f_c E_{l,n}^{[0]}}{c}\right) \exp\left(j\frac{4\pi f_c E_{l,n}^{[1]}}{c}t_m\right) \quad (9.58)$$
$$\times \exp\left(j\frac{4\pi f_c E_{l,n}^{[2]}}{c}t_m^2\right),$$

$$R_{2,cross}^{[3]}(\tau_2, t_m) = \sum_{i=1}^{K}\sum_{l=1}^{K} \sum_{n=1,n\neq l}^{K} \delta_{i,l,n}^{[3]} \mathrm{sinc}\left[B\left(\tau_2 + \frac{2F_{i,l,n}}{c}\right)\right]$$
$$\times \exp\left(j\frac{4\pi f_c F_{i,l,n}^{[0]}}{c}\right) \exp\left(j\frac{4\pi f_c F_{i,l,n}^{[1]}}{c}t_m\right) \quad (9.59)$$
$$\times \exp\left(j\frac{4\pi f_c F_{i,l,n}^{[2]}}{c}t_m^2\right) \exp\left(j\frac{4\pi f_c F_{i,l,n}^{[3]}}{c}t_m^3\right),$$

$$R_{2,cross}^{[4]}(\tau_2, t_m) = \sum_{i=1}^{K}\sum_{l=1}^{K} \sum_{n=1,n\neq l}^{K} \delta_{i,l,n}^{[4]} \mathrm{sinc}\left[B\left(\tau_2 + \frac{2G_{i,l,n}}{c}\right)\right]$$
$$\times \exp\left(j\frac{4\pi f_c G_{i,l,n}^{[0]}}{c}\right) \exp\left(j\frac{4\pi f_c G_{i,l,n}^{[1]}}{c}t_m\right) \quad (9.60)$$
$$\times \exp\left(j\frac{4\pi f_c G_{i,l,n}^{[2]}}{c}t_m^2\right) \exp\left(j\frac{4\pi f_c G_{i,l,n}^{[3]}}{c}t_m^3\right),$$

$$R_{2,cross}^{[5]}(\tau_2, t_m) = \sum_{l=1}^{K} \sum_{n=1,n\neq l}^{K} \sum_{d=1,d\neq l}^{K} \sum_{e=1,e\neq d,e\neq n}^{K} \delta_{l,n,d,e}^{[5]}$$

$$\times \operatorname{sinc}\left[B\left(\tau_2 + \frac{2H_{l,n,d,e}}{c}\right)\right]$$

$$\times \exp\left(j\frac{4\pi f_c H_{l,n,d,e}^{[0]}}{c}\right)$$

$$\times \exp\left(j\frac{4\pi f_c H_{l,n,d,e}^{[1]}}{c}t_m\right) \qquad (9.61)$$

$$\times \exp\left(j\frac{4\pi f_c H_{l,n,d,e}^{[2]}}{c}t_m^2\right)$$

$$\times \exp\left(j\frac{4\pi f_c H_{l,n,d,e}^{[3]}}{c}t_m^3\right),$$

where

$$D_{l,n} = D_{l,n}^{[0]} + D_{l,n}^{[1]}t_m + D_{l,n}^{[2]}t_m^2, \qquad (9.62a)$$

$$D_{l,n}^{[0]} = -N_{1,l} + N_{1,n} - N_{2,l}T_r - N_{3,l}T_r^2, \qquad (9.62b)$$

$$D_{l,n}^{[1]} = -N_{2,l} + N_{2,n} - 2N_{3,l}T_r, \qquad (9.62c)$$

$$D_{l,n}^{[2]} = -N_{3,l} + N_{3,n}, \qquad (9.62d)$$

$$E_{l,n} = E_{l,n}^{[0]} + E_{l,n}^{[1]}t_m + E_{l,n}^{[2]}t_m^2, \qquad (9.63a)$$

$$E_{l,n}^{[0]} = -C_{l,n}^{[1]}T_r - C_{l,n}^{[2]}T_r^2 - C_{l,n}^{[3]}T_r^3, \qquad (9.63b)$$

$$E_{l,n}^{[1]} = -2C_{l,n}^{[2]}T_r - 3C_{l,n}^{[3]}T_r^2, \qquad (9.63c)$$

$$E_{l,n}^{[2]} = -3C_{l,n}^{[3]}T_r, \qquad (9.63d)$$

$$F_{i,l,n} = F_{i,l,n}^{[0]} + F_{i,l,n}^{[1]}t_m + F_{i,l,n}^{[2]}t_m^2 + F_{i,l,n}^{[3]}t_m^3, \qquad (9.64a)$$

$$F_{i,l,n}^{[0]} = N_{1,k} - C_{l,n}^{[0]} - C_{l,n}^{[1]}T_r - C_{l,n}^{[2]}T_r^2 - C_{l,n}^{[3]}T_r^3, \qquad (9.64b)$$

$$F_{i,l,n}^{[1]} = N_{2,k} - C_{l,n}^{[1]} - 2C_{l,n}^{[2]}T_r - 3C_{l,n}^{[3]}T_r^2, \qquad (9.64c)$$

$$F_{i,l,n}^{[2]} = N_{3,k} - C_{l,n}^{[2]} - 3C_{l,n}^{[3]}T_r, \qquad (9.64d)$$

$$F_{i,l,n}^{[3]} = -C_{l,n}^{[3]}, \qquad (9.64e)$$

$$G_{i,l,n} = G_{i,l,n}^{[0]} + G_{i,l,n}^{[1]}t_m + G_{i,l,n}^{[2]}t_m^2 + G_{i,l,n}^{[3]}t_m^3, \tag{9.65a}$$

$$G_{i,l,n}^{[0]} = -N_{1,k} + C_{l,n}^{[0]} - N_{2,k}T_r - N_{3,k}T_r^2, \tag{9.65b}$$

$$G_{i,l,n}^{[1]} = -N_{2,k} + C_{l,n}^{[1]} - 2N_{3,k}T_r, \tag{9.65c}$$

$$G_{i,l,n}^{[2]} = -N_{3,k} + C_{l,n}^{[2]}, \tag{9.65d}$$

$$G_{i,l,n}^{[3]} = C_{l,n}^{[3]}, \tag{9.65e}$$

$$H_{l,n,d,e} = H_{l,n,d,e}^{[0]} + H_{l,n,d,e}^{[1]}t_m + H_{l,n,d,e}^{[2]}t_m^2 + H_{l,n,d,e}^{[3]}t_m^3, \tag{9.66a}$$

$$H_{l,n,d,e}^{[0]} = -C_{l,n}^{[0]} + C_{d,e}^{[0]} - C_{l,n}^{[1]}T_r - C_{l,n}^{[2]}T_r^2 - C_{l,n}^{[3]}T_r^3, \tag{9.66b}$$

$$H_{l,n,d,e}^{[1]} = -C_{l,n}^{[1]} + C_{d,e}^{[1]} - 2C_{l,n}^{[2]}T_r - 3C_{l,n}^{[3]}T_r^2, \tag{9.66c}$$

$$H_{l,n,d,e}^{[2]} = -C_{l,n}^{[2]} + C_{d,e}^{[2]} - 3C_{l,n}^{[3]}T_r, \tag{9.66d}$$

$$H_{l,n,d,e}^{[3]} = -C_{l,n}^{[3]} + C_{d,e}^{[3]}. \tag{9.66e}$$

From (9.55)–(9.66), we can see that $R_2(\tau_2, t_m)$ contains both self terms (i.e., $R_{2,self}(\tau_2, t_m)$) and cross terms (i.e., $R_{2,cross}^{[1]}(\tau_2, t_m)$, $R_{2,cross}^{[2]}(\tau_2, t_m)$, $\cdots$, $R_{2,cross}^{[5]}(\tau_2, t_m)$). Note that the self terms are all first-order phase functions of slow time t_m, whereas the cross terms are either quadratic phase functions or cubic phase functions of the slow time t_m. As a result, the energy of self terms can be integrated via FT; meanwhile, the energy of cross terms cannot be integrated due to the mismatch with FT, and thus the self terms can be highlighted and the cross terms could be suppressed via FT operation.

Applying FT on (9.55) with respect to t_m yields

$$R_2(\tau_2, f_{t_m}) = \sum_{k=1}^{K} A_{8,k}\text{sinc}\left[B\left(\tau_2 + \frac{2M_{0,k}}{c}\right)\right]$$
$$\times \text{sinc}\left[(N-2)T_r\left(f_{t_m} - \frac{2f_cM_{2,k}}{c}\right)\right] \tag{9.67}$$
$$\times \exp\left(j\frac{4\pi f_c M_{1,k}}{c}\right) + R_{2,cross}(\tau_2, f_{t_m}),$$

where $R_{2,cross}(\tau_2, f_{t_m})$ denotes the FT result of all the cross terms.

By (9.67), each target is completely accumulated and corresponding to a sole peak in the output of FT. Based on the peak location of (9.67), the target can be detected, and the motion parameters can be estimated. According to analyses above, an example is given below to show how the ACCF-based algorithm accumulates the self terms and suppresses the cross terms under multi-target.

Example 3: Two high maneuvering targets C and D are used in this example. Radar parameters are set as: the carrier frequency $f_c = 1$ GHz, the bandwidth $B = 10$ MHz, the sample frequency $f_s = 10$ MHz, pulse repetition frequency $f_p = 200$ Hz

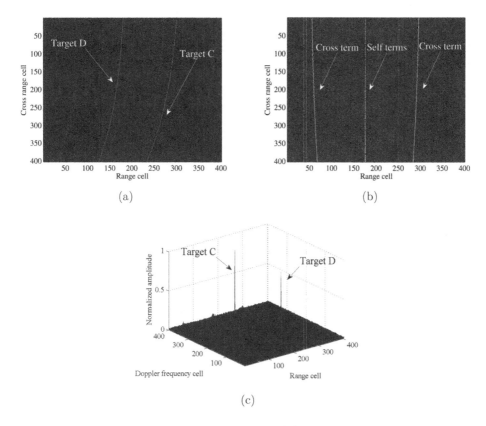

Figure 9.3 Simulation results of Example 3. (a) Result of pulse compression. (b) Result after ACCF operations. (c) Integration result of FT.

and the number of integration pulses $M = 403$. The motion parameters are set as: $r_{0,1} = 101.875$ km, $v_{0,1} = 180$ m/s, $a_{0,1} = 60$ m/s^2, $a_{1,1} = 50$ m/s^3 for target C; $r_{0,2} = 100$ km, $v_{0,2} = 150$ m/s, $a_{0,2} = 30$ m/s^2 and $a_{1,2} = 50$ m/s^3 for target D. Fig. 9.3 gives the simulation results of this example.

Fig. 9.3(a) shows the result of pulse compression, which implies that serious RM effect occurs (range walk and range curvature [6, 7]). Fig. 9.3(b) shows the envelope after ACCF operations. We can see that the self terms are located in the same range cell (i.e., RM has been corrected), while the cross terms are distributed in different range cells, which is helpful to the suppression of cross terms. The integration result via FT is given in Fig. 9.3(c). It can be seen that the target's energy is coherently accumulated as two obvious peaks in the output of FT, which indicates that the cross terms have been suppressed well.

9.4 ADVANTAGES OF THE PROPOSED METHODS

9.4.1 Low Computational Cost

Note that the ACCF can be achieved via complex multiplications, FFT and IFFT. Besides, LVD can be implemented by the scaled Fourier transform and IFFT, which

TABLE 9.1 Computational Complexity of GRFT and the
ACCF-Based Methods

Methods	Computational complexity
GRFT	$O\left(MM_rN_{v_0}N_{a_0}N_{a_1}\right)$
ACCF-LVD	$O(MM_r\log_2 M_r + 6M_rM_lM\log_2 M)$
ACCF iteratively	$O(3MM_r\log_2 M_r + 3M_rM\log_2 M)$

are interpolation free and only use the complex multiplications and FFT based on the scaling principle [8]. Therefore, the two ACCF-based algorithms can be easily implemented by using complex multiplications, FFT and IFTT. Compared with the GRFT method, the ACCF-based algorithms eliminate the brute-force searching procedure of unknown motion parameters, which will reduce the radar system complexity and make these two ACCF-based methods more practical for realistic applications. In the following, the computational complexities of GRFT and the ACCF-based algorithms are analyzed.

Denote the number of range cells, lag samples, searching jerk, searching acceleration and searching velocity by M_r, M_l, N_{a_1}, N_{a_0} and N_{v_0}, respectively. Then, the computational cost of ACCF-LVD is $O(MM_r\log_2 M_r + 6M_rM_lM\log_2 M)$ [8]. Besides, the computational burden of the method via ACCF iteratively is $O(3MM_r\log_2 M_r + 3M_rM\log_2 M)$. On the other hand, the computational complexity will be $O\left(\prod_{q=1}^{q=3} MM_rN_{a_q}\right)$ for GRFT [1]. Table 9.1 gives the computational costs of the three methods. Suppose that $M_r = M = M_l = N_{a_q}$ ($q = 1, 2, 3$), then the computational cost of the two ACCF-based methods are, respectively, $O(M_l^3\log_2 M_l)$ and $O(M_l^2\log_2 M_l)$, whereas the computational burden of GRFT is $O(M_l^5)$. Therefore, the two ACCF-based methods require a much lower computational load than GRFT. Below, we compare the time cost of the two ACCF-based methods and GRFT for coherent integration of a maneuvering target, where the simulation parameters of radar and moving target are the same as those in Example 1. Table 9.2 illustrates the time cost of GRFT and the two ACCF-based algorithms. We can see that the GRFT takes more time than the two ACCF-based methods due to the four-dimensional searching.

TABLE 9.2 Time Cost of GRFT and
the ACCF-Based Methods

Methods	Time cost (s)
GRFT	345.458175
ACCF-LVD	11.185624
ACCF iteratively	5.356537

Note: Main configuration of the computer: CPU: Intel Core i5-4430 3.0 GHz; RAM: 8.00G; Operating System: Windows 7; Software: Matlab 2012b.

9.4.2   Avoidance of Velocity Ambiguity

In general, most targets have a high speed in radar detection, and most existing radars adopt a high carrier frequency and a low pulse repetition frequency (PRF) to guarantee the accurate positioning and far-range target detection [11]. As a result, the velocity ambiguity appears, and the searching procedure of unknown velocity ambiguity integer is necessary [5, 12] or the blind speed sidelobe (BSSL) effect occurs [3, 13]. Fortunately, the ACCF-based methods can avoid the problem of velocity ambiguity. This is also another advantage of the ACCF-based algorithms. Suppose that the carrier frequency is f_c and the pulse repetition frequency is f_p, then the area of unambiguous velocity is $\left[0, \frac{\lambda f_p}{2}\right]$.

Note that the velocity of the kth target (i.e., $v_{0,k}$) is not directly estimated for the two methods based on ACCF. Instead, they firstly estimate the parameters of the ACCF, i.e., $N_{3,k}$, $N_{2,k}$ and $N_{1,k}$. Then, the estimation of $v_{0,k}$ can be achieved via the relationship between $N_{i,k}(i = 1, 2, 3)$ and $v_{0,k}$. Therefore, if the following condition

$$|N_{i,k}| \leq \frac{\lambda f_p}{2}, i = 1, 2, 3 \tag{9.68}$$

is satisfied, then the velocity ambiguity would not be existed in the ACCF-based methods. Generally speaking, the condition shown in (9.68) is satisfied. For example, based on the simulation parameters of Example 1, the $N_{i,k}$ ($i = 1, 2, 3$) can be calculated as $N_{1,k} = 8.0013$, $N_{2,k} = 0.5075$ and $N_{3,k} = 0.15$. In addition, the area of unambiguous velocity is $\left[0, \frac{\lambda f_p}{2}\right] = [0, 30]$. It can be seen that the condition shown in (9.68) is satisfied. As a result, the velocity ambiguity could be avoided for the ACCF-based algorithms.

9.5   NUMERICAL RESULTS

In this section, several numerical simulations are presented to demonstrate the effectiveness of the ACCF-based coherent integration algorithms for a maneuvering target with jerk motion, where the parameters of the radar and moving target are the same as those in Example 1. Comparisons with other popular coherent integration methods, i.e., moving target detection (MTD) [12], RFT [3], RFRFT [14] and GRFT [1], are also given, including the integration ability, detection probability and parameters estimation performance.

9.5.1   Coherent Integration for a Maneuvering Target

We first evaluate the coherent integration performance of the two ACCF-based algorithms for a maneuvering target in Fig. 9.4. Complex white Gaussian noise is added to the echoes, and the SNR is 6 dB after pulse compression. Fig. 9.4(a) shows the result of pulse compression, which indicates that serious RM occurs. Fig. 9.4(b)

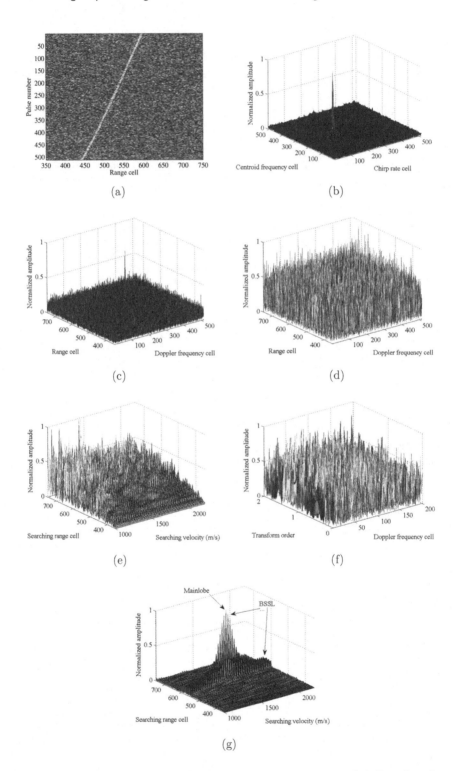

Figure 9.4 Coherent integration for a maneuvering target. (a) Result after pulse compression. (b) Integration result of ACCF-LVD. (c) Integration result of ACCF iteratively. (d) Integration result of MTD. (e) Integration result of RFT. (f) Integration result of RFRFT. (g) Integration result of GRFT.

and Fig. 9.4(c) show, respectively, the integration results of the two methods (i.e., ACCF-LVD and ACCF iteratively). It can be seen that the target energy is coherently accumulated as one peak in the output, which is helpful for the target detection and motion parameters estimation.

For the sake of comparison, Fig. 9.4(d)−Fig. 9.4(f) show, respectively, the integration result of MTD, RFT and RFRFT. Because of the high-order DFM induced by target's jerk motion, these three methods all become invalid. In addition. Fig. 9.4(g) shows the integration result of GRFT. Although GRFT is also able to achieve the coherent accumulation via four-dimensional searching, the BSSL with large peak value appears in the GRFT output, which may lead to serious false alarm and make it hard to obtain the number of target, especially for the scenario of multiple targets.

9.5.2 Maneuvering Target Detection Ability

The detection performances of moving target detection (MTD), Radon Fourier transform (RFT), Radon-fractional Fourier transform (RFRFT), generalized Radon Fourier transform (GRFT) and the two ACCF-based methods are investigated by Monte Carlo trials. We combine the constant false alarm (CFAR) detector [14] and the six methods as corresponding detectors. The Gaussian noises are added to the target echoes, and the false alarm ratio is set as $P_{fa} = 10^{-4}$. Fig. 9.5 shows the detection probability of the six detectors versus different SNR levels, and in each case, 1000 times of Monte Carlo simulations are done. The simulation results show that the probability of the detector based on the ACCF-based methods is superior to MTD, RFT and RFRFT thanks to their ability to deal with the high-order DFM. Moreover, it is worth pointing out that the ACCF-based methods suffer from some

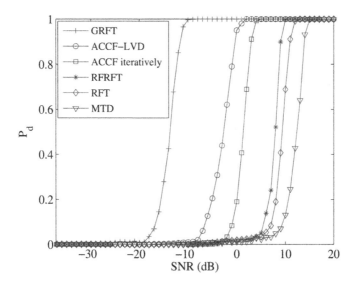

Figure 9.5 Detection probability of MTD, RFT, RFRFT, GRFT and the ACCF-based methods.

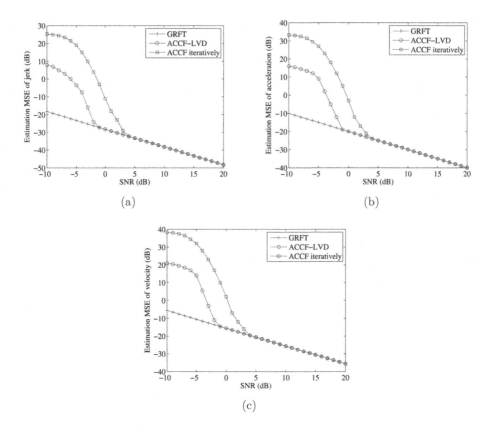

Figure 9.6 Motion parameters estimation performance. (a) Estimation MSE of jerk. (b) Estimation MSE of acceleration. (c) Estimation MSE of velocity.

detection performance loss in comparison with the GRFT algorithm, which is the cost of the reduced computational complexity.

9.5.3 Motion Parameters Estimation Performance

We also compare the motion parameters estimation performance of the ACCF-based algorithms and GRFT method via Monte Carlo trials in Fig. 9.6, where the SNR for the signal varied in [−10, 20 dB]. For each SNR value, 500 simulations were performed, and the measured mean square errors (MSEs) for the motion parameters estimation (i.e., jerk, acceleration and velocity) are, respectively, shown in Fig. 9.6(a), Fig. 9.6(b) and Fig. 9.6(c). It can be seen that the MSEs of the ACCF-based methods are in general close to those of the GRFT method at high SNR.

From the simulation results above (Fig. 9.4–Fig. 9.6 and Table 9.2), we can conclude that the coherent integration and detection performance of the two ACCF-based algorithms are superior to MTD, RFT and RFRFT for a maneuvering target with jerk motion. Besides, compared with the GRFT method, the ACCF-based algorithms can avoid the BSSL effect and achieve a good balance between the computational cost and detection ability as well as the motion parameters estimation performance.

9.6   SUMMARY

In this chapter, we have addressed the coherent integration problem for the detection and motion parameters estimation of a maneuvering target with jerk motion. Summarizing:

- We discussed a fast non-searching method based on ACCF and LVD, i.e., ACCF-LVD, to realize the coherent integration and parameters estimation. This method can effectively remove the migrations (range migration and Doppler frequency migration) and obtain the parameters estimation via adjacent correlation operation and Lv's transform.

- We introduced a fast coherent integration method via ACCF iteratively to further reduce the computational burden and achieve the motion parameters estimation. In this method, two adjacent correlation operations are employed to reduce the order of range migration and Doppler frequency migration. After that, the coherent integration and motion parameters estimation can be obtained via FT.

- We have evaluated the performance of the two ACCF-based algorithms by several numerical simulations. Compared with the GRFT method, the results highlighted that the ACCF-based methods can avoid the BSSL effect and have much lower computational cost. Besides, the integration and detection performances of the two ACCF-based algorithms are superior to the MTD, RFT and RFRFT thanks to their ability to deal with the high-order motions.

Bibliography

[1] J. Xu, X. G. Xia, S. B. Peng, J. Yu, Y. N. Peng, and L. C. Qian, "Radar maneuvering target motion estimation based on Generalized Radon-Fourier transform," *IEEE Transactions on Signal Processing*, vol. 60, no. 12, pp. 6190–6201, December 2012.

[2] Y. Wang and Y. C. Jiang, "ISAR imaging of maneuvering target based on the L-Class of fourth-order complex-lag PWVD," *IEEE Transactions on Geoscience and Remote Sensing*, vol. 48, no. 3, pp. 1518–1527, March 2010.

[3] J. Xu, J. Yu, Y. N. Peng, and X. G. Xia, "Radon-Fourier transform (RFT) for radar target detection (I): generalized Doppler filter bank processing," *IEEE Transactions on Aerospace and Electronic Systems*, vol. 47, no. 2, pp. 1186–1202, April 2011.

[4] X. L. Li, G. L. Cui, W. Yi, and L. J. Kong, "A fast maneuvering target motion parameters estimation algorithm based on ACCF," *IEEE Signal Processing Letters*, vol. 22, no. 3, pp. 270–274, March 2015.

[5] R. P. Perry, R. C. Dipietro, and R. L. Fante, "SAR imaging of moving targets," *IEEE Transactions on Aerospace and Electronic Systems*, vol. 35, no. 1, pp. 188–200, January 1999.

[6] D. Kirkland, "Imaging moving targets using the second-order keystone transform," *IET Radar Sonar and Navigation*, vol. 5, no. 8, pp. 902–910, October 2011.

[7] Y. Jungang, H. Xiaotao, J. Thompson, J. Tian, and Z. Zhimin, "Low-frequency ultra-wideband synthetic aperture radar ground moving target imaging," *IET Radar Sonar and Navigation*, vol. 5, no. 9, pp. 994–1001, December 2011.

[8] X. L. Lv, G. Bi, C. R. Wan, and M. D. Xing, "Lv's distribution: principle, implementation, properties, and performance," *IEEE Transactions on Signal Processing*, vol. 59, no. 8, pp. 3576–3591, August 2011.

[9] S. Luo, G. Bi, X. L. Lv, and F. Y. Hu, "Performance analysis on Lv distribution and its applications," *Digital Signal Processing*, vol. 23, no. 3, pp. 797–807, May 2013.

[10] S. Luo, X. Lv, and G. Bi, "Lv's distribution for time-frequency analysis," in *Proceedings of 2011 International Conference on Circuits, Systems, Control, Signals*, 2011, pp. 110–115.

[11] J. B. Zheng, T. Su, W. T. Zhu, X. H. He, and Q. H. Liu, "Radar high-speed target detection based on the scaled inverse fourier transform," *IEEE Journal of Selected Topics in Applied Earth Observations and Remote Sensing*, vol. 8, no. 3, pp. 1108–1119, March 2015.

[12] M. D. Xing, J. H. Su, G. Y. Wang, and Z. Bao, "New parameter estimation and detection algorithm for high speed small target," *IEEE Transactions on Aerospace and Electronic Systems*, vol. 47, no. 1, pp. 214–224, January 2011.

[13] J. Xu, J. Yu, Y. N. Peng, and X. G. Xia, "Radon-Fourier transform (RFT) for radar target detection (II): Blind speed sidelobe suppression," *IEEE Transactions on Aerospace and Electronic Systems*, vol. 47, no. 4, pp. 2473–2489, October 2011.

[14] X. L. Chen, J. Guan, N. B. Liu, and Y. He, "Maneuvering target detection via Radon-Fractional Fourier transform-based long-time coherent integration," *IEEE Transactions on Signal Processing*, vol. 62, no. 4, pp. 939–953, February 2014.

CHAPTER 10

STGRFT-Based Integration for Multiple Motion States

The previous chapters are all based on the assumption that the target is of single motion model (i.e., the target's motion model is fixed) during the long integration time. However, with increasing target's maneuverability and long observation time (needed to detect weak targets), the target's motion model is often changing (i.e., target is of multiple motion models) during the integration time. In this chapter, we address the coherent integration problem for detecting a maneuvering weak target with multiple motion models, where the target signal model is first established. Thereafter, the short-time generalized Radon Fourier transform (STGRFT)-based long-time coherent integration (LTCI) algorithm is introduced, including the transform definition, property and coherent integration processing procedure .

10.1   SIGNAL MODEL

Assume that the linear frequency modulation (LFM) is adopted as the baseband waveform [1–3]

$$s_{trans}(t, t_m) = \text{rect}\left(\frac{t}{T_p}\right) \exp\left(j\pi\mu t^2\right) \exp[j2\pi f_c(t + t_m)], \qquad (10.1)$$

where $\text{rect}\,(x) = \begin{cases} 1, & |x| \leq \frac{1}{2} \\ 0, & |x| > \frac{1}{2} \end{cases}$, μ, T_p, t and f_c denote the frequency modulated rate, pulse duration, fast time and carrier frequency, respectively. The pulse repetition time is denoted as T_r, and t_m is the slow-time. Suppose that within the radar detection area, a target of S motion models (corresponding to S motion stages) is presented during the long observation time. The target is supposed to be a nondispersive and isotropic point scattering object [4–7]. Neglecting the high-order motions, the instantaneous slant range of target during the ith motion stage satisfies

$$R_i(t_m) = R_{0,i} + V_i(t_m - T_{i-1}) + A_i(t_m - T_{i-1})^2, t_m \in [T_{i-1}, T_i], \qquad (10.2)$$

DOI: 10.1201/9781003529101-10

where $R_{0,i}$, V_i and A_i are, respectively, the initial slant range, radial velocity and acceleration for the target of ith motion stage and T_{i-1} and T_i denote, respectively, the beginning time and ending time of the ith motion stage ($i = 1, 2, \cdots, S$).

According to the radar principle, the received signal after PC during the ith motion stage could be stated as

$$s_i(t, t_m) = \sigma \text{sinc}\left[B\left(t - \frac{2R_i(t_m)}{c}\right)\right] \times \exp\left[-j4\pi\frac{R_i(t_m)}{\lambda}\right], t_m \in [T_{i-1}, T_i], \quad (10.3)$$

where σ, c, B and λ denote, respectively, the signal amplitude, the speed of light, the bandwidth and the wavelength.

Then, the received signal after PC during the total integration time (i.e., from T_0 to T_M) could be expressed as

$$\begin{aligned} s(t, t_m) &= \sum_{i=1}^{M} \sigma w_i(t_m)\text{sinc}\left[B\left(t - \frac{2R_i(t_m)}{c}\right)\right] \exp\left[-j4\pi\frac{R_i(t_m)}{\lambda}\right] \\ &= \sum_{i=1}^{M} w_i(t_m)s_i(t, t_m), \end{aligned}$$

$$(10.4)$$

where

$$w_i(t_m) = \text{rect}\left[\frac{t_m - 0.5(T_{i-1} + T_i)}{T_i - T_{i-1}}\right] = \begin{cases} 1, & T_{i-1} \leq t_m \leq T_i \\ 0, & \text{else} \end{cases}, \quad (10.5)$$

By (10.4), it could be seen that the target's motion parameters are $(R_{0,i}, V_i, A_i)$ during the ith motion stage (between T_{i-1} and T_i). Then, in the instantaneous time T_i, the target's motion model begins to change, and the target's motion parameters change into $(R_{0,i+1}, V_{i+1}, A_{i+1})$. Hence, we have

$$R_{0,i+1} = R_i(t_m)|_{t_m=T_i}, V_{i+1} = \frac{dR_i(t_m)}{dt}|_{t_m=T_i}. \quad (10.6)$$

10.2   SHORT TIME GRFT

10.2.1   Definition of STGRFT

The definition of STGRFT is

$$STGRFT_{g(t_m)}(r_{s0}, v, a) = \int_{\eta_0}^{\eta_1} s\left(2r(t_m)/c, t_m\right)g(t_m) \times \exp\left(j4\pi\frac{r(t_m)}{\lambda}\right)dt_m, \quad (10.7)$$

where $g(t_m)$ denotes the window function and η_0 and η_1 are, respectively, the beginning time and ending time of the window function's non-zero area, i.e.,

$$g(t_m) = \text{rect}\left[\frac{t_m - 0.5(\eta_1 + \eta_0)}{\eta_1 - \eta_0}\right] = \begin{cases} 1, & \eta_0 \leq t_m \leq \eta_1 \\ 0, & \text{else} \end{cases}$$

$$T_0 \leq \eta_0 \leq \eta_1 \leq T_S, \quad (10.8)$$

$r(t_m)$ represents the searching trajectory, which is determined by the searching motion parameters (r_{s0}, v, a), i.e.,

$$r(t_m) = r_{s0} + v(t_m - \eta_0) + a(t_m - \eta_0)^2, \tag{10.9}$$

where r_{s0}, v and a denote, respectively, the searching initial range, searching velocity and searching acceleration.

For comparison, the definition of GRFT [8] is also given, i.e.,

$$GRFT(r_{s0}, v, a) = \int_{T_0}^{T_S} s \left(2[r_{s0} + v(t_m - T_0) + a(t_m - T_0)^2]/c, t_m \right)$$
$$\times \exp \left(j4\pi \frac{v(t_m - T_0) + a(t_m - T_0)^2}{\lambda} \right) dt_m. \tag{10.10}$$

Compared with GRFT, the differences and improvements of STGRFT are as follows:

- GRFT extracts and integrates the target signal during the whole observation time (i.e., the beginning time and ending time of the integration operation is fixed), while STGRFT extracts and integrates the target signal within part of the observation time, which is determined by the window function $g(t_m)$ (i.e., the beginning time and ending time of the integration operation in STGRFT are adjustable in accordance with the target's motion model). Therefore, while it is difficult to match maneuvering target over the entire duration, STGRFT is able to match better within the shorter window time frame.

- Different from GRFT, STGRFT also compensates the phase term caused by the target's initial range. Thus, STGRFT is able to match and compensate the phase difference of the maneuvering target signal within multiple motion stages.

10.2.2 Property of STGRFT with Respect to Window Function

Substituting (10.4) into (10.7) yields

$$STGRFT_{g(t_m)}(r_{s0}, v, a) = \int_{\eta_0}^{\eta_1} \sum_{i=1}^{S} \sigma \text{sinc} \left[B \left(\frac{2(r(t_m) - R_i(t_m))}{c} \right) \right]$$
$$\times w_i(t_m) g(t_m) \exp \left[j4\pi \frac{r(t_m) - R_i(t_m)}{\lambda} \right] dt_m. \tag{10.11}$$

Without loss of generality, we focus on the analysis of STGRFT for target signal during the ith motion stage. Let

$$D = \{t|g(t_m) = 1\}, E = \{t|w_i(t_m) = 1\}, \tag{10.12}$$

where D and E indicate, respectively, the non-zero area of $g(t_m)$ and $w_i(t_m)$.

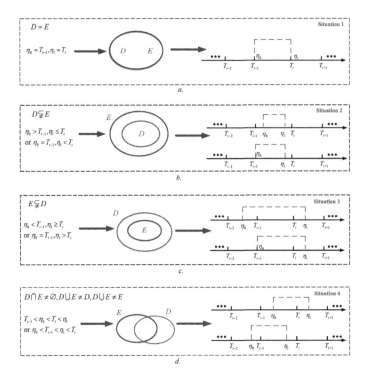

Figure 10.1 Sketch map of the four situations.

Based on the relationship between the set D and set E, the following conclusions could be arrived:

Situation 1: $D = E$, as shown in Fig. 10.1(a), i.e.,

$$\eta_0 = T_{i-1}, \eta_1 = T_i. \tag{10.13}$$

Then, we have

$$g(t_m)w_i(t_m) = w_i(t_m), \quad g(t_m)w_k(t_m) = 0 (k \neq i). \tag{10.14}$$

Note that (10.11) could be rewritten as

$$
\begin{aligned}
STGRFT_{g(t_m)}(r_{s0}, v, a) = &\int_{T_{i-1}}^{T_i} \sigma \mathrm{sinc}\left[B\left(\frac{2(r(t_m) - R_i(t_m))}{c}\right)\right] \\
&\times g(t_m)w_i(t_m)\exp\left[j4\pi\frac{r(t_m) - R_i(t_m)}{\lambda}\right]dt_m \\
&+ \int_{T_{i-1}}^{T_i}\sum_{k=1, k\neq i}^{S}\sigma\mathrm{sinc}\left[B\left(\frac{2(r(t_m) - R_k(t_m))}{c}\right)\right] \\
&\times g(t_m)w_k(t_m)\exp\left[j4\pi\frac{r(t_m) - R_k(t_m)}{\lambda}\right]dt_m,
\end{aligned}
\tag{10.15}
$$

where the first integral term of (10.15) denotes the STGRFT of target signal during the ith motion stage and the last integral term represents the STGRFT of target signal during the other motion stages.

Substituting (10.14) into (10.15) yields

$$
\begin{aligned}
STGRFT_{g(t_m)}&(r_{s0}, v, a) \\
&= \int_{T_{i-1}}^{T_i} \sigma \text{sinc}\left[B\left(\frac{2(r(t_m) - R_i(t_m))}{c}\right)\right] w_i(t_m) \\
&\quad \times \exp\left[j4\pi\frac{r(t_m) - R_i(t_m)}{\lambda}\right] dt_m + \int_{T_{i-1}}^{T_i} 0 dt_m \\
&= \int_{T_{i-1}}^{T_i} \sigma \text{sinc}\left[B\left(\frac{2(r(t_m) - R_i(t_m))}{c}\right)\right] \times \exp\left[j4\pi\frac{r(t_m) - R_i(t_m)}{\lambda}\right] dt_m.
\end{aligned}
\tag{10.16}
$$

When the searching motion parameters (r_{s0}, v, a) match with the target's true motion parameters during the ith motion stage, i.e., $r_{s0} = R_{0,i}, v = V_i, a = A_i$, (10.16) could be recast as

$$
\begin{aligned}
STGRFT_{g(t_m)}(r_{s0}, v, a) &= \int_{T_{i-1}}^{T_i} \sigma \text{sinc}\left[2B0t_m\right] \exp\left[j4\pi 0 t_m\right] dt_m \\
&= \sigma(T_i - T_{i-1}).
\end{aligned}
\tag{10.17}
$$

From (10.16) and (10.17), it could be seen that the target signal during the ith motion stage will be totally extracted and coherently integrated as a peak in the STGRFT domain (the location is corresponding to $r_{s0} = R_{0,i}$, $v = V_i$ and $a = A_i$), whereas the target signal during other motion stages all filter out.

Situation 2: $D \subsetneqq E$, as shown in Fig. 10.1(b). In this situation, η_0, T_{i-1}, η_1 and T_i satisfy the follow equation:

$$
\eta_0 > T_{i-1}, \eta_1 \le T_i \text{ or } \eta_0 = T_{i-1}, \eta_1 < T_i.
\tag{10.18}
$$

Then, we have

$$
w_i(t_m)g(t_m) = g(t_m), \ w_k(t_m)g(t_m) = 0(k \ne i).
\tag{10.19}
$$

Combining (10.19), the STGRFT in (10.11) could be written as

$$
\begin{aligned}
STGRFT_{g(t_m)}(r_{s0}, v, a) &= \int_{\eta_0}^{\eta_1} \sigma \text{sinc}\left[B\left(\frac{2(r(t_m) - R_i(t_m))}{c}\right)\right] \\
&\quad \times w_i(t_m)g(t_m) \exp\left[j4\pi\frac{r(t_m) - R_i(t_m)}{\lambda}\right] dt_m \\
&\quad + \int_{\eta_0}^{\eta_1} \sum_{k=1, k \ne i}^{S} \sigma \text{sinc}\left[B\left(\frac{2(r(t_m) - R_k(t_m))}{c}\right)\right] \\
&\quad \times w_k(t_m)g(t_m) \exp\left[j4\pi\frac{r(t_m) - R_k(t_m)}{\lambda}\right] dt_m \\
&= \int_{\eta_0}^{\eta_1} \sigma \text{sinc}\left[B\left(\frac{2(r(t_m) - R_i(t_m))}{c}\right)\right] \\
&\quad \times \exp\left[j4\pi\frac{r(t_m) - R_i(t_m)}{\lambda}\right] dt_m.
\end{aligned}
\tag{10.20}
$$

It should be pointed out that the coherent integration of (10.20) is obtained based on the fact that RM elimination and DFM compensation are accomplished when the searching motion parameters equal to the target's motion parameters [8], i.e.,

$$r(t_m) - R_i(t_m) \equiv 0, \text{ when } r_{s0} = R_{0,i}, v = V_i, a = A_i. \tag{10.21}$$

In other words, $r(t_m) - R_i(t_m)$ does not change with the time t_m and is equal to zero at every instantaneous time t_m.

(1) Case 1 of situation 2 (i.e., $\eta_0 > T_{i-1}, \eta_1 \leq T_i$), when $r_{s0} = R_{0,i}, v = V_i$ and $a = A_i$, we have

$$
\begin{aligned}
r(t_m) - R_i(t_m) =& (r_{s0} - R_{0,i}) + (v - V_i)t_m + (a - A_i)t_m^2 \\
& + 2(A_i T_{i-1} - a\eta_0)t_m + (a\eta_0^2 - A_i T_{i-1}^2) \\
& + (V_i T_{i-1} - v\eta_0) \\
=& 2A_i(T_{i-1} - \eta_0)t_m + A_i(\eta_0^2 - T_{i-1}^2) \\
& + V_i(T_{i-1} - \eta_0) \\
=& (T_{i-1} - \eta_0)(2A_i t_m - A_i(T_{i-1} + \eta_0) + V_i) \\
\not\equiv& 0.
\end{aligned}
\tag{10.22}
$$

From (10.22), we can see that $r(t_m) - R_i(t_m)$ still changes with the slow time and cannot be equivalent to zero at every instantaneous time t when $r_{s0} = R_{0,i}, v = V_i$ and $a = A_i$, since $A_i \neq 0$, $V_i \neq 0$ and $T_{i-1} \neq \eta_0$ in general. Hence, the RM elimination and DFM compensation would not be accomplished, and the coherent integration of the target signal during the ith motion stage would not be obtained via STGRFT.

In particular, based on the Cauchy-Shwartz inequality, we could obtain the upper bound for the integration term of target signal within the ith motion stage (i.e., the last integral term of (10.20)):

$$
\begin{aligned}
& \left| \int_{\eta_0}^{\eta_1} \sigma \text{sinc} \left[B \left(\frac{2(r(t_m) - R_i(t_m))}{c} \right) \right] \right. \\
& \left. \times \exp \left[j4\pi \frac{r(t_m) - R_i(t_m)}{\lambda} \right] dt_m \right| \\
\leq & \left[\int_{\eta_0}^{\eta_1} \left| \sigma \text{sinc} \left[B \left(\frac{2(r(t_m) - R_i(t_m))}{c} \right) \right] \right|^2 dt_m \right]^{0.5} \\
& \times \left[\int_{\eta_0}^{\eta_1} \left| \exp \left[j4\pi \frac{r(t_m) - R_i(t_m)}{\lambda} \right] \right|^2 dt_m \right]^{0.5} \\
< & \sqrt{[\sigma^2(\eta_1 - \eta_0)](\eta_1 - \eta_0)} \\
< & \sigma(\eta_1 - \eta_0) < \sigma(T_i - T_{i-1}).
\end{aligned}
\tag{10.23}
$$

Therefore, the integrated peak value in this case is less than that of situation 1.

(2) Case 2 of situation 2 (i.e., $\eta_0 = T_{i-1}, \eta_1 < T_i$), when $r_{s0} = R_{0,i}, v = V_i$ and $a = A_i$, we have

$$
\begin{aligned}
r(t_m) - R_i(t_m) =& 2A_i(T_{i-1} - \eta_0)t_m + A_i(\eta_0^2 - T_{i-1}^2) \\
& + V_i(T_{i-1} - \eta_0) \equiv 0.
\end{aligned}
\tag{10.24}
$$

However, in this case, the STGRFT shown in (10.20) could be recast as

$$STGRFT_{g(t_m)}(r_{s0}, v, a) = \int_{\eta_0}^{\eta_1} \sigma \text{sinc}\left[B\left(\frac{2(r(t_m) - R_i(t_m))}{c}\right)\right]$$

$$\times \exp\left[j4\pi\frac{r(t_m) - R_i(t_m)}{\lambda}\right] dt_m \quad (10.25)$$

$$= \sigma(\eta_1 - \eta_0) < \sigma(T_i - T_{i-1}),$$

$$\text{when } r_{s0} = R_{0,i}, v = V_i, a = A_i.$$

By (10.25), we could see that the integrated peak value for case 2 of situation 2 is still less than the peak value of situation 1 shown in (10.17). This is because in case 2 of situation 2, only part of the target signal during the ith motion stage is extracted and integrated, while the entire target signal during the ith motion stage is extracted and accumulated in situation 1.

Situation 3: $E \subsetneq D$, i.e.,

$$\eta_0 < T_{i-1}, \eta_1 \geq T_i \text{ or } \eta_0 = T_{i-1}, \eta_1 > T_i. \quad (10.26)$$

Then, we have

$$w_i(t_m)g(t_m) = w_i(t_m). \quad (10.27)$$

Without loss of generality, we consider the scene that shown in Fig. 10.1(c) for simplicity. Then, the STGRFT in (10.11) can be rewritten as

$$STGRFT_{g(t_m)}(r_{s0}, v, a) = \int_{T_{i-1}}^{T_i} \sigma \text{sinc}\left[B\left(\frac{2(r(t_m) - R_i(t_m))}{c}\right)\right]$$

$$\times \exp\left[j4\pi\frac{r(t_m) - R_i(t_m)}{\lambda}\right] dt_m$$

$$+ \int_{\eta_0}^{T_{i-1}} \sigma \text{sinc}\left[B\left(\frac{2(r(t_m) - R_{i-1}(t_m))}{c}\right)\right]$$

$$\times \exp\left[j4\pi\frac{r(t_m) - R_{i-1}(t_m)}{\lambda}\right] dt_m \quad (10.28)$$

$$+ \int_{T_i}^{\eta_1} \sigma \text{sinc}\left[B\left(\frac{2(r(t_m) - R_{i+1}(t_m))}{c}\right)\right]$$

$$\times \exp\left[j4\pi\frac{r(t_m) - R_{i+1}(t_m)}{\lambda}\right] dt_m,$$

where the first integral term of (10.28) represents the STGRFT of target signal during the ith motion stage, the second integral term is the STGRFT of target signal within the $(i-1)$th motion stage, and the third integral term represents the STGRFT of target signal within the $(i+1)$th motion stage.

(1) Case 1 of situation 3 (i.e., $\eta_0 < T_{i-1}, \eta_1 \geq T_i$), when $r_{s0} = R_{0,i}$, $v = V_i$ and $a = A_i$, we have

$$r(t_m) - R_i(t_m) = 2A_i(T_{i-1} - \eta_0)t_m + A_i(\eta_0^2 - T_{i-1}^2)$$

$$+ V_i(T_{i-1} - \eta_0) \not\equiv 0. \quad (10.29)$$

Hence, the RM and DFM of the target signal during the ith motion stage would not be totally removed. As a result, the coherent integration of the target signal during the ith motion stage would not be achieved.

Notably, based on the Cauchy-Shwartz inequality, we could obtain the upper bound for the integration term of target signal within the ith motion stage (i.e., the first integral term of (10.28)):

$$
\left| \int_{T_{i-1}}^{T_i} \sigma \text{sinc} \left[B \left(\frac{2(r(t_m) - R_i(t_m))}{c} \right) \right] \times \exp \left[j4\pi \frac{r(t_m) - R_i(t_m)}{\lambda} \right] dt_m \right|
$$

$$
\leq \left[\int_{T_{i-1}}^{T_i} \left| \sigma \text{sinc} \left[B \left(\frac{2(r(t_m) - R_i(t_m))}{c} \right) \right] \right|^2 dt_m \right]^{0.5}
$$

$$
\times \left[\int_{T_{i-1}}^{T_i} \left| \exp \left[j4\pi \frac{r(t_m) - R_i(t_m)}{\lambda} \right] \right|^2 dt_m \right]^{0.5}
$$

$$
< \sqrt{[\sigma^2(T_i - T_{i-1})](T_i - T_{i-1})}
$$

$$
< \sigma(T_i - T_{i-1}).
$$

Thus, the integrated peak value of this case is less than that of situation 1.

(2) Case 2 of situation 3 (i.e., $\eta_0 = T_{i-1}, \eta_1 > T_i$), then (10.28) could be described as

$$
STGRFT_{g(t_m)}(r_{s0}, v, a) = \int_{T_{i-1}}^{T_i} \sigma \text{sinc} \left[B \left(\frac{2(r(t_m) - R_i(t_m))}{c} \right) \right]
$$

$$
\times \exp \left[j4\pi \frac{r(t_m) - R_i(t_m)}{\lambda} \right] dt_m
$$

$$
+ \int_{T_i}^{\eta_1} \sigma \text{sinc} \left[B \left(\frac{2(r(t_m) - R_{i+1}(t_m))}{c} \right) \right]
$$

$$
\times \exp \left[j4\pi \frac{r(t_m) - R_{i+1}(t_m)}{\lambda} \right] dt_m.
$$

When $r_{s0} = R_{0,i}$, $v = V_i$ and $a = A_i$, we have

$$
r(t_m) - R_i(t_m) = 2A_i(T_{i-1} - \eta_0)t_m + A_i(\eta_0^2 - T_{i-1}^2)
$$

$$
+ V_i(T_{i-1} - \eta_0) \equiv 0.
$$

In this case, the coherent integration of target signal during the ith motion stage is obtained. However, it should be pointed out that in case 2 of situation 3, not only that the target signal during the ith motion stage is extracted and integrated, but also part of the target signal during the $(i+1)$th motion stage is extracted and integrated (corresponding to the second integral term of (10.31)). On the one hand, the target signal within different motion stages is with different motion parameters, and thus the second integral term of (10.31) would hardly provide benefit for the first integral term of (10.31). On the other hand, the partially extraction of the target signal during the $(i+1)$th motion stage means that we can no longer extract and integrate the target signal within the $(i+1)$th motion stage, via next STGRFT operation, which is detrimental to the LTCI of the target signal in all the motion stages.

Situation 4: $D \cap E \neq \varnothing$ and $D \cup E \neq D$ and $D \cup E \neq E$, i.e.,

$$T_{i-1} < \eta_0 < T_i < \eta_1, \text{ or } \eta_0 < T_{i-1} < \eta_1 < T_i. \tag{10.33}$$

Without loss of generality, for simplicity we consider the scene that shown in Fig. 10.1(d).

(1) Case 1 of situation 4 (i.e., $T_{i-1} < \eta_0 < T_i < \eta_1$), the STGRFT shown in (10.11) could be rewritten as

$$
\begin{aligned}
STGRFT_{g(t_m)}(r_{s0}, v, a) = &\int_{\eta_0}^{T_i} \sigma \text{sinc} \left[B \left(\frac{2(r(t_m) - R_i(t_m))}{c} \right) \right] \\
&\times \exp \left[j4\pi \frac{r(t_m) - R_i(t_m)}{\lambda} \right] dt_m \\
&+ \int_{T_i}^{\eta_1} \sigma \text{sinc} \left[B \left(\frac{2(r(t_m) - R_{i+1}(t_m))}{c} \right) \right] \\
&\times \exp \left[j4\pi \frac{r(t_m) - R_{i+1}(t_m)}{\lambda} \right] dt_m,
\end{aligned} \tag{10.34}
$$

where the first integral term of (10.34) denotes the STGRFT of target signal within the ith motion stage and the second integral term represents the STGRFT of target signal within the $(i+1)$th motion stage. It could be seen from (10.34) that only part of the signal during the ith motion stage is extracted and accumulated.

When $r_{s0} = R_{0,i}$, $v = V_i$ and $a = A_i$, we have

$$
\begin{aligned}
r(t_m) - R_i(t_m) = &2A_i(T_{i-1} - \eta_0)t_m + A_i(\eta_0^2 - T_{i-1}^2) \\
&+ V_i(T_{i-1} - \eta_0) \not\equiv 0.
\end{aligned} \tag{10.35}
$$

Hence, the coherent integration of the extracted target signal during the ith motion stage would not be obtained.

In particular, with the Cauchy-Shwartz inequality, we could obtain the upper bound for the integration term of the target signal within the ith motion stage (i.e., the first integral term of (10.34)):

$$
\begin{aligned}
&\left| \int_{\eta_0}^{T_i} \sigma \text{sinc} \left[B \left(\frac{2(r(t_m) - R_i(t_m))}{c} \right) \right] \times \exp \left[j4\pi \frac{r(t_m) - R_i(t_m)}{\lambda} \right] dt_m \right| \\
&\leq \left[\int_{\eta_0}^{T_i} \left| \sigma \text{sinc} \left[B \left(\frac{2(r(t_m) - R_i(t_m))}{c} \right) \right] \right|^2 dt_m \right]^{0.5} \\
&\times \left[\int_{\eta_0}^{T_i} \left| \exp \left[j4\pi \frac{r(t_m) - R_i(t_m)}{\lambda} \right] \right|^2 dt_m \right]^{0.5} \\
&< \sqrt{[\sigma^2(T_i - \eta_0)](T_i - \eta_0)} \\
&< \sigma(T_i - T_{i-1}).
\end{aligned} \tag{10.36}
$$

Therefore, the integrated peak value of this case is less than that of situation 1.

(2) Case 2 of situation 4 (i.e., $\eta_0 < T_{i-1} < \eta_1 < T_i$), the STGRFT of (10.11) could be rewritten as

$$
\begin{aligned}
STGRFT_{g(t_m)}&(r_{s0}, v, a) \\
&= \int_{T_{i-1}}^{\eta_1} \sigma \mathrm{sinc} \left[B \left(\frac{2(r(t_m) - R_i(t_m))}{c} \right) \right] \times \exp \left[j4\pi \frac{r(t_m) - R_i(t_m)}{\lambda} \right] dt_m \\
&+ \int_{\eta_0}^{T_{i-1}} \sigma \mathrm{sinc} \left[B \left(\frac{2(r(t_m) - R_{i-1}(t_m))}{c} \right) \right] \times \exp \left[j4\pi \frac{r(t_m) - R_{i-1}(t_m)}{\lambda} \right] dt_m,
\end{aligned}
$$
(10.37)

where the first integral term of (10.37) represents the STGRFT of target signal within the ith motion stage and the second integral term denotes the STGRFT of target signal within the $(i-1)$th motion stage. By (10.37), we can see that only part of the signal during the ith motion stage is extracted and integrated.

When $r_{s0} = R_{0,i}$, $v = V_i$ and $a = A_i$, we have

$$
\begin{aligned}
r(t_m) - R_i(t_m) =& 2A_i(T_{i-1} - \eta_0)t_m + A_i(\eta_0^2 - T_{i-1}^2) \\
& + V_i(T_{i-1} - \eta_0) \not\equiv 0.
\end{aligned}
$$
(10.38)

Therefore, the coherent integration of the extracted target signal during the ith motion stage would not be obtained.

Similarly, with the Cauchy-Shwartz inequality, we could obtain the upper bound for the integration term of the target signal within the ith motion stage (i.e., the first integral term of (10.37)):

$$
\begin{aligned}
\Bigg| \int_{T_{i-1}}^{\eta_1} & \sigma \mathrm{sinc} \left[B \left(\frac{2(r(t_m) - R_i(t_m))}{c} \right) \right] \\
& \times \exp \left[j4\pi \frac{r(t_m) - R_i(t_m)}{\lambda} \right] dt_m \Bigg| \\
\leq & \left[\int_{T_{i-1}}^{\eta_1} \left| \sigma \mathrm{sinc} \left[B \left(\frac{2(r(t_m) - R_i(t_m))}{c} \right) \right] \right|^2 dt_m \right]^{0.5} \\
& \times \left[\int_{T_{i-1}}^{\eta_1} \left| \exp \left[j4\pi \frac{r(t_m) - R_i(t_m)}{\lambda} \right] \right|^2 dt_m \right]^{0.5} \\
< & \sqrt{[\sigma^2(\eta_1 - T_{i-1})](\eta_1 - T_{i-1})} \\
< & \sigma(T_i - T_{i-1}).
\end{aligned}
$$
(10.39)

Based on the discussions and analysis of the four situations above, we could conclude that the integration outputs could be obtained via STGRFT with respect to the different window functions. Only when the window function of STGRFT matches the target signal during the ith motion stage (i.e., $g(t_m) = w_i(t_m)$), the target signal of the ith motion stage could then be coherently integrated and reach its maximum value and would also not affect the coherent integration of the target signal during the other motion stages.

A simulation example is given below to show how the STGRFT performs with different window functions (corresponding to different $g(t)$ of the four situations), where the window functions of the four situations are given in Table 10.1. The radar parameters are set as: carrier frequency $f_c = 0.15$ GHz, bandwidth $B = 10$ MHz, sample frequency $f_s = 50$ MHz, pulse duration $T_p = 10$ μs, pulse repetition frequency $f_p = 200$ Hz. The whole integration time is $T = 3.825$ s, where the total pulse number is 765. The target moves with uniform motion (i.e., initial range cell is 600, radial velocity $V_1 = 100$ m/s) in the first stage (between 0 and 1.275 s) and moves with uniformly accelerated motion (i.e., radial acceleration $A_2 = 60$ m/s^2) during the second stage (between 1.275 and 2.55 s). Thereafter, the target moves with uniformly decelerated motion (radial acceleration $A_3 = -30$ m/s^2) within the third stage (between 2.55 and 3.825 s).

Fig. 10.2 shows the simulation results, where Fig. 10.2(a) gives the result of PC. From this figure, we could see that the target has three motion models (i.e., the uniform motion model in the first stage, the uniformly accelerated motion model in the second stage and the uniformly decelerated motion model in the third stage) during the whole integration time. Fig. 10.2(b)−Fig. 10.2(h) give, respectively, the STGRFT outputs under four situations (corresponding to different window functions of Table 10.1). From Fig. 10.2(b) to Fig. 10.2(h), several observations are noted:

- In situation 1, the window function of STGRFT matches the target signal during the second motion stage, and thus the target signal in this stage will be totally extracted and coherently integrated, where the peak location is corresponding to target's motion parameters during second motion stage, as shown in Fig. 10.2(b).

- For case 1 of situation 2, the RM and DFM of the extracted target signal within the second motion stage could not be totally compensated. Thus, the extracted signal could not be coherently accumulated, which makes the integrated peak value of Fig. 10.2(c) less than that of Fig. 10.2(b). On the other hand, for case 2 of situation 2, although the RM and DFM of extracted target's signal within the second motion stage are eliminated, the extracted target's signal is only part of the second motion stage. Hence, the corresponding integrated peak value of this case (Fig. 10.2(d)) is still less than that of situation 1.

- As to case 1 of situation 3, the RM and DFM of target signal during the second stage are still present and cannot be totally removed. Hence, the coherent integration of target signal during the second stage could not be obtained, which implies that the integrated peak value of Fig. 10.2(e) is less than that of Fig. 10.2(b). For case 2 of situation 3, the target signal during the second stage will be totally extracted and coherently accumulated, and the integrated peak value of Fig. 10.2(f) is close to that of Fig. 10.2(b). However, note that in this case, part of target signal during the third stage is also extracted, which indicates that we could not extract and integrate the entire target signal of the third stage simultaneously.

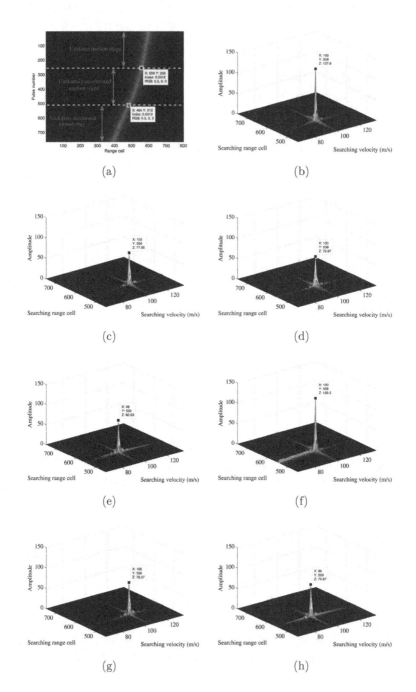

Figure 10.2 STGRFT outputs with different situations. (a) PC. (b) Situation 1. (c) Case 1 of situation 2. (d) Case 2 of situation 2. (e) Case 1 of situation 3. (f) Case 2 of situation 3. (g) Case 1 of situation 4. (h) Case 2 of situation 4.

TABLE 10.1　Different Widow Functions of STGRFT under Four Situations

Different situations	Window function of STGRFT
Situation 1	$g(t_m) = \begin{cases} 1, & t_m \in [1.28, 2.55 \text{ s}] \\ 0, & \text{otherwise} \end{cases}$
Case 1 of situation 2	$g(t_m) = \begin{cases} 1, & t_m \in [1.325, 2.5 \text{ s}] \\ 0, & \text{otherwise} \end{cases}$
Case 2 of situation 2	$g(t_m) = \begin{cases} 1, & t_m \in [1.28, 2 \text{ s}] \\ 0, & \text{otherwise} \end{cases}$
Case 1 of situation 3	$g(t_m) = \begin{cases} 1, & t_m \in [1.25, 2.575 \text{ s}] \\ 0, & \text{otherwise} \end{cases}$
Case 2 of situation 3	$g(t_m) = \begin{cases} 1, & t_m \in [1.28, 2.8 \text{ s}] \\ 0, & \text{otherwise} \end{cases}$
Case 1 of situation 4	$g(t_m) = \begin{cases} 1, & t_m \in [1.325, 2.575 \text{ s}] \\ 0, & \text{otherwise} \end{cases}$
Case 2 of situation 4	$g(t_m) = \begin{cases} 1, & t_m \in [1.25, 2.5 \text{ s}] \\ 0, & \text{otherwise} \end{cases}$

- For case 1 and case 2 of situation 4, due to the mismatch between the window function of STGRFT and target's signal during the second stage, the RM and DFM still exist. Therefore, the integrated peak value of this situation is much less than that of situation 1, as shown in Fig. 10.2(g) and Fig. 10.2(h).

Table 10.2 lists the values for the amplitude and location of the peaks in Fig. 10.2.

TABLE 10.2　Values for Amplitude and Location of the Peaks in Fig. 10.2

Different situations	Figures	Peak value	Peak location
Situation 1	Fig. 10.2(b)	127.8	(100,558)
Case 1 of situation 2	Fig. 10.2(c)	77.52	(105,556)
Case 2 of situation 2	Fig. 10.2(d)	72.67	(100,558)
Case 1 of situation 3	Fig. 10.2(e)	80.63	(96,559)
Case 2 of situation 3	Fig. 10.2(f)	129.2	(100,558)
Case 1 of situation 4	Fig. 10.2(g)	78.57	(105,556)
Case 2 of situation 4	Fig. 10.2(h)	79.67	(96,559)

10.3 COHERENT INTEGRATION VIA STGRFT

Motivated by the property of STGRFT with respect to different window functions, the LTCI method based on STGRFT is proposed in this section.

10.3.1 Definition of LTCI Based on STGRFT

For the received signal of a target with S motion models (as shown in (10.4)), the LTCI based on STGRFT is as follows:

$$LTCI(\boldsymbol{\eta_0}, \boldsymbol{\eta_1}, \mathbf{r}, \mathbf{v}, \mathbf{a}) = \sum_{k=1}^{S} STGRFT_{g_l(t_m)}(r_{0,l}, v_l, a_l), \tag{10.40}$$

where

$$\boldsymbol{\eta_0} = \begin{bmatrix} \eta_{1,0} & \eta_{2,0} & \cdots & \eta_{S,0} \end{bmatrix}, \tag{10.41}$$

$$\boldsymbol{\eta_1} = \begin{bmatrix} \eta_{1,1} & \eta_{2,1} & \cdots & \eta_{S,1} \end{bmatrix}, \tag{10.42}$$

$$\mathbf{r} = \begin{bmatrix} r_{0,1} & r_{0,2} & \cdots & r_{0,S} \end{bmatrix}, \tag{10.43}$$

$$\mathbf{v} = \begin{bmatrix} v_1 & v_2 & \cdots & v_S \end{bmatrix}, \tag{10.44}$$

$$\mathbf{a} = \begin{bmatrix} a_1 & a_2 & \cdots & a_S \end{bmatrix}, \tag{10.45}$$

$STGRFT_{g_l(t_m)}(r_{0,l}, v_l, a_l)$ denotes the STGRFT with respect to window function $g_l(t_m)$, i.e.,

$$STGRFT_{g_l(t_m)}(r_{0,l}, v_l, a_l) = \int_{\eta_{l,0}}^{\eta_{l,1}} s\left(2r_l(t_m)/c, t_m\right) g_l(t_m)$$
$$\times \exp\left(j4\pi\frac{r_l(t_m)}{\lambda}\right) dt_m, \tag{10.46}$$

$$g_l(t_m) = \text{rect}\left[\frac{t_m - 0.5(\eta_{l,1} + \eta_{l,0})}{\eta_{l,1} - \eta_{l,0}}\right] = \begin{cases} 1, & \eta_{l,0} \le t_m \le \eta_{l,1} \\ 0, & \text{else} \end{cases} \tag{10.47}$$

$$r_l(t_m) = r_{0,l} + v_l(t_m - \eta_{l,0}) + a_l(t_m - \eta_{l,0})^2, \tag{10.48}$$

$r_{0,l}$, v_l and a_l represent, respectively, the searching initial range, searching radial velocity and searching radial acceleration with respect to $STGRFT_{g_l(t_m)}(r_{0,l}, v_l, a_l)$. $\eta_{l,0}$ and $\eta_{l,1}$ denote, respectively, the beginning time and ending time of non-zero area for $g_l(t_m)$. By using (10.40), the LTCI output can be achieved with respect to different searching parameters. Only when the searching motion parameters match the actual target's motion parameters (i.e., $\eta_{l,0} = T_{i-1}$, $\eta_{l,1} = T_i$, $r_{0,l} = R_{0,i}$, $v_l = V_i$ and $a_l = A_i$), the target energy distributed in multiple motion stages (corresponding to different motion models) could then be totally extracted and coherently accumulated.

More specifically, when $\eta_{l,0} = T_{i-1}$ and $\eta_{l,1} = T_i$, the $LTCI(\boldsymbol{\eta_0}, \boldsymbol{\eta_1}, \mathbf{r}, \mathbf{v}, \mathbf{a})$ of (10.40) could be expressed as

$$
\begin{aligned}
LTCI(\boldsymbol{\eta_0}, \boldsymbol{\eta_1}, \mathbf{r}, \mathbf{v}, \mathbf{a}) &= \sum_{l=1}^{S} \int_{\eta_{l,0}}^{\eta_{l,1}} s\left(2r_l(t_m)/c, t_m\right) g_l(t_m) \\
&\quad \times \exp\left(j4\pi \frac{r_l(t_m)}{\lambda}\right) dt_m \\
&= \sum_{l=1}^{S} \int_{\eta_{l,0}}^{\eta_{l,1}} \sigma \mathrm{sinc}\left[B\left(\frac{2(r_l(t_m) - R_i(t_m))}{c}\right)\right] \\
&\quad \times \exp\left[j4\pi \frac{r_l(t_m) - R_i(t_m)}{\lambda}\right] dt_m.
\end{aligned}
\tag{10.49}
$$

For the case that $r_{0,l} = R_{0,i}$, $v_l = V_i$ and $a_l = A_i$, the RM and DFM of the extracted target signal during the ith motion stage will be eliminated, and coherent integration could be obtained, i.e.,

$$
\begin{aligned}
LTCI(\boldsymbol{\eta_0}, \boldsymbol{\eta_1}, \mathbf{r}, \mathbf{v}, \mathbf{a}) &= \sum_{l=1}^{S} \int_{\eta_{l,0}}^{\eta_{l,1}} \sigma \mathrm{sinc}\left[2B0t_m\right] \exp\left[j4\pi 0 t_m\right] dt_m \\
&= \sum_{l=1}^{S} \sigma(\eta_{l,1} - \eta_{l,0}) = \sigma(T_S - T_0).
\end{aligned}
\tag{10.50}
$$

Thus, the target signal among multiple motion stages (corresponding to multiple motion models) is coherently accumulated and reached its maximum value, which is good for the signal-to-noise ratio (SNR) improvement and target detection thereafter.

Furthermore, it should be pointed out that there are two constraints for the window function, i.e.,

- 1) The sum of the length of all the window functions' non-zero area should equal to the target's signal length during the total integration time, i.e.,

$$
\sum_{l=1}^{S} (\eta_{l,1} - \eta_{l,0}) = \sum_{i=1}^{S} (T_i - T_{i-1}).
\tag{10.51}
$$

- 2) The non-zero area of two arbitrary different window functions (e.g., $g_l(t)$ and $g_p(t)$, $l \neq p$) should not be overlapped, i.e.,

$$
\int_{T_0}^{T_S} g_l(t_m) g_p(t_m) dt_m = 0.
\tag{10.52}
$$

10.3.2 Detailed Procedure of STGRFT-Based LTCI

The detailed processing procedure of LTCI method based on STGRFT is given in this section.

Step 1: Necessary parameters preparation for the STGRFT.

According to the radar observation time and the relative prior information such as varieties and maneuverability as well as the moving status of targets to be detected, the expected motion model number, searching area of initial range, velocity and acceleration could be determined, which are, respectively, denoted as S, $[R_{\min}, R_{\max}]$, $[V_{\min}, V_{\max}]$ and $[A_{\min}, A_{\max}]$. The beginning time of $g_l(t_m)$ is $\eta_{l,0}$, and the ending time of $g_l(t_m)$ is $\eta_{l,1}$. The searching interval of $\eta_{l,0}$ and $\eta_{l,1}$ is denoted as $\Delta\eta$. Searching motion parameters pair is $(r_{0,l}, v_l, a_l)$, where $l = 1, 2, 3, \cdots, S$.

Step 2: Determine the searching interval and discrete values of the motion parameters $(r_{0,l}, v_l, a_l)$.

Based on the radar system parameters, the searching interval of the initial range, radial velocity and acceleration (i.e., ΔR, ΔV and ΔA) could be determined as [5]

$$\Delta R = c/2B, \tag{10.53}$$

$$\Delta V = \lambda/2(T_S - T_0), \tag{10.54}$$

$$\Delta A = \lambda/2(T_S - T_0)^2. \tag{10.55}$$

Thus, the discrete values of the searching range, searching velocity and searching acceleration are respectively:

$$r_{0,l} = R_{\min} : \Delta R : R_{\max}, \tag{10.56}$$

$$v_l = V_{\min} : \Delta V : V_{\max}, \tag{10.57}$$

$$a_l = A_{\min} : \Delta A : A_{\max}. \tag{10.58}$$

Step 3: Determine the searching interval and discrete values of the beginning/ending time for window function.

First, note that the beginning time of the first motion stage is T_0 and the ending time of the last motion stage is T_S. Therefore, we have

$$\eta_{1,0} = T_0, \eta_{S,1} = T_S. \tag{10.59}$$

Second, it should be pointed out that the lth motion stage and the $(l+1)$th motion stage of target are adjacent, i.e., the difference between the ending time of the lth motion stage and the beginning time of the $(l+1)$th motion stage is one pulse time. In this regard, the searching interval of beginning time or ending time could be set as: $\Delta\eta = T_r$. Thus, during the beginning-time/ending-time searching process of the window function for STGRFT, the following relationship could be employed:

$$\eta_{l,1} = \eta_{l+1,0} - T_r(l = 1, 2, 3, \cdots, S - 1), \tag{10.60}$$

$$\Delta\eta = T_r. \tag{10.61}$$

Then, the discrete values of the beginning time/ending time could be expressed as:

$$\eta_{1,1} = T_0 : \Delta\eta : T_S, \tag{10.62}$$

$$\eta_{l,0} = \eta_{l-1,1} + T_r : \Delta\eta : T_S - T_r, l = 2, 3, \cdots, S - 1, \tag{10.63}$$

$$\eta_{M,0} = \eta_{S-1,1} + T_r : \Delta\eta : T_S. \tag{10.64}$$

Combining equations $(10.59)-(10.64)$, we could obtain the searching values of $\eta_{l,0}$ and $\eta_{l,1}, l = 1, 2, \cdots, S$.

Third, note that the instantaneous slant range of the lth motion stage in its ending time is the initial slant range of the $(l + 1)$th motion stage. Meanwhile, the instantaneous velocity of the lth motion stage in its ending time is the initial velocity of the $(l + 1)$th motion stage. Thus, during the motion parameters searching, the following relationship could be applied:

$$r_{0,l+1} = r_{0,l} + v_l(\eta_{l,1} - \eta_{l,0}) + a_l(\eta_{l,1} - \eta_{l,0})^2, \tag{10.65}$$

$$v_{l+1} = v_l + 2a_l(\eta_{l,1} - \eta_{l,0}). \ l = 1, 2, \cdots, S - 1. \tag{10.66}$$

Step 4: Apply the STGRFT to achieve the LTCI among different motion stages.

Firstly, we determine the window function, and the moving target trajectory to be searched for based on the searching parameters.

$$g_l(t_m) = \text{rect} \left[\frac{t_m - 0.5(\eta_{l,1} + \eta_{l,0})}{\eta_{l,1} - \eta_{l,0}} \right], \tag{10.67}$$

$$r_l(t_m) = r_{0,l} + v_l(t_m - \eta_{l,0}) + a_l(t_m - \eta_{l,0})^2. \tag{10.68}$$

Secondly, perform the STGRFT operation on the received signal after PC with respect to $g_l(t_m), l = 1, 2, \cdots, S$, i.e.,

$$STGRFT_{g_l(t_m)}(r_{0,l}, v_l, a_l) = \int_{\eta_{l,0}}^{\eta_{l,1}} s\left(2r_l(t_m)/c, t_m\right) g_l(t_m)$$
$$\times \exp\left(j4\pi \frac{r_l(t_m)}{\lambda}\right) dt_m. \tag{10.69}$$

Thirdly, apply the addition operation to achieve the LTCI of target signal among different motion stages.

$$LTCI(\boldsymbol{\eta_0}, \boldsymbol{\eta_1}, \mathbf{r}, \mathbf{v}, \mathbf{a}) = \sum_{l=1}^{S} STGRFT_{g_l(t_m)}(r_{0,l}, v_l, a_l). \tag{10.70}$$

When the $STGRFT_{g_l(t_m)}(r_{0,l}, v_l, a_l)$ matches the target signal during the ith motion stage, i.e.,

$$\eta_{l,0} = T_{i-1}, \eta_{l,1} = T_i, \tag{10.71}$$

$$r_{0,l} = R_{0,i}, v_l = V_i, a_l = A_i. \tag{10.72}$$

Then, the target signal distributed in different motion stages would be coherently integrated, and $LTCI(\boldsymbol{\eta_0}, \boldsymbol{\eta_1}, \mathbf{r}, \mathbf{v}, \mathbf{a})$ reaches its maximum value. Hence, the target motion parameters could be estimated by

$$(\hat{\boldsymbol{\eta}}_0, \hat{\boldsymbol{\eta}}_1, \hat{\mathbf{r}}, \hat{\mathbf{v}}, \hat{\mathbf{a}}) = \underset{(\boldsymbol{\eta_0}, \boldsymbol{\eta_1}, \mathbf{r}, \mathbf{v}, \mathbf{a})}{\arg\max} |LTCI(\boldsymbol{\eta_0}, \boldsymbol{\eta_1}, \mathbf{r}, \mathbf{v}, \mathbf{a})|, \quad (10.73)$$

where $\hat{\boldsymbol{\eta}}_0 = \begin{bmatrix} \hat{\eta}_{1,0} & \hat{\eta}_{2,0} & \cdots & \hat{\eta}_{S,0} \end{bmatrix}$, $\hat{\boldsymbol{\eta}}_1 = \begin{bmatrix} \hat{\eta}_{1,1} & \hat{\eta}_{2,1} & \cdots & \hat{\eta}_{S,1} \end{bmatrix}$, $\hat{\mathbf{r}} = \begin{bmatrix} \hat{r}_{0,1} & \hat{r}_{0,2} & \cdots & \hat{r}_{0,S} \end{bmatrix}$, $\hat{\mathbf{v}} = \begin{bmatrix} \hat{v}_1 & \hat{v}_2 & \cdots & \hat{v}_S \end{bmatrix}$ and $\hat{\mathbf{a}} = \begin{bmatrix} \hat{a}_1 & \hat{a}_2 & \cdots & \hat{a}_S \end{bmatrix}$.

Step 5: Perform the constant false alarm ratio (CFAR) detection based on the LTCI output.

The amplitude of $LTCI(\boldsymbol{\eta_0}, \boldsymbol{\eta_1}, \mathbf{r}, \mathbf{v}, \mathbf{a})$ output in step 4 is taken as the test statistic and compared with the adaptive threshold of a given false alarm probability

$$|LTCI(\boldsymbol{\eta_0}, \boldsymbol{\eta_1}, \mathbf{r}, \mathbf{v}, \mathbf{a})|, \quad (10.74)$$

where γ is the detection threshold [5], and it is obtained by the reference unit after LTCI via STGRFT. If the test statistic is larger than the threshold, target detection is declared.

10.3.3 Computational Complexity Analysis

According to the analysis presented in Section 10.3.2, we can see that the STGRFT-based method involves the searching of $\eta_{l,0}$, $\eta_{l,1}$, $r_{0,l}$, v_l and $a_l(l = 1, 2, \cdots, S)$, which needs $5S$ dimensions searching. However, based on the relationship shown in (10.59)−(10.66), it could be seen that we only need to search for $a_l(l = 1, 2, \cdots, S)$, $\eta_{l,0}(l = 2, 3, \cdots, S)$, $r_{0,1}$ and v_1, which only requires $2S + 1$ dimensions searching. Other $3S - 1$ searching parameters (i.e., $\eta_{1,0}$, $\eta_{1,1}$, $\eta_{l,1}$, $r_{0,l}$ and v_l, $l = 2, 3, \cdots, S$) could be determined via (10.59)−(10.66).

Let S represent the pulse number, and suppose that N_{a_l} denotes the number of searching parameter $a_l(l = 1, 2, \cdots, S)$ and $N_{\eta_{l,0}}$ represents the number of searching parameter $\eta_{l,0}(l = 2, \cdots, S)$. $N_{r_{0,1}}$ and N_{v_1} denote, respectively, the number of searching parameters $r_{0,1}$ and v_1. Then, $(S + 4)N_{a_1}N_{r_{0,1}}N_{v_1}\prod_{l=2}^{S} N_{a_l}N_{\eta_{l,0}}$ complex multiplications and $(S + 6)N_{a_1}N_{r_{0,1}}N_{v_1}\prod_{l=2}^{S} N_{a_l}N_{\eta_{l,0}}$ complex additions are required for the STGRFT-based method. Hence, the computational cost of STGRFT-based algorithm would be higher than GRFT. Fortunately, on the one hand, intelligent optimization techniques, such as particle swarm optimization (PSO), may be used to realize the multiple-dimension searching processing to save computational time. On the other hand, the prior information of the target or environment could be used to reduce the searching area.

10.4 SIMULATION ANALYSIS

10.4.1 LTCI in Different SNR Background

Simulation experiments in this section are given to verify the LTCI performance of the STGRFT-based method for maneuvering target under a different SNR background. A

TABLE 10.3   Radar Parameters

Parameters	Values
Carrier frequency	1.5 GHz
Bandwidth	20 MHz
Sample frequency	100 MHz
Pulse repetition frequency	200 Hz
Pulse duration	10 μs
Pulse number	510

maneuvering target is considered in this scene, where the target moves with constant acceleration (i.e., velocity is 100 m/s and acceleration is 60 m/s^2) in the first stage (between 0 and 1.275 s) and moves with constant velocity in the second stage (between 1.275 and 2.55 s). Table 10.3 lists the radar parameters.

10.4.1.1   *SNR = 6 dB After PC*

Fig. 10.3(a) shows the compressed signal. The integration results of STGRFT method are given in Fig. 10.3(b)–Fig. 10.3(f). More specifically, Fig. 10.3(b) gives the slice of initial range cell and velocity during the first stage, where the estimations of target's initial range cell and velocity of the first stage could be obtained. Fig. 10.3(c) shows the slice of velocity and acceleration during the first stage, where the peak indicates the estimations of target's velocity and acceleration within the first stage. Fig. 10.3(d) gives the slice of acceleration and ending time during the first stage, where the peak implies the estimations of target's acceleration and ending time of the first stage. Fig. 10.3(e) shows the slice of ending time during the first stage and acceleration in the second stage, where the peak indicates the estimations of target's ending time during the first stage and acceleration in the second stage. Fig. 10.3(f) gives the slice of acceleration in the second stage and initial range cell during the first stage, where the peak implies the estimations of target's acceleration within the second stage and the initial range cell of the first stage. Combining the results of Fig. 10.3(b)–Fig. 10.3(f), we could obtain the estimations of target's motion parameters.

For comparison, the integration results of GRFT [8], KTMFP [4], RFT [9] and MTD [10] are also given in Fig. 10.4. We could see from Fig. 10.4(a) and Fig. 10.4(b) that the integrated peak values of GRFT and KTMFP are about half of that of the STGRFT method. This is because that GRFT and KTMFP could only realize the coherent accumulation of target signal during the first stage (between 0 and 1.275 s) and their effective integrated pulse number is 255. However, the STGRFT method could achieve the coherent integration of target signal distributed in the first and second stages (between 0 and 2.55 s), and its effective integrated pulse number is 510, which indicates that the STGRFT algorithm could achieve higher integration gain than GRFT and KTMFP. Besides, RFT and MTD both become invalid and could not achieve the coherent integration because of the DFM, RM and changing motions of the maneuvering target, as shown in Fig. 10.4(c) and Fig. 10.4(d).

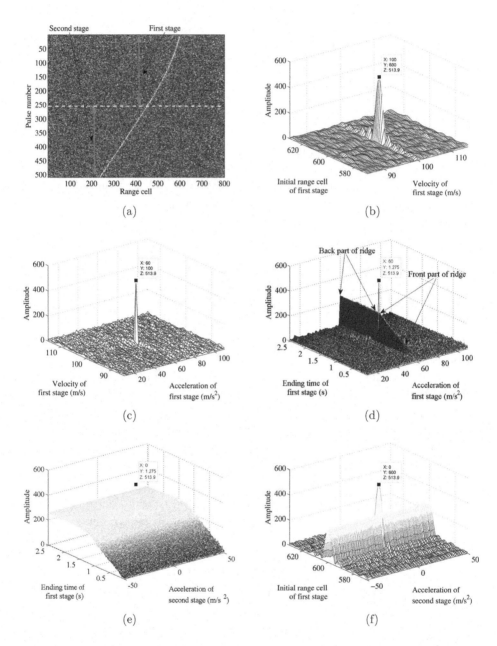

Figure 10.3 LTCI via STGRFT for SNR = 6 dB. (a) PC. (b) Initial range cell and velocity of first stage (slice). (c) Velocity and acceleration of first stage (slice). (d) Acceleration and ending time of first stage (slice). (e) Ending time of first stage and acceleration of second stage (slice). (f) Acceleration of second stage and initial range cell of first stage (slice).

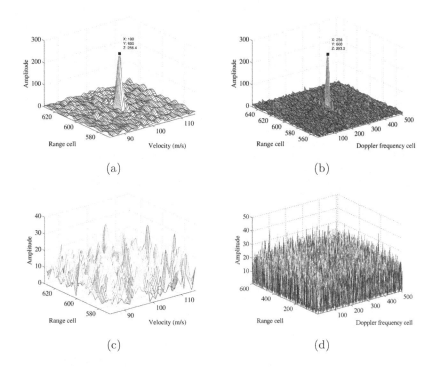

Figure 10.4 LTCI results of GRFT, KTMFP, RFT and MTD for SNR = 6 dB. (a) GRFT. (b) KTMFP. (c) RFT. (d) MTD.

10.4.1.2 *SNR = 0 dB after PC*

Fig. 10.5(a) shows the compressed signal. We can see that the target signal is almost buried in noise and that the target's moving trajectory is indefinable because of the low SNR. After LTCI via STGRFT, the integration results are given in Fig. 10.5(b)−Fig. 10.5(f). With the integration gain via STGRFT, the target is clearly located in the output. In particular, Fig. 10.5(b) gives the slice of initial range cell and velocity during the first stage. Fig. 10.5(c) shows the slice of velocity and acceleration during the first stage. Fig. 10.5(d) gives the slice of acceleration and ending time during the first stage. Fig. 10.5(e) shows the slice of ending time during the first stage and acceleration in the second stage. Fig. 10.5(f) gives the slice of acceleration in the second stage and initial range cell during the first stage.

For comparison, Fig. 10.6 also shows the integration result of GRFT, KTMFP, RFT and MTD. It could be seen from Fig. 10.6(a) and Fig. 10.6(b) that the integrated peak values of GRFT and KTMPF are about half of that of the STGRFT-based method, which means that the integration gain of GRFT and KTMFP is 3 dB less than that of the STGRFT-based algorithm. In addition, the RFT and MTD both become ineffective and could not obtain the coherent integration. Hence, the target energy is still unfocused in Fig. 10.6(c) and Fig. 10.6(d).

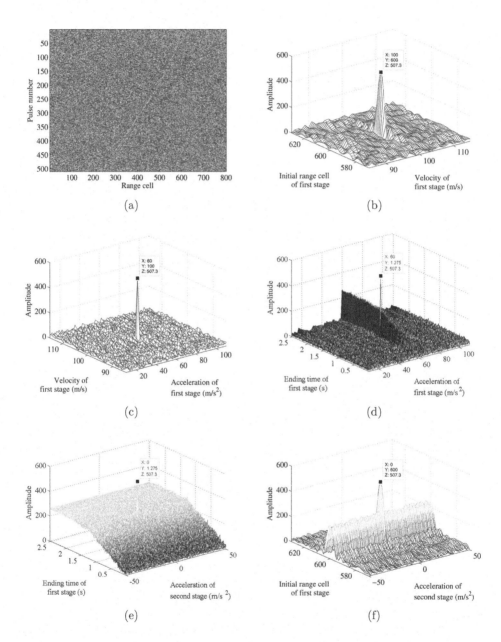

Figure 10.5 LTCI via STGRFT for SNR = 0 dB. (a) PC. (b) Initial range cell and velocity of the first stage (slice). (c) Velocity and acceleration of the first stage (slice). (d) Acceleration and ending time of the first stage (slice). (e) Ending time of the first stage and acceleration of the second stage (slice). (f) Acceleration of the second stage and initial range cell of the first stage (slice).

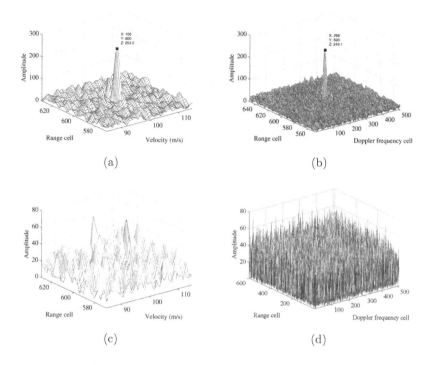

Figure 10.6 LTCI results of GRFT, KTMFP, RFT and MTD for SNR = 0 dB. (a) GRFT. (b) KTMFP. (c) RFT. (d) MTD.

10.4.1.3 *SNR = −13 dB After PC*

Fig. 10.7(a) shows the compressed signal. We could see that the target's signal is totally buried in noise, and it is impossible to see the target's moving trajectory because of the extremely low SNR, which implies that it is difficult to accomplish target detection via Fig. 10.7(a). Fortunately, with the LTCI via STGRFT, the target's signal distributed in different pulses within the whole observation time is coherently accumulated as a peak in the output, as shown in Fig. 10.7(b)−Fig. 10.7(f). Thereafter, target detection could be declared. For comparison, the integration results of Radon-Lv's distribution (RLVD) [7], generalized Radon Fourier transform (GRFT), keystone transform match filtering process (KTMFP) and segmented keystone transform and Doppler Lv's transform (SKTDLVT) are also given in Fig. 10.8(a)−Fig. 10.8(d). Because of the extremely low SNR and the target is of multiple motion models, the target signal is still buried in noise after integration via RLVD or GRFT or KTMFP or SKTDLVT, and target detection based on Fig. 10.8 is not possible.

Based on the experimental results of Fig. 10.3−Fig. 10.8, we can see that the STGRFT method could obtain higher integration gain than RLVD, GRFT, KTMFP, SKTDLVT, RFT and MTD for the target with multiple motion models. Thus, the STGRFT algorithm could improve the detection ability of the target with low SNR.

Note that there are two large ridges in Fig. 10.3(d) (or Fig. 10.5(d)) and Fig. 10.3(f) (or Fig. 10.5(f)). For Fig. 10.3(d) or Fig. 10.5(d), the ridge is caused by the STGRFT's property shown in (10.25) and (10.31) (as shown in case 2 of situation

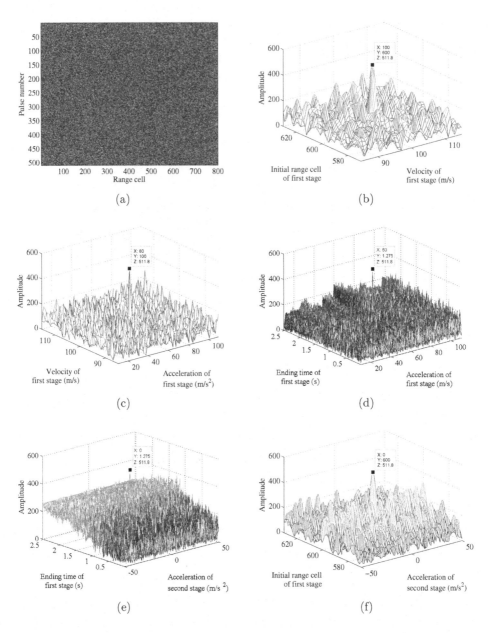

Figure 10.7 LTCI via STGRFT for SNR = −13 dB. (a) PC. (b) Initial range cell and velocity of the first stage (slice). (c) Velocity and acceleration of the first stage (slice). (d) Acceleration and ending time of the first stage (slice). (e) Ending time of the first stage and acceleration of the second stage (slice). (f) Acceleration of the second stage and initial range cell of the first stage (slice).

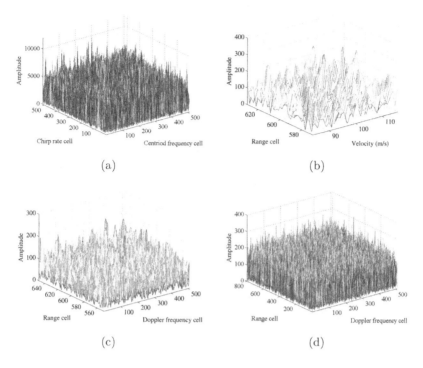

Figure 10.8 LTCI results of RLVD, GRFT, KTMFP and SKTDLVT for SNR = −13 dB. (a) RLVD. (b) GRFT. (c) KTMFP. (d) SKTDLVT.

2 and case 2 of situation 3 in Section 10.2.2). In particular, the front part of the ridge (where the searching value of the ending time is less than the target's true ending time of first stage) is induced by the STGRFT's response shown in (10.25), where part of the target signal during the first stage is extracted and integrated. Besides, the back part of the ridge (where the searching value of the ending time is bigger than the target's true ending time of the first stage) is induced by the STGRFT's response shown in (10.31). In this case, the entire target signal during the first stage is extracted and integrated. As to Fig. 10.3(f) or Fig. 10.5(f), the ridge is caused by the STGRFT's property shown in (10.31), where the target signal during the first stage is totally extracted and integrated as a ridge in the output.

Furthermore, in order to make clearer comparisons between different SNR scenarios, the ratios of the integrated peaks with respect to the noise floor and ridges of Fig. 10.3, Fig. 10.5 and Fig. 10.7 are quantitatively calculated. More specifically, we compare the ratio between the integrated peak value of the target signal and the ridge, which is defined as:

$$\Gamma_1 = 10\log 10(P_s^2/P_r^2), \qquad (10.75)$$

where P_s represents the integrated peak value of the target signal and P_r denotes the peak value of the ridge. In addition, we define the SNR in STGRFT domain as follows

$$\Gamma_2 = 10\log 10(P_s^2/\xi^2), \qquad (10.76)$$

where ξ^2 denotes the noise power after STGRFT. Table 10.4 shows the calculated results. We could notice that the ratio between the integrated peak value of the target

TABLE 10.4 The Ratio of the Peak with Respect to the Noise Floor and Ridge

Figures	Input SNR (dB)	Γ_1 (dB)	Γ_2 (dB)	SNR gain (dB)
Fig. 10.3	SNR = 6	6.08	33.05	27.05
Fig. 10.5	SNR = 0	6.05	27.02	27.02
Fig. 10.7	SNR = −13	6.04	13.96	26.96

signal and ridge is stable under different SNR backgrounds, while Γ_2 reduces with the decrease in input SNR. However, the SNR gains (i.e., the difference between Γ_2 and the input SNR) of Fig. 10.3, Fig. 10.5 and Fig. 10.7 are all approximately equal to 27 dB (which is close to the integration gain of 510 pulses).

10.4.2 LTCI of Multi-Target

In Fig. 10.9, the LTCI performance of the STGRFT-based algorithm for multi-target is evaluated, where the radar parameters are the same as those in Section 10.4.1. Eight maneuvering targets (i.e., targets 1–8) are considered, where all the targets are of two motion stages during the observation time, and the SNR is set as 6 dB after PC. The detailed parameters of the targets are shown in Table 10.5.

Firstly, targets 1 and 2 are considered, which have different initial range cells and radial velocities within the first motion stage. Fig. 10.9(a) gives the PC results, and Fig. 10.9(b) shows the focused results of targets 1 and 2 (slice of initial range cell and velocity of the first motion stage). Secondly, we consider targets 3 and 4, which are different in radial velocities and accelerations during the first motion stage. The PC results are given in Fig. 10.9(c), and the LTCI results of targets 3 and 4 are shown in Fig. 10.9(d) (slice of velocity and acceleration of the first motion stage). Thirdly, targets 5 and 6 are considered, which have different ending times and radial accelerations during the first motion stage. The PC results are shown in Fig. 10.9(e), and the LTCI results of targets 5 and 6 are given in Fig. 10.9(f) (slice of acceleration and ending time of the first motion stage). Fourthly, we consider targets 7 and 8, which are different in ending times of the first motion stage and accelerations of the second motion stage. Fig. 10.9(g) gives the PC results. Noting that targets 7 and 8 have the same initial range cell, radial velocity and acceleration during the first motion stage. Hence, the moving trajectories of these two targets are partly overlapped (as shown in Fig. 10.9(g)). Fig. 10.9(h) shows the LTCI results of targets 7 and 8 (slice of ending time in the first stage and acceleration in the second stage).

Based on the experimental results of Fig. 10.9, it could be seen that the STGRFT-based method can effectively realize the LTCI of multiple targets with multiple motion models.

10.4.3 Anti-Noise Performance Analysis

In this section, the input-output SNR performance and detection ability are examined to evaluate the anti-noise performance of the STGRFT method. Fig. 10.10(a) and

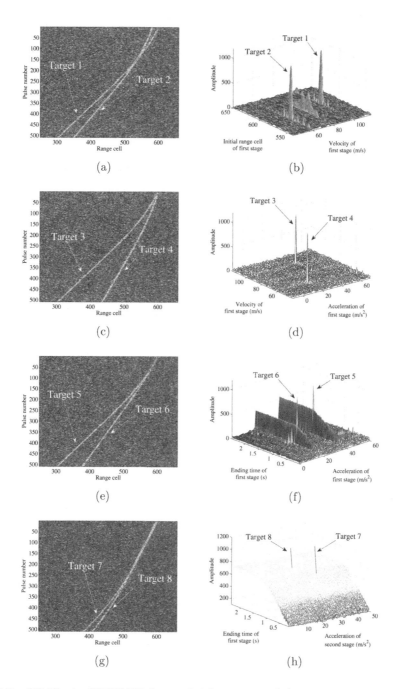

Figure 10.9 LTCI via STGRFT for multiple targets. (a) PC result of targets 1 and 2. (b) LTCI result of targets 1 and 2. (c) PC result of targets 3 and 4. (d) LTCI result of targets 3 and 4. (e) PC result of targets 5 and 6. (f) LTCI result of targets 5 and 6. (g) PC result of targets 7 and 8. (h) LTCI result of targets 7 and 8.

TABLE 10.5 Parameters of the Multiple Targets

Targets	Initial range cell	Time of first stage (s)	Motion of first stage	Time of second stage (s)	Motion of second stage
1	600	0−1.275	100 m/s, 38 m/s^2	1.275−2.55	Constant velocity
2	580	0−1.275	60 m/s, 38 m/s^2	1.275−2.55	Constant velocity
3	603	0−1.250	100 m/s, 40 m/s^2	1.250−2.55	Constant velocity
4	603	0−1.250	60 m/s, 20 m/s^2	1.250−2.55	Constant velocity
5	601	0−1.270	100 m/s, 41 m/s^2	1.270−2.55	Constant velocity
6	601	0−1.000	100 m/s, 20 m/s^2	1.000−2.55	Constant velocity
7	600	0−1.275	100 m/s	1.275−2.55	30 m/s^2
8	600	0−1.000 s	100 m/s	1.000−2.55	20 m/s^2

Fig. 10.10(b) show, respectively, the input-output SNR performance and the detection performance, where the radar parameter and target's motion parameters are the same as those in Section 10.4.1.

For the input-output SNR performance, the input SNRs tested are $\text{SNR}_{\text{in}} = [-30 \text{ dB} : 20 \text{ dB}]$, and 1000 trials are performed for each SNR_{in} value. Results of the STGRFT, GRFT and KTMFP are shown in Fig. 10.10(a), where the result of ideal MTD (which supposes that the RM and DFM of target during the observation time have been fully eliminated and target signal is coherently accumulated) is also given. As to the detection performance, the corresponding false alarm ratio is $P_{fa} = 10^{-4}$, and the detection probabilities versus SNRs are calculated by 10^6 times of Monte Carlo trials. From Fig. 10.10(a) and 10.10(b), we could see that the STGRFT-based algorithm could obtain almost identical input-output SNR performance and detection ability with the ideal MTD. Besides, several points may be noted:

1) Compared with RLVD, KTMFP and GRFT, the STGRFT-based algorithm could obtain a better input-output SNR performance (3 dB or better) and target detection performance. This is because that the STGRFT-based algorithm could achieve the LTCI of the target signal during the whole integration time (including

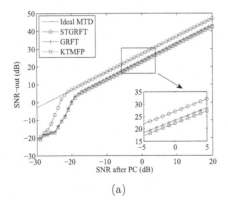

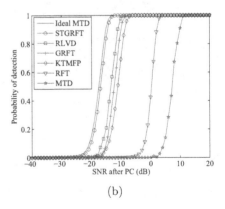

(a)   (b)

Figure 10.10 Anti-noise performance. (a) Input-output SNR performance. (b) Detection ability.

the first and second stages), while the RLVD/GRFT/KTMFP could only obtain the LTCI of target signal during the first stage (i.e., the effective integrated pulse number of RLVD or GRFT or KTMFP is 255, while the effective integrated pulse number of the STGRFT algorithm is 510).

2) Compared with the STGRFT-based method, RFT has a 16-dB detection performance loss. The reason is that RFT could not compensate for the DFM and could not realize the LTCI of target with acceleration and changing motion model. Besides, compared with the MTD method, the STGRFT method obtains a better detection ability (23-dB gain). This is because MTD could not remove the RM/DFM, and it could only achieve the LTCI of target signal within the same range cell and Doppler frequency cell. Hence, it suffers serious integration performance loss for the target with RM/DFM and multiple motion models.

10.4.4 Real Scene Processing Results

In the following, the real radar data are applied to validate the STGRFT algorithm. The real data were recorded by a S-band radar in the city of Chengdu, Sichuan, China, in 2017. An unmanned aerial vehicle (UAV) is utilized as the experimented target. The radar parameters are bandwidth $B = 50$ MHz, sample frequency $f_s = 100$ MHz, pulse repetition frequency $f_p = 500$ Hz and pulse duration $T_p = 4$ μs. During the experiment, the UAV flight consisted of two motion modes (acceleration motion mode and uniform motion mode).

Fig. 10.11(a) shows the PC result of 2000-pulses echo of target, where we can see that the target signal is partially buried in the clutter and noise. Fig. 10.11(b)−Fig. 10.11(f) give the LTCI result of the STGRFT-based algorithm. With the help of LTCI among 2000 pulses, the target's energy distributed in different motion models is coherently accumulated as a peak in the output, where the noise and clutter are suppressed well and the target signal is enhanced. Thereafter, the target detection could be accomplished via the LTCI output.

10.5 SUMMARY

In this chapter, the STGRFT-based coherent integration method is discussed for the signal accumulation and detection of high-speed targets with multiple motion models. The STGRFT-based method could not only deal with range migration and Doppler frequency migration, but also obtain the estimation of model-changing points and realize the LTCI of target signal distributed in different motion stages (corresponding to different motion models). Simulation and real data verifications show that the STGRFT-based algorithm could achieve higher integration gain and input-output SNR performance and better detection ability than the existing representative LTCI methods.

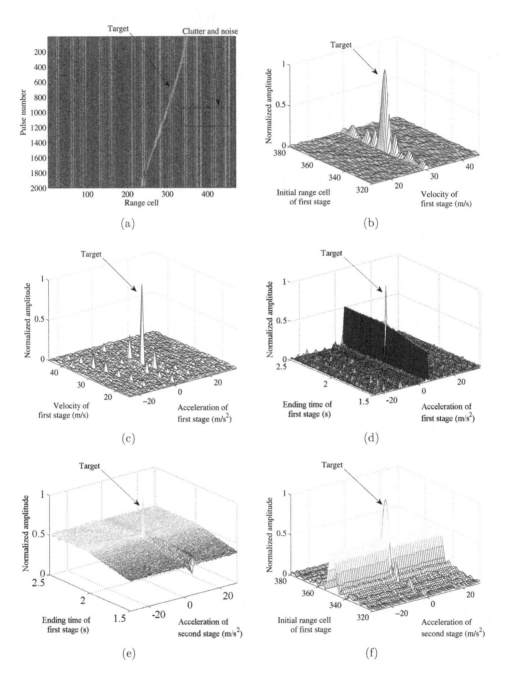

Figure 10.11 Real data processing. (a) PC. (b) Slice of initial range cell and velocity during the first stage. (c) Slice of velocity and acceleration during the first stage. (d) Slice of acceleration and ending time during the first stage. (e) Slice of ending time during the first stage and acceleration during the second stage. (f) Slice of acceleration during the second stage and initial range cell within the first stage.

Bibliography

[1] X. L. Li, G. L. Cui, W. Yi, and L. J. Kong, "A fast maneuvering target motion parameters estimation algorithm based on ACCF," *IEEE Signal Processing Letters*, vol. 22, no. 3, pp. 270–274, March 2015.

[2] X. L. Li, G. L. Cui, L. J. Kong, and W. Yi, "Fast non-searching method for maneuvering target detection and motion parameters estimation," *IEEE Transactions on Signal Processing*, vol. 64, no. 9, pp. 2232–2244, May 2016.

[3] X. L. Li, L. J. Kong, G. L. Cui, and W. Yi, "A fast detection method for maneuvering target in coherent radar," *IEEE Sensors Journal*, vol. 15, no. 11, pp. 6722–6729, November 2015.

[4] Z. Sun, X. L. Li, W. Yi, G. L. Gui, and L. J. Kong, "Detection of weak maneuvering target based on keystone transform and matched filtering process," *Signal Processing*, vol. 140, pp. 127–138, November 2017.

[5] X. L. Chen, J. Guan, N. B. Liu, and Y. He, "Maneuvering target detection via Radon-Fractional Fourier transform-based long-time coherent integration," *IEEE Transactions on Signal Processing*, vol. 62, no. 4, pp. 939–953, February 2014.

[6] X. L. Chen, J. Guan, N. B. Liu, W. Zhou, and Y. He, "Detection of a low observable sea-surface target with micromotion via the Radon-linear canonical transform," *IEEE Geoscience and Remote Sensing Letters*, vol. 11, no. 7, pp. 1225–1229, July 2014.

[7] X. L. Li, G. L. Cui, W. Yi, and L. J. Kong, "Coherent integration for maneuvering target detection based on Radon-Lv's distribution," *IEEE Signal Processing Letters*, vol. 22, no. 9, pp. 1467–1471, September 2015

[8] J. Xu, X. G. Xia, S. B. Peng, J. Yu, Y. N. Peng, and L. C. Qian, "Radar maneuvering target motion estimation based on generalized Radon-Fourier transform," *IEEE Transactions on Signal Processing*, vol. 60, no. 12, pp. 6190–6201, December 2012.

[9] J. Xu, J. Yu, Y. N. Peng, and X. G. Xia, "Radon-Fourier transform (RFT) for radar target detection (I): generalized Doppler filter bank processing," *IEEE Transactions on Aerospace and Electronic Systems*, vol. 47, no. 2, pp. 1186–1202, April 2011.

[10] M. D. Xing, J. H. Su, G. Y. Wang, and Z. Bao, "New parameter estimation and detection algorithm for high speed small target," *IEEE Transactions on Aerospace and Electronic Systems*, vol. 47, no. 1, pp. 214–224, January 2011.

CHAPTER 11

Multi-Frame Coherent Integration Processing

The previous chapters focus on the coherent integration with the intra-frame, i.e., only one single frame with multi-pulse is considered. Intuitively, the best way for performance improvement should be to combine both the inter-frame coherent integration and intra-frame coherent integration. In this chapter, we consider the multi-frame coherent integration (including intra-frame integration and inter-frame integration) problem for the detection of high-speed weak targets. The signal model is established, and the modified Radon Fourier transform (RFT) is presented to realize the intra-frame coherent integration for every frame of echo signals. After that, the RFT-domain integration algorithm is presented to achieve the inter-frame coherent integration based on the characteristics of RFT outputs. The explicit expressions and analysis for various performance measures, such as integration output response, probability of detection, input-output signal-to-noise ratio (SNR) and blind speed sidelobe (BSSL) response are derived and given.

11.1 SIGNAL MODELING AND MODIFIED RFT

11.1.1 Signal Model

Suppose that the linear frequency modulation (LFM) is used as the baseband waveform [1]

$$s_{trans}(t, t_m) = \text{rect}\left(\frac{t + t_m}{T_p}\right) \exp\left(j\pi\mu(t + t_m)^2\right) \exp(j2\pi f_c(t + t_m)), \qquad (11.1)$$

where $\text{rect}(x) = \begin{cases} 1, & |x| \leq \frac{1}{2} \\ 0, & |x| > \frac{1}{2} \end{cases}$, μ, T_p, t and f_c denote the frequency modulated rate, pulse duration, fast time and carrier frequency, respectively, and $t_m = mT_r (m = 1, 2, \cdots, M)$ represents the slow time.

Consider a moving target with constant velocity v_0 and the total observation time of radar is T_{total}, during which L-frame echoes are received, while the slant

DOI: 10.1201/9781003529101-11

range between radar and target in the lth$(l = 1, 2, \cdots, L)$ frame varies as:

$$R_l(t_m) = r_{0,1} + v_0[t_m + (l - 1)T_{sum}] \qquad (11.2)$$
$$= r_{0,l} + v_0 t_m,$$

where $r_{0,l} = r_{0,1} + v_0(l-1)T_{sum}$ denotes the target's initial range during the lth frame, T_r is the pulse repetition time M is the pulse number of each frame and $T_{sum} = MT_r$ denotes the time of each frame. One frame in the chapter is corresponding to the radar's collected two-dimensional (fast time-slow time) data within one observation period, and the frame interval is corresponding to the time duration of each frame.

The radar's received baseband echo signal of target within the lth frame could be expressed as [2]:

$$s_l(t, t_m) = A_0 \text{rect}\left(\frac{t - 2R_l(t_m)/c}{T_p}\right) \exp\left[j\pi\mu(t - \frac{2R_l(t_m)}{c})\right]$$
$$\times \exp\left[-j\frac{4\pi R_l(t_m)}{\lambda}\right], \qquad (11.3)$$

where A_0 denotes the received signal's amplitude.

After pulse compression, the compressed signal within the lth frame is:

$$s_l(t, t_m) = A_1 \text{sinc}\left[\pi B\left(t - \frac{2R_l(t_m)}{c}\right)\right] \exp\left[-j\frac{4\pi R_l(t_m)}{\lambda}\right], \qquad (11.4)$$

where A_1 denotes the signal amplitude after compression and B is the bandwidth.

Let $r = ct/2$, then equation (11.4) could be recast as

$$s_l(r, t_m) = A_1 \text{sinc}\left[\pi\left(\frac{r - R_l(t_m)}{\rho_r}\right)\right] \exp\left[-j\frac{4\pi R_l(t_m)}{\lambda}\right], \qquad (11.5)$$

where $\rho_r = \frac{c}{2B}$ denotes the range resolution.

Based on the radar system's pulse repetition time T_r and range sampling frequency f_s, the discrete compressed signal in the lth frame (corresponding to (11.5)) could be rewritten as

$$s_l(n, m) = A_1 \text{sinc}\left[\frac{\pi[(n - i_T^l)\Delta_r - v_0 m T_r]}{\rho_r}\right]$$
$$\times \exp\left[-j\frac{4\pi}{\lambda}(i_T^l \Delta_r + v_0 m T_r)\right] \qquad (11.6)$$
$$m = 1, 2, \cdots, M, n = 1, 2, \cdots, N_g,$$

where $n = \text{round}(r/\Delta_r)$ denotes the range sampling index; round$(\cdot)$ represents the round up to an integer operation; $\Delta_r = c/(2f_s)$, i_T^l is the range sampling index corresponding to the target's initial range $r_{0,l}$ and N_g denotes the sampling number of range.

11.1.2 Intra-Frame Integration via Modified RFT

According to the relative prior information (e.g., moving status and varieties) of the target to be detected, the expected search scope of range and initial radial velocity for the target could be preset, which are, respectively, denoted as $[-R_{\max}, R_{\max}]$ and $[-V_{\max}, V_{\max}]$. The searching intervals of range and initial radial velocity are $\Delta_r = c/2f_s$ and $\Delta_v = \lambda/2T_{sum}$, while the target's radial velocity could be approximately represented as $v_0 \approx q_T\Delta_v$ [3]. Thus, the searching range number and the searching velocity number could be calculated as:

$$N_r = \text{round}\left(\frac{2R_{\max}}{\Delta_r}\right), \tag{11.7}$$

$$N_v = \text{round}\left(\frac{2V_{\max}}{\Delta_v}\right). \tag{11.8}$$

Then, the discrete searching range series and searching velocity series could be determined, i.e.,

$$r(i) = i\Delta_r \in [-R_{\max}, R_{\max}], i = 1, 2, \cdots, N_r, \tag{11.9}$$

$$v(q) = q\Delta_v \in [-V_{\max}, V_{\max}], q = 1, 2, \cdots, N_v. \tag{11.10}$$

With the discrete searching range and searching velocity, the modified RFT of (11.6) could be written as

$$G_l(i, q) = \sum_{m=1}^{M} s\left(m, \frac{\text{round}\left(r(i) + mT_r v(q)\right)}{\Delta_r}\right) \\ \times \exp\left(j\frac{4\pi}{\lambda}(r(i) + mT_r v(q))\right). \tag{11.11}$$

Substituting (11.6) into (11.11), we have

$$G_l(i, q) \approx \sum_{m=1}^{M} A_1 \text{sinc}\left[\frac{\pi}{\rho_r}[(i - i_T^l)\Delta_r + (q - q_T)\Delta_v m T_r]\right] \\ \times \exp\left[j\frac{4\pi(q - q_T)\Delta_v m T_r}{\lambda}\right]\exp\left[j\frac{4\pi}{\lambda}(i - i_T^l)\Delta_r\right], \tag{11.12}$$

where the equation is obtained by neglect the round() term.

After some mathematical operations, the modified RFT result of (11.12) could be expressed as [2, 4, 5]

$$G_l(i, q) \approx A_1 M \text{sinc}\left[\pi\frac{(i - i_T^l)\Delta_r}{\rho_r}\right]\delta(q - q_T) \\ \times \exp\left[j\frac{4\pi}{\lambda}(i - i_T^l)\Delta_r\right]. \tag{11.13}$$

It could be noticed from (11.13) that when the searching parameters match with the target's motion parameters, i.e., $i = i_T^l, q = q_T$ (corresponding to

$r(i) = r_{0,l}, v(q) = v_0$), the intra-frame accumulation of the target energy within the lth frame could be realized via the modified RFT. That is to say, with the modified RFT operation, we could obtain the intra-frame coherent integration output of each frame in the RFT-domain, as shown in (11.13), from which several points could be made.

1) After the modified RFT, the target signal energy of the lth frame is coherently accumulated as a peak in the range-velocity plane, where the peak location (i_l, q_l) is corresponding to

$$i_l = i_T^l, \tag{11.14}$$

$$q_l = q_T. \tag{11.15}$$

2) The moving state of target signal energy of adjacent frames (e.g., lth frame and $l+1$th frame) after modified RFT satisfies

$$i_{l+1} = i_l + \text{round}(v_0 * T_{sum}/\Delta_r), \tag{11.16}$$

$$q_{l+1} = q_l. \tag{11.17}$$

3) The position (i.e, i and q) of each grid point for the modified RFT output indicates the corresponding searching range and searching velocity. In particular, for the grid point (i, q), we have

$$i \to r(i), q \to v(q), \tag{11.18}$$

where $\to$ implies one-to-one correspondence.

Remark: Compared with the original RFT [2], the modified RFT also compensates the phase term induced by the target's initial range. As a result, the modified RFT could also match and compensate the phase difference of the moving target signal distributed in different frames, which is helpful to the inter-frame coherent integration thereafter.

11.2 RFT-DOMAIN INTER-FRAME INTEGRATION ALGORITHM

11.2.1 Algorithm Description

Based on the characteristics of intra-frame integration outputs after the modified RFT operation, the RFT-domain inter-frame integration algorithm is proposed to achieve the coherent integration of target energy distributed in multiple frames, which could be expressed as

$$
\begin{aligned}
I_L(i, q) =& G_L(i, q) + G_{L-1}\left(i - \frac{v(q)T_{sum}}{\Delta_r}, q\right) \\
&+ G_{L-2}\left(i - \frac{2v(q)T_{sum}}{\Delta_r}, q\right) + \cdots \\
&+ G_1\left(i - \frac{(K-1)v(q)T_{sum}}{\Delta_r}, q\right) \\
=& \sum_{l=1}^{L} G_{L+1-l}\left(i - \frac{(l-1)v(q)T_{sum}}{\Delta_r}, q\right).
\end{aligned}
\tag{11.19}
$$

More specifically, the RFT-domain inter-frame integration algorithm could be described in detail as follows.

(1) Input: The modified RFT results of L-frames compressed echoes, i.e., $G_l(i, q), 1 \leq l \leq L$, discrete searching ranges $r(i)$ and discrete searching velocities $v(q)$, $i \in [1, N_r], j \in [1, N_v]$.

(2) Initialization:

$$I_1(i, q) = G_1(i, q). \tag{11.20}$$

(3) Recursion: For $2 \leq l \leq L$, $1 \leq i \leq N_r$ and $1 \leq q \leq N_v$, perform

$$I_l(i, q) = G_l(i, q) + [I_{l-1}(i_{trans}, q_{trans})], \tag{11.21}$$

where

$$i_{trans} = i - \text{round}[v(q) * T_{sum}/\Delta_r], \tag{11.22}$$

$$q_{trans} = q, \tag{11.23}$$

i_{trans} denotes the target's range position change between two adjacent frames and q_{trans} denotes the target's velocity position change between two adjacent frames.

(4) Output: Multi-frame integration output $I_L(\cdot)$.

The detailed procedures of the RFT-domain inter-frame integration method are also given in **Algorithm 1**.

Algorithm 1 : RFT-domain Inter-frame Integration Algorithm

Require: The modified RFT results of L-frames compressed echoes, i.e., $G_l(i, q), 1 \leq l \leq L$, discrete searching ranges $r(i)$ and discrete searching velocities $v(q)$, $i \in [1, N_r], j \in [1, N_v]$

Initialization: $I_1 = G_1$

  Recursion:

for frame number $l \in [2, L]$ **do**,

    for each range-index $1 \leq i \leq N_r$ **do**

      for each velocity-index $1 \leq q \leq N_v$ **do**

        $i_{trans} = i - \text{round}[v(q) * T_{sum}/\Delta_r]$;

        $q_{trans} = q$;

        $I_l(i, q) = I_{l-1}(i_{trans}, q_{trans}) + G_l(i, q)$;

      end for

    end for

end for

  *3) **Termination:*** For threshold γ find,

  $\{\hat{\mathbf{x}}_l\} = \{\mathbf{x}_l : I(\mathbf{x}_L) > \gamma\}$

Ensure: Multi-frame integration output $I_L(\cdot)$

11.2.2 Integration Output Response

Considering the target signal without noise, we could obtain the inter-frame integration output response for l-frame as:

$$
\begin{aligned}
I_l(i,q) \approx & kA_1 M \mathrm{sinc}\left[\pi \frac{(i - i_T^l)\Delta_r}{\rho_r}\right]\delta(q - q_T) \\
& \times \exp\left[j\frac{4\pi}{\lambda}(i - i_T^l)\Delta_r\right].
\end{aligned}
\tag{11.24}
$$

The proof for (11.24) under various conditions are given as follows:
(i) When $l = 1$, we have

$$
I_1(i,q) = G_1(i,q).
\tag{11.25}
$$

Thus, equation (11.24) is obviously satisfied.
(ii) Suppose that when $l = n$,

$$
\begin{aligned}
I_n(i,q) \approx & nA_1 M \mathrm{sinc}\left[\pi \frac{(i - i_T^n)\Delta_r}{\rho_r}\right]\delta(q - q_T) \\
& \times \exp\left[j\frac{4\pi}{\lambda}(i - i_T^n)\Delta_r\right].
\end{aligned}
\tag{11.26}
$$

Then, for $l = n + 1$, we have

$$
I_{n+1}(i,q) = I_n(i_{trans}, q_{trans}) + G_{n+1}(i,q),
\tag{11.27}
$$

where

$$
\begin{aligned}
G_{n+1}(i,q) \approx & A_1 M \mathrm{sinc}\left[\pi \frac{(i - i_T^{n+1})\Delta_r}{\rho_r}\right]\delta(q - q_T) \\
& \times \exp\left[j\frac{4\pi}{\lambda}(i - i_T^{n+1})\Delta_r\right],
\end{aligned}
\tag{11.28}
$$

$$
\begin{aligned}
I_n(i_{trans}, q_{trans}) \approx & nA_1 M \mathrm{sinc}\left[\pi \frac{(i_{trans} - i_T^n)\Delta_r}{\rho_r}\right] \\
& \times \delta(q_{trans} - q_T) \\
& \times \exp\left[j\frac{4\pi}{\lambda}(i_{trans} - i_T^n)\Delta_r\right].
\end{aligned}
\tag{11.29}
$$

Note that

$$
i_{trans} = i - \mathrm{round}(v(q) * T_{sum}/\Delta_r),
\tag{11.30}
$$

$$
q_{trans} = q.
\tag{11.31}
$$

Thus, we have

$$
\begin{aligned}
r(i_{trans}) &= r(i) - \text{round}(v(q) * T_{sum}/\Delta_r) * \Delta_r \\
&= r(i) - v(q) * T_{sum},
\end{aligned} \tag{11.32}
$$

$$
v(q_{trans}) = v(q). \tag{11.33}
$$

Case 1: When $q = q_T$, we have: $q_{trans} = q_T, v(q_{trans}) = v_0, r_{trans} = r(i) - v_0 * T_{sum}$. Then:

$$
\begin{aligned}
I_n(i_{trans}, q_{trans}) &= nA_1 M \text{sinc}\left[\pi \frac{(i_{trans} - i_T^n)\Delta_r}{\rho_r}\right] \\
&\times \exp\left[j\frac{4\pi}{\lambda}(i_{trans} - i_T^n)\Delta_r\right],
\end{aligned} \tag{11.34}
$$

$$
\begin{aligned}
G_{n+1}(i, q) &\approx A_1 M \text{sinc}\left[\pi \frac{(i - i_T^{n+1})\Delta_r}{\rho_r}\right] \\
&\times \exp\left[j\frac{4\pi}{\lambda}(i - i_T^{n+1})\Delta_r\right].
\end{aligned} \tag{11.35}
$$

Note that

$$
\begin{aligned}
i_{trans} - i_T^n &= \text{round}\left[\frac{r(i_{trans}) - R_{0,n}}{\Delta_r}\right] \\
&= \text{round}\left[\frac{r(i) - v_0 * T_{sum} - R_{0,n}}{\Delta_r}\right] \\
&= \text{round}\left[\frac{r(i) - (R_{0,n} + v_0 * T_{sum})}{\Delta_r}\right] \\
&= \text{round}\left[\frac{r(i) - R_{0,n+1}}{\Delta_r}\right] \\
&= i - i_T^{n+1}.
\end{aligned} \tag{11.36}
$$

Thus, we have

$$
\begin{aligned}
I_{n+1}(i, q) &= I_n(i_{trans}, q_{trans}) + G_{n+1}(i, q) \\
&= nA_1 M \text{sinc}\left[\pi \frac{(i - i_T^{n+1})\Delta_r}{\rho_r}\right] \\
&\times \exp\left[j\frac{4\pi}{\lambda}(i - i_T^{n+1})\Delta_r\right] \\
&+ A_1 M \text{sinc}\left[\pi \frac{(i - i_T^{n+1})\Delta_r}{\rho_r}\right] \\
&\times \exp\left[j\frac{4\pi}{\lambda}(i - i_T^{n+1})\Delta_r\right] \\
&= (n+1)A_1 M \text{sinc}\left[\pi \frac{(i - i_T^{n+1})\Delta_r}{\rho_r}\right] \\
&\times \exp\left[j\frac{4\pi}{\lambda}(i - i_T^{n+1})\Delta_r\right].
\end{aligned} \tag{11.37}
$$

Case 2: When $q \neq q_T$, we have

$$G_{n+1}(i, q) \approx 0, \tag{11.38}$$

$$q_{trans} \neq q_T, \tag{11.39}$$

$$I_n(i_{trans}, q_{trans}) \approx 0. \tag{11.40}$$

Hence, the inter-frame integration response in this case is

$$\begin{aligned} I_{n+1}(i, q) &= I_n(i_{trans}, q_{trans}) + G_{n+1}(i, q) \\ &\approx 0. \end{aligned} \tag{11.41}$$

Combining (11.34)−(11.41), we have

$$\begin{aligned} I_{n+1}(i, q) \approx &(n + 1)A_1 M \mathrm{sinc} \left[\pi \frac{(i - i_T^{n+1})\Delta_r}{\rho_r} \right] \\ &\times \delta(q - q_T) \exp \left[j \frac{4\pi}{\lambda}(i - i_T^{n+1})\Delta_r \right]. \end{aligned} \tag{11.42}$$

Thus, equation (11.24) is validated. It is shown that the integration peak amplitude after the multi-frame integration method linearly increases with M and frame number l.

11.2.3   Calculation of P_d and P_{fa}

Assume that the additive noise of lth frame after pulse compression is $n_{1,l}(n, m) \sim CN(0, \sigma_n^2)$ and independent identically distributed (IID). Then, after modified RFT, the noise component (denoted as $w_l(i, q)$) satisfies $w_l(i, q) \sim CN(0, \sigma_n^2 M)$ [4], which means the noise after modified RFT in each frame still satisfies the Gaussian distribution and is also IID. Note that the inter-frame integration process shown in (11.21)−(11.23) is a linear operation and the inter-frame integration process of the noise of the L frames is actually the sum of L independent Gaussian random variables (which are IID). Hence, the noise $n_2(i, q)$ after multi-frame integration process satisfies

$$n_2(i, q) \sim CN(0, \sigma_n^2 ML). \tag{11.43}$$

Thus, detecting moving targets after coherent integration via the RFT domain method could be represented as a binary hypothesis test in the background of complex white Gaussian noise [6], which could be stated as

$$\begin{aligned} H_1 : Z(i, q) &= I_L(i, q) + n_2(i, q) \\ H_0 : Z(i, q) &= n_2(i, q), \end{aligned} \tag{11.44}$$

where $n_2(i, q) \sim CN(0, \sigma_n^2 ML)$ and

$$\begin{aligned} I_l(i, q) \approx &KA_1 M \mathrm{sinc} \left[\pi \frac{(i - i_T^l)\Delta_r}{\rho_r} \right] \delta(q - q_T) \\ &\times \exp \left[j \frac{4\pi}{\lambda}(i - i_T^l)\Delta_r \right]. \end{aligned} \tag{11.45}$$

$Z(i, q)$ is a complex vector decomposed into real $\text{Re}[Z(i, q)]$ and imaginary part $\text{Im}[Z(i, q)]$. The amplitude of $Z(i, q)$ is

$$M = \sqrt{\text{Re}^2[Z(i, q)] + \text{Im}^2[Z(i, q)]} = |Z(i, q)|. \tag{11.46}$$

Under the hypothesis H_0, the random vectors $\text{Re}[Z(i, q)]|H_0$ and $\text{Im}[Z(i, q)]|H_0$ are consistent with normal distributions with zero mean value and variance $\sigma_n^2 ML/2$, and the probability density function (PDF) can be represented as

$$p(\text{Re}[Z(i, q)]|H_0) = \frac{1}{\sqrt{2\pi}\sigma} \exp\left(-\frac{|\text{Re}[Z(i, q)]|^2}{2\sigma^2}\right), \tag{11.47}$$

$$p(\text{Im}[Z(i, q)]|H_0) = \frac{1}{\sqrt{2\pi}\sigma} \exp\left(-\frac{|\text{Im}[Z(i, q)]|^2}{2\sigma^2}\right), \tag{11.48}$$

where $\sigma_n^2 ML/2 = \sigma^2$.

Obviously, the envelope $M|H_0$ conforms to Rayleigh distribution, and its PDF can be acquired [7]

$$p(M|H_0) = \frac{M}{\sigma^2} \exp\left(-\frac{M^2}{2\sigma^2}\right). \tag{11.49}$$

Under the signal plus noise condition (i.e., hypothesis H_1), $\text{Re}[Z(i, q)]|H_1$ and $\text{Im}[Z(i, q)]|H_1$ are also Gaussian distributed, whose expectations and PDF are

$$E(\text{Re}[Z(i, q)]|H_1) = LA_1 M \cos(\theta), \tag{11.50}$$

$$E(\text{Im}[Z(i, q)]|H_1) = LA_1 M \sin(\theta) \tag{11.51}$$

$$p(\text{Re}[Z(i, q)]|H_1),$$
$$= \frac{1}{\sqrt{2\pi}\sigma} \exp\left(-|\text{Re}[Z(i, q)] - E(\text{Re}[Z(i, q)]|H_1|^2/2\sigma^2\right) \tag{11.52}$$

$$p(\text{Im}[Z(i, q)]|H_1),$$
$$= \frac{1}{\sqrt{2\pi}\sigma} \exp\left(-|\text{Im}[Z(i, q)] - E(\text{Im}[Z(i, q)]|H_1|^2/2\sigma^2\right), \tag{11.53}$$

where θ denotes the phase angle of the peak value of $l_L(i, q)$.

Then, the envelope $M|H_1$ obeys the Rice distribution [7], i.e.,

$$p(M|H_1) = \frac{M}{\sigma^2} \exp\left(-\frac{M^2 + (LMA_1)^2}{2\sigma^2}\right) I_0\left(\frac{LMA_1}{\sigma^2}M\right), \tag{11.54}$$

where $I_0(u)$ denotes the modified Bessel function of the first kind. As a result, the likelihood ratio test (LRT) detector can be obtained

$$L(M) = \frac{p(M|H_1)}{p(M|H_0)}$$
$$= I_0\left(\frac{LMA_1}{\sigma^2}M\right) \exp\left[-\frac{(LMA_1)^2}{2\sigma^2}\right]. \tag{11.55}$$

Note that the modified Bessel function $I_0(u)$ is a monotonically increasing function, and thus the variable M can be taken as the test statistic. In this regard, the target can be declared according to the principle

$$M = |Z(i,q)| \overset{\overset{H_1}{>}}{\underset{\underset{H_0,}{<}}{}} \gamma, \tag{11.56}$$

where γ is the LRT threshold, which is determined by the given false alarm and noise power.

Moreover, since the random $M|H_0$ and $M|H_1$ obey the Rayleigh distribution and Rice distribution, respectively, the detection probability and the false alarm probability could be calculated as :

$$
\begin{aligned}
P_d &= \int_\gamma^{+\infty} p(M|H_1)dM \\
&= \int_\gamma^{+\infty} \frac{M}{\sigma^2} \exp\left(-\frac{M^2 + (LMA_1)^2}{2\sigma^2}\right) \\
&\quad \times I_0\left(\frac{LMA_1}{\sigma^2}M\right) dM,
\end{aligned} \tag{11.57}
$$

$$
\begin{aligned}
P_{fa} &= \int_\gamma^{+\infty} p(M|H_0)dM, \\
&= \exp(-\gamma^2/2\sigma^2).
\end{aligned} \tag{11.58}
$$

11.2.4 Input-Output SNR

From (11.24) and (11.43), the ultimate SNR of the integration output of the multi-frame integration method is

$$
\begin{aligned}
SNR_{output} &= 10 * \log10\left(\frac{(LMA_1)^2}{\sigma_n^2 ML}\right) \\
&= 10 * \log10\left(\frac{A_1^2}{\sigma_n^2}\right) + 10 * \log10(M) \\
&\quad + 10 * \log10(L) \\
&= SNR_{input} + Gain_{intra} + Gain_{inter},
\end{aligned} \tag{11.59}
$$

where SNR_{input} denotes the input SNR of the radar compressed echo of each frame, $Gain_{intra}$ denotes the SNR gain of modified RFT and $Gain_{inter}$ denotes the SNR gain of the inter-frame integration process, i.e.,

$$SNR_{input} = 10 * \log10\left(\frac{A_1^2}{\sigma_n^2}\right), \tag{11.60}$$

$$Gain_{intra} = 10 * \log10(M), \tag{11.61}$$

$$Gain_{inter} = 10 * \log10(L). \tag{11.62}$$

It is noted from (11.59)−(11.62) that the output SNR increases with the integrated frame number L and pulse number M.

11.3 NUMERICAL EXPERIMENTS AND PERFORMANCE ANALYSIS

In this section, numerical experiments are performed to evaluate the performance of the multi-frame coherent integration method on several aspects, including integration output response, integration for weak target, integration for multiple targets, input-output SNR performance and detection capability. The radar parameters are carrier frequency $f_c = 0.15$ GHz, bandwidth $B = 20$ MHz, sampling frequency $f_s = 40$ MHz, pulse repetition frequency $f_p = 200$ Hz, pulse duration $T_p = 5$ μs and pulse number of each frame is $M = 200$.

11.3.1 Integration Output Response

In this subsection, the integration output response of the multi-frame processing method is evaluated. Target motion parameters are initial range cell (corresponding to the target's initial range) is 270, radial velocity $v_0 = 150$ m/s and frame number is $L = 6$. The target's compressed echoes during the sixth frame are given in Fig. 11.1(a), where range walk effect (RWE) occurs. After the integration via the multi-frame processing, the integration result is given in Fig. 11.1(b), from which we could see that the target signal is focused as a peak. To show the integration output more clearly, the range response slice and the velocity response slice are, respectively, given in Fig. 11.1(c) and Fig. 11.1(e), where the theoretical response curves (as shown in (11.24)) are also plotted for comparison. Furthermore, Fig. 11.1(d) and Fig. 11.1(f) show, respectively, the local enlarged views of Fig. 11.1(c) and Fig. 11.1(e). It is noticed from Fig. 11.1 that the experiment results agree well with the theoretical curves.

11.3.2 Integration for Weak Target

In this subsection, simulation experiment is carried out to verify the integration performance of the multi-frame integration method for a weak target. The target's motion parameters are the same as those of Section 11.3.1, while the SNR of the target's raw data is −37 dB before pulse compression. Because the time-bandwidth product $D = 100$, the SNR gain of pulse compression is $G_{pc} = 10\log10(D) = 20$ dB. With the pulse compression SNR gain, the outputs (i.e., the compressed echoes of target) are still below noise level (SNR = −17 dB) and unobservable as in Fig. 11.2(a). Fig. 11.2(b) shows the MTD result of target echoes of Fig. 11.2(a). Due to the target's RWE, MTD could not obtain the coherent integration and thus the target is still buried in noise as in Fig. 11.2(b).

After coherent accumulation via RFT, the integration result of the first frame echo is given in Fig. 11.2(c). It could be noticed that it is still not easy to achieve target detection based on Fig. 11.2(c). This is because with the 200-pulse integration (SNR gain is $10\log10(200) = 23$ dB), the output SNR of RFT is about 6 dB, which is still not enough to make the target stands out for detection. The integration results

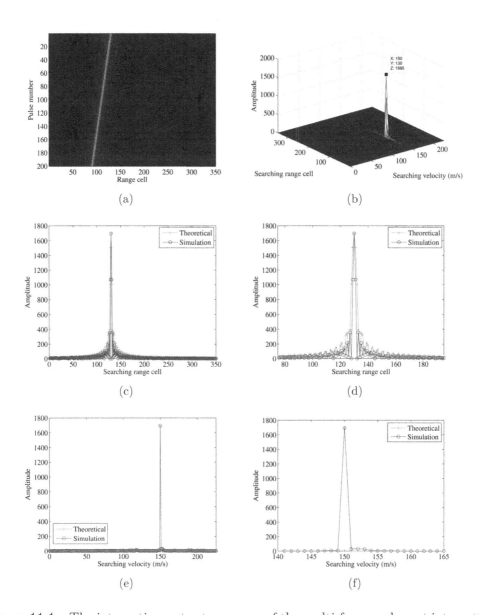

Figure 11.1 The integration output response of the multi-frame coherent integration method. (a) Compressed echoes of the sixth frame. (b) Two-dimensional response. (c) Range response slice. (d) Local enlarged view of range response slice. (e) Velocity response slice. (f) Local enlarged view of velocity response slice.

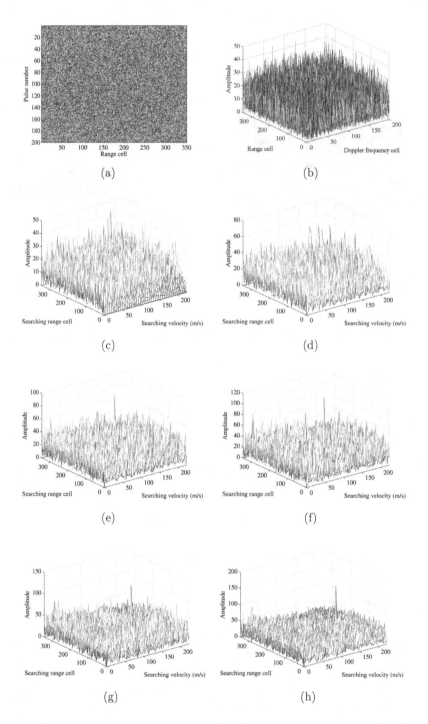

Figure 11.2 Integration of a weak target. (a) Compressed echoes of the first frame. (b) MTD result. (c) RFT result. (d) Integration result of two frames. (e) Integration result of three frames. (f) Integration result of four frames. (g) Integration result of five frames. (h) Integration result of six frames.

TABLE 11.1   Parameters of the Multiple Targets

Targets	Target 1	Target 2	Target 3	Target 4
Initial range cell	270	260	270	240
Radial velocity (m/s)	150	75	75	150
SNR after PC (dB)	6	6	6	6

of the RFT domain method with different frame numbers ($l = 2, 3, 4, 5, 6$) are shown in Fig 11.2(d)−Fig. 11.2(h), where we could see that with the increase in integrated frame number, the integrated peak of target energy gradually becomes prominent. In particular, after integration with six frames echoes, the target signal is focused as an obvious peak (which is significantly higher than the surrounding noise level) in the output, as shown in Fig. 11.2(h), from which target detection can be easily accomplished. The results in Fig. 11.2 show that the multi-frame integration algorithm can improve the target SNR through the joint accumulation of multi-frame echoes.

11.3.3   Integration for Multiple Targets

Where multiple targets exist, there are normally three cases according to the targets' motion parameters, i.e., **case 1:** targets with different initial ranges and different radial velocities; **case 2:** targets with the same initial range but with different radial velocities; **case 3:** targets with different initial ranges but the same radial velocity. Table 11.1 lists the motion parameters of four different moving targets. First, targets 1 and 2 (satisfy case 1 condition) are considered, and Fig. 11.3(a) shows their compressed echoes (first frame), while Fig. 11.3(b) gives the integration result. Second, targets 1 and 3 (satisfy case 2 condition) are used, and their compressed echoes (first frame) are shown in Fig. 11.3(c). After integration via the RFT domain method, the focused result is given in Fig. 11.3(d). Finally, targets 1 and 4 (satisfy case 3 condition) are chosen, while Fig. 11.3(e) shows their compressed echoes (first frame) and Fig. 11.3(f) shows the integration result of the. It could be noticed from Fig. 11.3 that the multi-frame integration method performs well in multi-target scenarios.

11.3.4   Input-Output SNR Performance

In order to validate the SNR gain of the multi-frame integration method, the input-output SNR performance is evaluated in this subsection. The target motion parameters are the same as those of Section 11.3.1, and the input SNRs of the target's compressed echo for each frame vary from −23 to 20 dB, while the step size is 1 dB. Fig. 11.4 shows the experimental input-output SNR curve of the multi-frame integration method, where the theoretical curve using equation (11.59) is also given as a benchmark. From Fig. 11.4, we could see that the experimental input-output SNR curve matches with the theoretical one very well, which means that the multi-frame integration method could obtain integration gain close to the theoretical value.

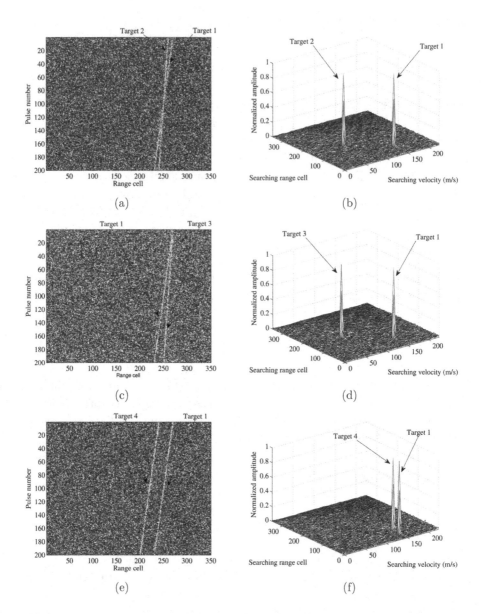

Figure 11.3 Integration of multi-target. (a) Compressed echoes of the first frame for targets 1 and 2. (b) Integration results of targets 1 and 2. (c) Compressed echoes of the first frame for targets 1 and 3. (d) Integration results of targets 1 and 3. (e) Compressed echoes of the first frame for targets 1 and 4. (f) Integration results of targets 1 and 4.

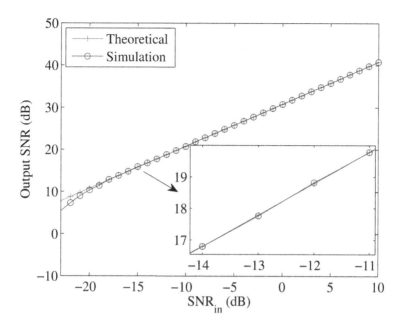

Figure 11.4 Input-output SNR performance.

11.3.5 Detection Performance

In this section, the detection performances of the multi-frame integration method are assessed. The motion parameters of the target are the same as those of Section 11.3, and the false alarm rate is $P_{fa} = 10^{-4}$. Since there is no study in the literature that considers both intra-frame integration and inter-frame integration of a target with RWE, we combine two existing intra-frame integration methods (HT [10] and RFT [2]) and an existing inter-frame integration method (DP [11]), denoted as HT+DP and RFT+DP, respectively, to compare with the RFT domain multi-frame coherent integration algorithm. The results are shown in Fig. 11.5, from which we could see:

- Compared with RFT+DP, the multi-frame coherent integration method can obtain about 3.5 dB of SNR improvement for the detection probability of 0.9. This is because although the multi-frame coherent integration algorithm and the RFT+DP method perform intra-frame accumulation in a similar way, the inter-frame integration of the multi-frame coherent integration method is coherent, while the inter-frame accumulation of RFT+DP is incoherent, i.e., the multi-frame coherent integration method could obtain higher SNR gain within the inter-frame integration processing.

- Compared with HT+DP that carries out intra-frame incoherent integration and inter-frame incoherent integration, the multi-frame coherent integration method obtains a higher detection ability (about 11 dB gain) since it could realize coherent integrations within both intra-frame processing and inter-frame processing.

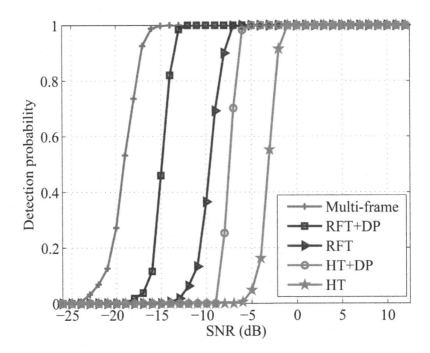

Figure 11.5   Detection performance comparison.

- Compared with RFT and HT, the multi-frame coherent integration method can obtain, respectively, 8 and 15 dB of SNR improvement for the detection probability of 0.9. This is because the RFT and HT algorithms can only realize the intra-frame integration (coherent for RFT but incoherent for HT), while the multi-frame coherent integration method could also realize the inter-frame integration (six frames), resulting in better SNR gain and detection performance.

In addition, the detection ability of the multi-frame coherent integration algorithm with different integration frame number L is shown in Fig. 11.6. From this figure, it could be noticed that the detection probability is better with increasing integration frame number L. In particular, with $P_d = 0.9$, compared with two-frame detection, the performance gains are 1.65, 3.02, 4.01 and 4.75 dB for the multi-frame coherent integration method with frame numbers L=3, 4, 5 and 6, respectively.

11.4   BSSL RESPONSE ANALYSIS

For some special cases, such as target with high velocity and/or radar with low repetition frequency, the BSSL term may appear in the RFT-based intra-frame integration output [4], while the RFT result shown in (11.12) should be further expressed as

$$G_l(i,q) \approx E_{0,l}(i,q) + \sum_{p=1,p\neq s}^{P} E_{p,l}(i,q), \qquad (11.63)$$

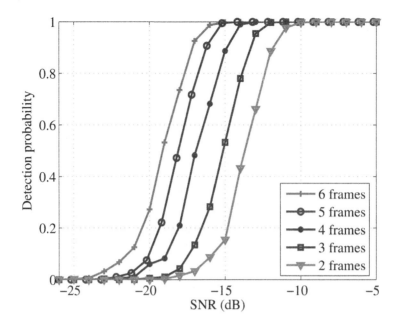

Figure 11.6 Detection probability of the multi-frame coherent integration method for different frame numbers L.

where E_0 denotes the main peak term, with its location corresponding to the target's motion parameters, and E_p denotes the BSSL term, i.e.,

$$
\begin{aligned}
E_0 =& A_1 M \mathrm{sinc}\left[\pi \frac{(i - i_T^l)\Delta_r}{\rho_r}\right] \delta(q - q_T) \\
& \times \exp\left[\frac{4\pi}{\lambda}(i - i_T^l)\Delta_r\right],
\end{aligned}
\tag{11.64}
$$

$$
\begin{aligned}
E_p =& \frac{2A_1 \rho_r}{\lambda(p - s)} \mathrm{sinc}\left[\pi \frac{2\rho_r}{\lambda(p - s)M}(q' - q_T')\right], \\
& \times \mathrm{rect}\left(\frac{2(i - i_T^l)\Delta_r}{\lambda M(p - s)}\right) \exp\left[\frac{4\pi}{\lambda}(i - i_T^l)\Delta_r\right],
\end{aligned}
\tag{11.65}
$$

$$
p = \mathrm{round}(q/M),
\tag{11.66}
$$

$$
s = \mathrm{round}\left(\frac{2v_0}{\lambda f_p}\right),
\tag{11.67}
$$

$$
q' = q - pM,
\tag{11.68}
$$

$$
q_T' = q_T - \mathrm{round}\left(\frac{2v_0}{\lambda f_p}\right)M.
\tag{11.69}
$$

It could be seen from (11.64)–(11.65) that the main lobe-to-BSSL ratio (MBR) after modified RFT is

$$\text{MBR}_1 = \frac{A_1 M}{\left(\frac{2A_1\rho_r}{\lambda(p-s)}\right)} = \frac{\lambda(p-s)M}{2\rho_r}. \tag{11.70}$$

Furthermore, it could be noticed that after the modified RFT, the location of the main peak for the lth frame is (i_T^l, q_T), while the location of the BSSL satisfies

$$q' = q_T'$$
$$\left| \frac{2(i - i_T^l)\Delta_r}{\lambda M(p-s)} \right| \leq 1/2. \tag{11.71}$$

Instituting (11.63) into (11.19), we have

$$I_L(i, q) = \sum_{l=1}^{L} E_{0,L+1-l}\left(i - \frac{(l-1)v(q)T_{sum}}{\Delta_r}, q\right)$$
$$+ \sum_{l=1}^{L} \sum_{p=1, p\neq s}^{P} E_{p,L+1-l}\left(i - \frac{(l-1)v(q)T_{sum}}{\Delta_r}, q\right), \tag{11.72}$$

where the first summation term of (11.72) denotes the integration of the main peak and the last summation term of (11.72) denotes the integration of the BSSL. According to the derivation of Section 11.3.1, it could be obtained that the first summation term can be approximatively expressed as

$$\sum_{l=1}^{L} E_{0,L+1-l}\left(i - \frac{(l-1)v(q)T_{sum}}{\Delta_r}, q\right)$$
$$\approx lA_1 M\text{sinc}\left[\pi\frac{(i - i_T^l)\Delta_r}{\rho_r}\right]\delta(q' - q_T') \tag{11.73}$$
$$\times \exp\left[\frac{4\pi}{\lambda}(i - i_T^l)\Delta_r\right].$$

In the following, we will demonstrate that the BSSL among different frames could not be integrated together.

11.4.1 BSSL Response

As shown in Fig. 11.7, without loss of generality, let A and A' denote, respectively, the peak locations of the main peaks for the lth frame and $l + h$th ($h \geq 1$) frame after the modified RFT operation, while their coordinates are, respectively, denoted as (i_T^l, q_T^l) and (i_T^{l+h}, q_T^{l+h}). Similarly, let B and B' denote, respectively, the locations of the BSSL for the lth frame and $l+h$th frame after the modified RFT operation, while their coordinates are, respectively, denoted as (i_b^l, q_b^l) and (i_b^{l+h}, q_b^{l+h}). According to the integration principle of the inter-frame integration method, several points could be made regarding to the location relationship of the main peak and the BSSL:

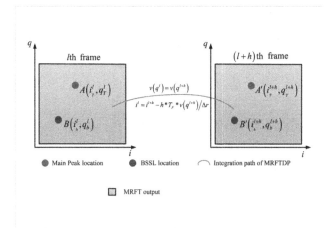

Figure 11.7 Sketch map of accumulate path for RFT-domain integration algorithm.

1) As to the relationship between the main peak location and the BSSL position shown in (11.64)−(11.69), we have

$$|i_b^l - i_T^l| \le \frac{\lambda M|p-s|}{4\Delta_r}$$
$$|q_b^l - q_T^l| = M|p-s|, \tag{11.74}$$

$$|i_b^{l+h} - i_T^{l+h}| \le \frac{\lambda M|p-s|}{4\Delta_r}$$
$$|q_b^{l+h} - q_T^{l+h}| = M|p-s|, \tag{11.75}$$

2) Based on the integration path of RFT-domain inter-frame integration algorithm, the following equations could be obtained for the main peak locations between the lth frame and the $l+h$th frame

$$i_T^l = i_T^{l+h} - v_0 * h * T_{sum}/\Delta_r, \tag{11.76}$$

$$q_T^l = q_T^{l+h} = q_T, \tag{11.77}$$

$$v(q_T^l) = v(q_T^{l+h}) = v_0. \tag{11.78}$$

To make the BSSL between the lth frame and $l+h$th frame accumulate together within the RFT-domain inter-frame integration process, the following condition should be satisfied:

$$i_b^l = i_b^{l+h} - v(q_b^{l+h}) * h * T_{sum}/\Delta_r$$
$$q_b^l = q_b^{l+h}. \tag{11.79}$$

However, the condition shown in (11.79) could not be satisfied. The proof is given in detailed as follows:

First, when $q_b^l = q_b^{l+h}$, the relationship shown in (11.74)−(11.75) could be divided into two cases:

Case 1:

$$0 \le i_b^l - i_T^l \le \frac{\lambda M |p - s|}{4\Delta_r}$$
$$q_b^l - q_T^l = -M|p - s|, \tag{11.80}$$

$$0 \le i_b^{l+h} - i_T^{l+h} \le \frac{\lambda M |p - s|}{4\Delta_r}$$
$$q_b^{l+h} - q_T^{l+h} = -M|p - s|. \tag{11.81}$$

Then, we have

$$i_b^l \le i_T^l + \frac{\lambda M |p - s|}{4\Delta_r},$$
$$v(q_b^l) - v(q_T^l) = -\frac{\lambda f_p |p - s|}{2} \tag{11.82}$$

$$i_T^{l+h} \le i_b^{l+h}. \tag{11.83}$$

Thus, we could have the following inequality:

$$i_b^l - i_b^{l+h} + v(q_b^{l+h}) * h * T_{sum}/\Delta_r$$
$$\le \left(i_T^l + \frac{\lambda M |p - s|}{4\Delta_r} \right) - i_b^{l+h} + v(q_b^{l+h}) * h * T_{sum}/\Delta_r$$
$$\le \left(i_T^l + \frac{\lambda M |p - s|}{4\Delta_r} \right) - i_T^{l+h} + v(q_b^{l+h}) * h * T_{sum}/\Delta_r \tag{11.84}$$
$$\le (i_T^l - i_T^{l+h}) + \frac{\lambda M |p - s|}{4\Delta_r} + v(q_b^{l+h}) * h * T_{sum}/\Delta_r.$$

Substituting (11.76) into (11.84) yields

$$i_b^l - i_b^{l+h} + v(q_b^{l+h}) * h * T_{sum}/\Delta_r$$
$$\le \frac{\lambda M |p - s|}{4\Delta_r} + [v(q_b^{l+h}) - v_0] * h * T_{sum}/\Delta_r$$
$$\le \frac{\lambda M |p - s|}{4\Delta_r} - \frac{\lambda |p - s| f_p}{2} * h * T_{sum}/\Delta_r \tag{11.85}$$
$$\le \frac{\lambda M |p - s|}{4\Delta_r} (1 - 2 * h)$$
$$< 0.$$

Therefore, the condition shown in (11.79) could not be satisfied.

Case 2:

$$-\frac{\lambda M|p-s|}{4\Delta_r} \leq i_b^l - i_T^l < 0$$

$$q_b^l - q_T^l = M|p-s|,$$

$$(11.86)$$

$$-\frac{\lambda M|p-s|}{4\Delta_r} \leq i_b^{l+h} - i_T^{l+h} < 0$$

$$q_b^{l+h} - q_T^{l+h} = M|p-s|.$$

$$(11.87)$$

Then, we have

$$i_T^l - \frac{\lambda M|p-s|}{4\Delta_r} \leq i_b^l$$

$$v(q_b^l) - v(q_T^l) = \frac{\lambda f_p|p-s|}{2}$$

$$(11.88)$$

$$i_b^{l+h} < i_T^{l+h}.$$

$$(11.89)$$

Thus, the following inequality could be obtained:

$$i_b^l - i_b^{l+h} + v(q_b^{l+h}) * h * T_{sum}/\Delta_r$$

$$\geq \left(i_T^l - \frac{\lambda M|p-s|}{4\Delta_r}\right) - i_b^{l+h} + v(q_b^{l+h}) * h * T_{sum}/\Delta_r$$

$$\geq \left(i_T^l - \frac{\lambda M|p-s|}{4\Delta_r}\right) - i_T^{l+h} + v(q_b^{l+h}) * h * T_{sum}/\Delta_r$$

$$\geq (i_T^l - i_T^{l+h}) - \frac{\lambda M|p-s|}{4\Delta_r} + v(q_b^{l+h}) * h * T_{sum}/\Delta_r.$$

$$(11.90)$$

Substituting (11.76) into (11.90) yields

$$i_b^l - i_b^{l+h} + v(q_b^{l+h}) * h * T_{sum}/\Delta_r$$

$$\geq -\frac{\lambda M|p-s|}{4\Delta_r} + [v(q_b^{l+h}) - v_0] * h * T_{sum}/\Delta_r$$

$$\geq -\frac{\lambda M|p-s|}{4\Delta_r} + \frac{\lambda|p-s|f_p}{2} * h * T_{sum}/\Delta_r$$

$$\geq \frac{\lambda M|p-s|}{4}(2h-1) > 0.$$

$$(11.91)$$

Therefore, the condition shown in (11.79) could not be satisfied in case 2.

Through the analysis of cases 1 and 2, it can be concluded that in the process of accumulating multiple frames of signals, the BSSL of any two frames will not be accumulated together. However, the main peak term of all the frames will be coherently accumulated. That is to say, the integration amplitude of the main peak term linearly increases with the frame number l, while the integration amplitude of the BSSL term approximately unchanged with the frame number l. As a result, the main peak term is enhanced and the BSSL term is suppressed in a sense, while the MBR with L frames could be approximately expressed as

$$\mathrm{MBR}_2 \approx \frac{LA_1M}{\left(\frac{2A_1\rho_r}{\lambda(p-s)}\right)} = L\mathrm{MBR}_1.$$

$$(11.92)$$

11.4.2 Simulation Example

In this subsection, a simulation example is provided to show how the BSSL performs within the RFT-domain inter-frame integration process. The system parameters of the radar are radar carrier $f_c = 0.75$ GHz, signal bandwidth $B = 15$ MHz, sampling frequency $f_s = 2\ B$, pulse duration $T_p = 10\ \mu s$, pulse number of each frame $M = 200$, frame number $L = 6$ and pulse repetition frequency $f_p = 200$ Hz. The target moves with constant velocity $v_0 = 150$ m/s, and its initial range location is 200 km (corresponding to the 330th range cell). The integration results are given in Fig. 11.8, where Fig. 11.8(a) shows the integration result of the first frame echo and Fig. 11.8(b)−Fig. 11.8(f) are, respectively, the outputs of the multi-frame integration method with frame number $L = 2, 3, 4, 5$ and 6. It could be seen from Fig. 11.8 that the BSSL term becomes lower with the increase of the frame number L, i.e., the BSSL suppression performance improves with the increase of L.

To show the BSSL suppression response of the RFT-domain integration method more clear, we also calculate the MBR of Fig. 11.8(a)–Fig. 11.8(f), which could well reflect the change of BSSL suppression within the integration processing. The results are listed in Table 11.2. In addition, both the simulated MBR curve and the theoretical MBR curve versus frame number are illustrated in Fig. 11.9, from which we could notice that the simulation curve agrees with the theoretical results well.

11.5 SUMMARY

In this chapter, a multi-frame coherent integration method is introduced to realize the intra-frame integration and inter-frame integration of the high-speed moving target. The multi-frame integration algorithm firstly accomplish the intra-frame integration via the modified RFT operation. Thereafter, based on the signal characteristics of the modified RFT outputs, the RFT-domain inter-frame integration algorithm is presented to achieve the accumulation of target signal energy distributed in multiple frames. Also, this chapter provides detailed derivations and analyses on four aspects, i.e., integration output response, probability of detection, input-output SNR performance and BSSL response. Finally, the theoretically derivation and analysis are validated with detailed numerical experiments.

TABLE 11.2 The MBR within the Integration Processing

Output	MBR
Integration with first frame (Fig. 11.8(a))	0.263
Integration with two frames (Fig. 11.8(b))	0.131
Integration with three frames (Fig. 11.8(c))	0.087
Integration with four frames (Fig. 11.8(d))	0.065
Integration with five frames (Fig. 11.8(e))	0.052
Integration with six frames (Fig. 11.8(f))	0.0434

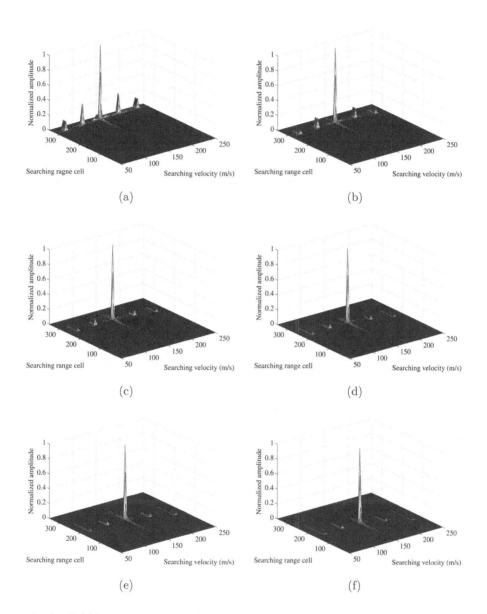

Figure 11.8   BSSL suppression response of the multi-frame integration method with respect to frame number. (a) Integration result of the first frame echo. (b) Integration result with two-frame echoes. (c) Integration result with three-frame echoes. (d) Integration result with four frame echoes. (e) Integration result with five-frame echoes. (f) Integration result with six-frame echoes.

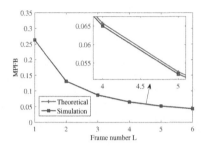

Figure 11.9   MBR curve versus frame number.

Bibliography

[1] X. L. Li, Z. Sun, T. S. Yeo, T. X. Zhang, W. Yi, G. L. Cui, and L. J. Kong, "STGRFT for detection of maneuvering weak target with multiple motion models," *IEEE Transactions on Signal Processing*, vol. 67, no. 7, pp. 1902–1917, April 2019.

[2] J. Xu, J. Yu, Y. N. Peng, and X. G. Xia, "Radon-fourier transform (RFT) for radar target detection (I): generalized doppler filter bank processing," *IEEE Transactions on Aerospace and Electronic Systems*, vol. 47, no. 2, pp. 1186–1202, April 2011.

[3] J. Xu, J. Yu, Y. N. Peng, X. G. Xia, and T. Long "Space-time Radon-Fourier transform and applications in radar target detection," *IET Radar, Sonar and Navigation*, vol. 6, no. 9, pp. 846–857, December 2012.

[4] J. Xu, J. Yu, Y. N. Peng, and X. G. Xia, "Radon-fourier transform (RFT) for radar target detection (II): blind speed sidelobe suppression," *IEEE Transactions on Aerospace and Electronic Systems*, vol. 47, no. 4, pp. 2473–2489, October 2011.

[5] J. Yu, J. Xu, Y. N. Peng, and X. G. Xia, "Radon-fourier transform (RFT) for radar target detection (III): optimality and fast implementations," *IEEE Transactions on Aerospace and Electronic Systems*, vol. 48, no. 2, pp. 991–1004, April 2012.

[6] J. Xu, X. G. Xia, S. B. Peng, J. Yu, Y. N. Peng, and L. C. Qian, "Radar maneuvering target motion estimation based on generalized radon-fourier transform," *IEEE Transactions on Signal Processing*, vol. 60, no. 12, pp. 6190–6201, December 2012.

[7] S. M. Kay, *Fundamentals of Statistical of Signal Processing Volume II: Detection Theory*. Englewood Cliffs, NJ, USA: Prentice Hall PTR, 1998, pp. 141–165.

[8] S. B. Peng, J. Xu, X. G. Xia, F. Liu, T. Long, J. Yang, and Y. N. Peng, "Multi-aircraft formation identification for narrowband coherent radar in a long coherent integration time," *IEEE Transactions on Aerospace and Electronic Systems*, vol. 51, no. 3, pp. 2121–2137, July 2015.

[9] J. Xu, Y. N. Peng, X. G. Xia, and A. Farina, "Focus-before-detection radar signal processing: part i-challenges and methods," *IEEE Aerospace and Electronic Systems Magazine*, vol. 32, no. 9, pp. 48–59, September 2017.

[10] B. D. Carlson, E. D. Evans, and S. L. Wilson, "Search radar detection and track with the hough transform part I: system concept," *IEEE Transactions on Aerospace and Electronic Systems*, vol. 30, no. 1, pp. 102–108, January 1994.

[11] L. A. Johnston and V. Krishnamurthy, "Performance analysis of a dynamic programming track before detect algorithm," *IEEE Transactions on Aerospace and Electronic Systems*, vol. 38, no. 1, pp. 228–242, January 2002.

CHAPTER 12

SCRFT-Based Coherent Integration Processing

With the increase of target's velocity and the time-bandwidth product of the transmitted signal, the scale effect will appear in the echo signal, which influences the performance of pulse compression (PC) and the subsequent coherent integration. This chapter addresses the coherent integration problem for the hypersonic target with scale effect. The scaled Radon Fourier transform (SCRFT)-based coherent integration method is discussed, including the signal model, transform definition and computational complexity analysis.

12.1   SIGNAL MODEL

Suppose the coherent radar emits the linear frequency modulated (LFM) signal [1–3], i.e.,

$$s_{trans}(t, t_m) = \text{rect}\left(\frac{t}{T_p}\right) \exp\left(j\pi\mu t^2\right) \exp\left[j2\pi f_c(t + t_m)\right], \qquad (12.1)$$

where

$$\text{rect}\left(\frac{t}{T_p}\right) = \begin{cases} 1 & \left|\frac{t}{T_p}\right| \leq \frac{1}{2}, \\ 0 & \left|\frac{t}{T_p}\right| > \frac{1}{2}, \end{cases}$$

t is the fast time and t_m indicates the slow time. $t_m = mT_r$ $(m = 0, \ldots, M - 1)$, where T_r is the pulse repetition interval and M is the number of total pulses. Besides, f_c and T_p, respectively, indicate the carrier frequency and pulse duration. Furthermore, μ denotes the frequency modulation rate.

Assume that the target moves away from radar with velocity v_0 and the initial slant range between radar and the target is r_0. When the target begins to move, the pulses will also be transmitted. Therefore, before the mth pulse is emitted, the target has moved for the period of $t + t_m$. Whereafter, the transmitted mth pulse will hit

                                                      DOI: 10.1201/9781003529101-12

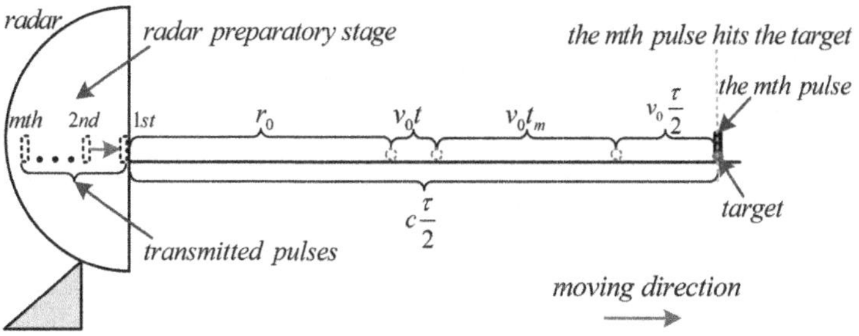

Figure 12.1 Diagram of the target and pulses motions.

the target after moving for the time of $\frac{\tau}{2}$ (τ denotes the time delay). The diagram of the target and pulses motions is shown in Fig. 12.1, from which we can obtain

$$\frac{c\tau}{2} = r_0 + v_0 t + v_0 t_m + \frac{v_0 \tau}{2}, \tag{12.2}$$

where c is the light speed.

According to equation (12.2) and considering the intra-pulse and inter-pulse motions, the instantaneous slant range of the target can be stated as [4, 5]

$$r(t, t_m) = r_0 + v_0(t + t_m). \tag{12.3}$$

On the basis of equations (12.2) and (12.3), the time delay τ can be written as

$$\tau = \frac{2r(t, t_m)}{c - v_0}. \tag{12.4}$$

After the demodulation, the echo signal could be presented as

$$s_r(t, t_m) = A_0 \text{rect} \left[\frac{\rho}{T_p}(t - \tau_m) \right] \exp \left[j\pi\mu\rho^2 (t - \tau_m)^2 \right]$$
$$\times \exp \left(-j2\pi f_c \rho \tau_m \right) \exp \left(-j2\pi\alpha t \right), \tag{12.5}$$

where A_0 indicates the complex amplitude of the echo signal. $\rho = \frac{c - 3v_0}{c - v_0}$ denotes the scale coefficient, and $\tau_m = \frac{2(r_0 + v_0 t_m)}{c - 3v_0}$ is the time delay of the inter-pulses. $\alpha = \frac{2v_0 f_c}{c - v_0}$. Applying FT on (12.5) with respect to t and utilizing the principle of stationary phase (PSP) [6, 7], we can get the target signal in the range frequency domain, i.e.,

$$S_r(f, t_m) = A_1 \text{rect} \left(\frac{f + \alpha}{\rho B} \right) \exp \left(-j\pi \frac{f^2}{\mu\rho^2} \right)$$
$$\times \exp \left(-j2\pi \frac{f\alpha}{\mu\rho^2} \right) \exp \left(-j\pi \frac{\alpha^2}{\mu\rho^2} \right) \tag{12.6}$$
$$\times \exp \left[-j2\pi (f + f_c) \tau_m \right],$$

where f denotes the range frequency with respect to t and $B = \mu T_p$ is the radar bandwidth. A_1 is the complex amplitude after the FT operation.

In the next section, we will discuss the traditional pulse compression (TPC) and coherent integration (CI) of the typical traditional method (i.e., Radon Fourier transform [RFT]) to evaluate the influence of scale effect.

12.2 TPC AND CI OF RFT

Usually, the TPC can be realized in frequency domain [8, 9], i.e.,

$$s(t, t_m) = \mathop{\text{IFT}}_{f} (S_r(f, t_m) H_t(f)),\tag{12.7}$$

where $\mathop{\text{IFT}}_{f}(\cdot)$ represents the inverse Fourier transform (IFT) on f. $H_t(f)$ is the traditional matched filter with the following form [8]:

$$H_t(f) = \text{rect}\left(\frac{f}{B}\right) \exp\left(j\pi \frac{f^2}{\mu}\right).\tag{12.8}$$

Substituting (12.6) and (12.8) into (12.7), the TPC in the time domain can be further written as

$$
\begin{aligned}
s(t, t_m) &= \int_{-\infty}^{+\infty} A_1 \exp\left(-j\pi \frac{\alpha^2}{\mu\rho^2}\right) \exp\left(-j2\pi f_c \tau_m\right) \\
&\times \text{rect}\left(\frac{f+\alpha}{\rho B}\right) \text{rect}\left(\frac{f}{B}\right) \exp\left[j\pi \frac{f^2}{\mu}\left(1 - \frac{1}{\rho^2}\right)\right] \\
&\times \exp\left(-j2\pi f \tau_m\right) \exp\left(-j2\pi \frac{f\alpha}{\mu\rho^2}\right) \exp\left(j2\pi f t\right) df \\
&= \begin{cases} A_{x1} \cdot \chi_1^*, & (\rho^2 < 1) \\ A_{x1} \cdot \chi_2, & (\rho^2 > 1) \end{cases},
\end{aligned}\tag{12.9}
$$

where

$$
\begin{aligned}
A_{x1} &= A_1 \sqrt{\frac{1}{2|\mu\triangle q|}} \exp\left(-j2\pi \frac{f\alpha}{\mu\rho^2}\right) \exp\left(-j2\pi f_c \tau_m\right) \\
&\times \exp\left\{-j\pi \left[\frac{1}{\mu\triangle q}\left(t - \frac{\alpha}{\mu\rho^2} - \tau_m\right)\right]^2\right\},
\end{aligned}\tag{12.10}
$$

where $\triangle q = \frac{1}{\mu^2}\left(1 - \frac{1}{\rho^2}\right)$. Besides, χ_1^* and χ_2 indicate the Fresnel functions. The proof of (12.9) is given in Section 12.7, including the concrete expressions of χ_1^* and χ_2.

In (12.9), $\exp\left[j\pi \frac{f^2}{\mu}\left(1 - \frac{1}{\rho^2}\right)\right]$ will cause the sinc envelope deformation (i.e., the envelope peak descent and extension of mainlobe). Besides, $\exp\left(-j2\pi \frac{f\alpha}{\mu\rho^2}\right)$ will lead to the sinc envelop center offset. As a result, the TPC result is a combination of Fresnel function rather than a sinc function, which is called as scale effect.

TABLE 12.1 System Parameters of Radar

Parameters	Values
Carrier frequency f_c	1.5 GHz
Bandwidth B	200 MHz
Sampling frequency f_s	400 MHz
Pulse repetition frequency PRF	250 Hz
Pulse duration T_p	1.5 ms
Number of pulses M	256

Below, an example (Example 1) is given to compare the performance of ideal PC and TPC when the scale effect occurs.

Example 1: A hypersonic target is considered, and the radar parameters are listed in Table 12.1. Besides, the motion parameters are set as: $r_0 = 300$ km (corresponding to the 300th range cell) and $v_0 = 3400$ m/s. The ideal PC and TPC results of a single transmitted pulse are, respectively, shown in Fig. 12.2(a) and Fig. 12.2(b), where the envelope of TPC is no longer a sinc function due to scale effect. From these figures, we can see that the TPC suffers from obvious performance loss and that the peak location shifts from the range cell 300 to 410, which are caused by the scale effect.

To discuss the relationship between the TPC result and target's velocity, another example (Example 2) is given.

Example 2: The radar parameters are the same as those of Example 1, and the hypersonic target moves with different velocities (i.e., $0, 2.5, 5, 7.5, 10, 12.5$, and 15 Mach). Fig. 12.3 gives the TPC with respect to different target's velocities, where the ideal PC refers to the PC result for a target with zero speed and is shown as a benchmark. From the experiment results, we can see that with the increase of the target's velocity, the TPC performance will become worse. Besides, the center and peak location will shift more. That is, the target's velocity is higher, and the scale effect becomes more obvious.

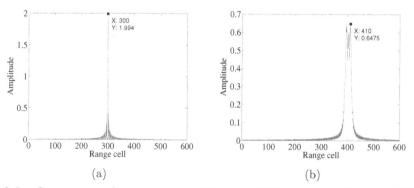

Figure 12.2 Comparison between ideal PC and TPC. (a) Ideal PC. (b) TPC.

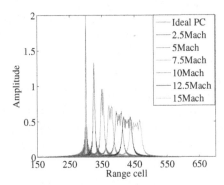

Figure 12.3 TPC with different velocities.

Furthermore, the RFT output for (12.9) can be written as

$$
S_{\mathrm{RFT}}\left(r', v'\right) = \begin{cases} \displaystyle\int_0^{T_{sum}} A_{x1} \cdot \chi_1^* \exp\left(\frac{j4\pi v' t_m}{\lambda}\right) dt_m, (\rho^2 < 1) \\ \displaystyle\int_0^{T_{sum}} A_{x1} \cdot \chi_2 \exp\left(\frac{j4\pi v' t_m}{\lambda}\right) dt_m, (\rho^2 > 1) \end{cases}, \tag{12.11}
$$

where $T_{sum} = MT_r$ represents the coherent processing interval.

In (12.11), we can see that the RFT is the integral result of the Fresnel function rather than sinc function because of the scale effect, which will cause CI performance loss.

To obtain good CI performance for a target with scale effect and RW, we introduce the SCRFT in the next section.

12.3 INTRODUCTION OF SCRFT

This section proposes the SCRFT to realize the CI for the hypersonic target with scale effect and RW, which includes the SCRFT for a single target, the SCRFT for multi-targets and analysis of computational complexity.

12.3.1 SCRFT for a Single Target

With $t = 2r'/(c - 3v_0)$, the SCRFT, which can successively realize the PC and CI, is defined as follows:

$$
S_{\mathrm{SCRFT}}\left(r', v'\right) = \int_0^{T_{sum}} s_c\left(\frac{2r'}{c - 3v_0}, t_m\right) H_2\left(v', t_m\right) dt_m, \tag{12.12}
$$
$$
\left(r' \in [r'_{\min}, r'_{\max}], \ v' \in [v'_{\min}, v'_{\max}]\right),
$$

where

$$s_c \left(\frac{2r'}{c - 3v_0}, t_m \right) = \underset{f}{\mathrm{IFT}} \left[S_r \left(f, t_m \right) H_1 \left(f, v' \right) \right], \tag{12.13}$$

$$
\begin{aligned}
H_1 \left(f, v' \right) =& \mathrm{rect} \left(\frac{f + \alpha'}{\rho' B} \right) \exp \left(j\pi \frac{f^2}{\mu \rho'^2} \right) \\
& \times \exp \left(j2\pi \frac{f\alpha'}{\mu \rho'^2} \right) \exp \left(j\pi \frac{\alpha'^2}{\mu \rho'^2} \right),
\end{aligned}
\tag{12.14}
$$

$$H_2 \left(v', t_m \right) = \exp \left(j4\pi f_c \frac{v' t_m}{c - 3v'} \right), \tag{12.15}$$

where r' and v' represent the searching radial range and velocity, respectively. $[r'_{min}, r'_{max}]$ and $[v'_{min}, v'_{max}]$ denote the searching regions of r' and v'. $\rho' = \frac{c - 3v'}{c - v'}$ is the searching scale coefficient and $\alpha' = \frac{2v' f_c}{c - v'}$.

The SCRFT in (12.12) mainly includes two implementation processes: one is the process of matched PC and the other is the process of extraction and CI, which are, respectively, introduced in the next two subsections.

12.3.1.1 *The Process of Matched PC*

The process of matched PC realizes the intra-pulse energy integration by velocity searching, as shown in (12.13), which can be further written as

$$
\begin{aligned}
s_c \left(\frac{2r'}{c - 3v_0}, t_m \right) =& \underset{f}{\mathrm{IFT}} \left[S_r \left(f, t_m \right) H_1 \left(f, v' \right) \right] \\
=& \int_{-\infty}^{+\infty} A_1 \mathrm{rect} \left(\frac{f + \alpha}{\rho B} \right) \mathrm{rect} \left(\frac{f + \alpha'}{\rho' B} \right) \\
& \times \exp \left[\frac{-j\pi f^2}{\mu} \left(\frac{1}{\rho^2} - \frac{1}{\rho'^2} \right) \right] \exp \left[\frac{-j2\pi f}{\mu} \left(\frac{\alpha}{\rho^2} - \frac{\alpha'}{\rho'^2} \right) \right] \\
& \times \exp \left[\frac{-j\pi}{\mu} \left(\frac{\alpha^2}{\rho^2} - \frac{\alpha'^2}{\rho'^2} \right) \right] \exp \left[-j4\pi \left(f + f_c \right) \tau_m \right] \\
& \times \exp \left(j\pi f \frac{4r'}{c - 3v_0} \right) df \\
=&
\begin{cases}
A_{z1} \cdot \chi_3^*, & (v' < v_0) \\
A_2 \mathrm{sinc} \left\{ \frac{2\pi B}{c - v_0} \left[\left(r' - r_0 \right) - v_0 t_m \right] \right\} \\
\quad \times \exp \left[-j4\pi f_c \frac{\left(r_0 + v_0 \right) t_m}{c - 3v_0} \right], & (v' = v_0) \\
A_{z1} \cdot \chi_4, & (v' > v_0)
\end{cases},
\end{aligned}
\tag{12.16}
$$

where

$$A_{z1} = A_1 \sqrt{\frac{1}{|2\mu\triangle q_0|}} \exp\left(-j\pi\mu\triangle q_2\right) \exp\left(-j2\pi f_c \tau_m\right)$$

$$\times \exp\left\{-j\pi\left[\frac{1}{\mu\triangle q_0}\left(t - \mu\triangle q_1 - \tau_m\right)\right]^2\right\}, \tag{12.17}$$

$\triangle q_0 = \frac{1}{\mu^2 \rho'^2} - \frac{1}{\mu^2 \rho^2}$. $\triangle q_1 = \frac{\alpha}{\mu^2 \rho^2} - \frac{\alpha'}{\mu^2 \rho'^2}$. $\triangle q_2 = \frac{\alpha^2}{\mu^2 \rho^2} - \frac{\alpha'^2}{\mu^2 \rho'^2}$. χ_3^* and χ_4 denote the Fresnel functions corresponding to the situation of $v' < v_0$ and $v' > v_0$, respectively. The proof of (12.16) and the specific expressions of χ_3^* and χ_4 are given in Section 12.8.

In (12.16), if the searching velocity is matched with the actual one, i.e., $v' = v_0$, the PC result is a strict sinc function, else the PC output is a combination of Fresnel function.

12.3.1.2 *The Process of Extraction and CI*

After the matched PC process shown in (12.16) (i.e., $v' = v_0$), the signal extraction and CI process are employed to realize the inter-pulse accumulation. With the matched PC result of (12.16), this process can be stated as

$$S_{\text{SCRFT}}\left(r', v'\right) = \int_0^{T_{sum}} A_2 \text{sinc}\left\{\frac{2\pi B}{c - v_0}\left[\left(r' - r_0\right) + \left(v' - v_0\right)t_m\right]\right\}$$

$$\times \exp\left(-j4\pi f_c \frac{v_0 t_m}{c - 3v_0}\right)$$

$$\times \exp\left(j4\pi f_c \frac{v' t_m}{c - 3v'}\right) \exp\left(-j4\pi f_c \frac{r_0}{c - 3v_0}\right) dt_m$$

$$\approx \begin{cases} A_2 T_{sum} \text{sinc}\left[\frac{2\pi B}{c - v_0}\left(r' - r_0\right)\right] \\ \times \text{sinc}\left[2\pi T_{sum} f_c\left(\frac{v'}{c - 3v'} - \frac{v_0}{c - 3v_0}\right)\right] \\ \times \text{rect}\left[\frac{2T_{sum}B}{c - v_0}\left(v' - v_0\right)\right], \\ \left(v' = v_0 \text{ or } r' = r_0\right) \\ 0, \\ \left(v' \neq v_0 \text{ and } r' \neq r_0\right) \end{cases} \tag{12.18}$$

From (12.18), we can see that when $v' = v_0$ and $r' = r_0$, the CI of target signal is obtained and there will appear an energy peak in the SCRFT output.

The flowchart of the SCRFT-based algorithm is shown in Fig. 12.4, and the implementation procedures of SCRFT are summarized in Algorithm 2.

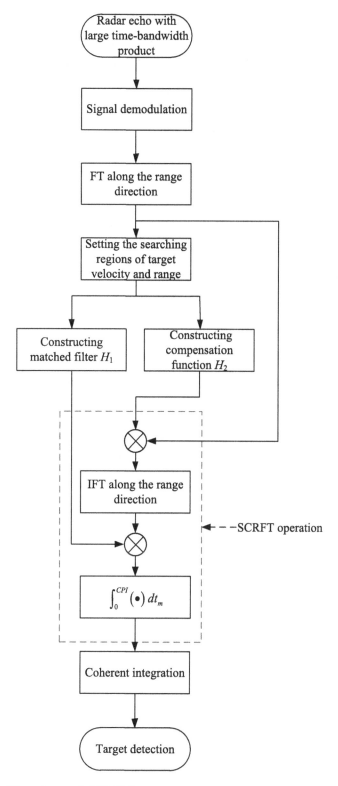

Figure 12.4　Flowchart of SCRFT.

Algorithm 2 Summarize: primary procedures of SCRFT

Input: raw data of transmitting LFM signal, i.e., $s(t, t_m)$.

Signal demodulation and FT: Receiving the echo signal and performing the demodulation, we can obtain $s_r(t, t_m)$ in (12.5). Then, performing the FT with respect to t, we have $S_r(f, t_m)$ in (12.6).

Setting searching regions of r' and v': Parameter searching regions are set as $[r'_{\min}, r'_{\max}]$ and $[v'_{\min}, v'_{\max}]$. Parameter searching steps are set as ρ_r and ρ_v. **Constructing H_1 and H_2:** Constructing the matched filter H_1 in (12.14) and compensation function H_2 in (12.15).

SCRFT operation: Go through each r' in $[r'_{\min}, r'_{\max}]$ with the interval ρ_r and v' in $[v'_{\min}, v'_{\max}]$ with the interval ρ_v when performing SCRFT operation.

   for $r' = r'_{\min} : \rho_r : r'_{\max}$ **do**
      for $v' = v'_{\min} : \rho_v : v'_{\max}$ **do**
         Multiplying the matched filtering function $H_1(f, v')$ with $S_r(f, t_m)$ and applying IFT along the f direction. Then multiplying the compensation function $H_2(v', t_m)$ with the IFT output and integrating with respect to t_m.
      end for
   end for

Output: CI result $S_{\text{SCRFT}}(r', v')$.

Target detection: Applying the constant false alarm rate (CFAR) detection based on the CI result, i.e., comparing the peak value of the CI with the self-adaptive threshold for a certain false alarm probability, namely [10]

$$|S_{\text{SCRFT}}(r', v')| > \xi, \tag{12.19}$$

where ξ denotes the threshold level, which is determined by the reference unit of CI. The test statistic is higher than the threshold ξ means the target can be detected. Otherwise, the target cannot be detected [10].

Parameter estimation: Finding the maximal peak value. The corresponding r' and v' are the estimated initial range and velocity.

12.3.2 SCRFT for Multi-Target

Suppose that there are K targets moving with a constant velocity in the scene. The received signals of the K targets in the range frequency domain can be expressed as

$$
\begin{aligned}
S_{r,k}(f, t_m) = \sum_{k=1}^{K} & A_{2,k} \text{rect}\left(\frac{f + \alpha_k}{\rho_k B}\right) \exp\left(-j\pi \frac{f^2}{\mu \rho_k^2}\right) \\
& \times \exp\left(-j2\pi \frac{f \alpha_k}{\mu \rho_k^2}\right) \exp\left(-j\pi \frac{\alpha_k^2}{\mu \rho_k^2}\right) \\
& \times \exp\left[-j2\pi (f + f_c) \tau_{m,k}\right],
\end{aligned}
\tag{12.20}
$$

where $A_{2,k}$ is the complex amplitude of the kth target's received signal after the FT operation. $\rho_k = \frac{c - 3v_{0,k}}{c - v_{0,k}}$ is the kth target's scale coefficient and $\tau_{m,k} = \frac{2(r_{0,k} + v_{0,k} t_m)}{c - 3v_{0,k}}$

denotes the inter-pulse time delay of the kth target. Besides, $\alpha_k = \frac{2v_{0,k}f_c}{c-v_{0,k}}$. $r_{0,k}$ and $v_{0,k}$, respectively, indicate the initial slant range and velocity of the kth target.

With $t = \frac{2r'_k}{c-3v_{0,k}}$, the SCRFT for the K targets can be given as

$$
S_{\text{SCRFT}}\left(r'_k, v'_k\right) = \int_0^{T_{sum}} s_{c,k}\left(\frac{2r'_k}{c-3v_{0,k}}, t_m\right)
$$
$$
\times H_{2,k}\left(v'_k, t_m\right) dt_m, \tag{12.21}
$$

where

$$
s_{c,k}\left(\frac{2r'_k}{c-3v_{0,k}}, t_m\right) = \underset{f}{\text{IFT}}\left[S_{r,k}\left(f, t_m\right) H_{1,k}\left(f, v'_k\right)\right], \tag{12.22}
$$

$$
H_{1,k}\left(f, v'_k\right) = \text{rect}\left(\frac{f+\alpha'_k}{\rho'_k B}\right) \exp\left(j\pi\frac{f^2}{\mu\rho'^2_k}\right)
$$
$$
\times \exp\left(j2\pi\frac{f\alpha'_k}{\mu\rho'^2_k}\right) \exp\left(j\pi\frac{\alpha'^2_k}{\mu\rho'^2_k}\right), \tag{12.23}
$$

$$
H_{2,k}\left(v'_k, t_m\right) = \exp\left(j4\pi f_c\frac{v'_k t_m}{c-3v'_k}\right), \tag{12.24}
$$

r'_k and v'_k are the searching slant range and velocity of the kth target, separately. $\rho'_k = \frac{c-3v'_k}{c-v'_k}$ denotes the searching scaled coefficient of the kth target and $\alpha'_k = \frac{2v'_k f_c}{c-v'_k}$.

Based on the analysis in Section 12.3.1, it can be noticed from (12.21) that if the searching velocity and initial range match with the kth target's motion parameters, and the matched PC and CI of kth target could be obtained. Consequently, according to the relationship of different targets' motion parameters, three cases can be presented within the SCRFT of multiple targets.

Case 1: The K targets are with different velocities and initial ranges. In this case, the matched PC of the K targets is corresponding to different searching velocities, and the focused peak locations in SCRFT domain for the K targets are also different, i.e., the peak location of each target is different in both dimensions of searching velocity and range.

Case 2: The K targets are with the same initial range but different velocities. In this case, the matched PC of the K targets is corresponding to different searching velocities. The focused peak locations in SCRFT domain for the K targets are only different in the searching velocity dimension, but the same in the searching range dimension. In other words, the K targets could be separated because of their different velocities.

Case 3: The K targets are with the same velocities but different initial ranges. In this case, the matched PC of the K targets is corresponding to the same searching velocity. The focused peak locations in SCRFT domain for the K targets are only different in the searching range dimension, but the same in the searching velocity dimension. That is to say, the K targets could be separated because of their different initial ranges.

TABLE 12.2 Motion Parameters of Targets A, B and C

Targets	Initial range cells	Radial velocities (m/s)
A	120	3000
B	240	3000
C	240	2800

Based on the three cases discussed above, a simulation example (Example 3) is provided to show how SCRFT performs in these three cases.

Example 3: In this example, three targets, i.e., targets A, B and C, are considered, and their motion parameters are listed in Table 12.2. Additionally, the input-SNR is −49 dB, and we still use the radar parameters in Table 12.1 (just change the pulse number as 128). The SCRFT results of multi-target under the mentioned three cases are shown in Fig. 12.5. Particularly, targets A and C are considered for case 1, and Fig. 12.5(a)–12.5(b) show their SCRFT results, where the integrated peak locations of targets A and C are different in both dimensions of searching velocity and range. Targets B and C are considered for case 2, and Fig. 12.5(c)–12.5(d) give the corresponding SCRFT results. The peak locations of targets B and C are the same along the searching range axis but different in the searching, making them can be separated and detected well along the velocity axis. Targets A and B are selected for case 3, and the corresponding SCRFT results are given in Fig. 12.5(e) and 12.5(f). Although these two targets have the same velocity, they can be well distinguished in the range direction based on the SCRFT outputs.

12.3.3 Analysis of Computational Complexity

We analyze the computational complexity of the traditional methods, including RFT [11], modified location rotation transform (MLRT) [8] and scaled inverse Fourier transform (SCIFT) [12], and the SCRFT-based algorithm in this section. Define the number of total pulses, searching range cells, searching angles and searching velocities by M, K_r, K_θ and K_v, respectively.

Essentially, except for SCIFT, the rest methods (including SCRFT) all need two-dimensional (2-D) searching. For RFT, it requires to search in range-velocity domain, and its computational complexity is $O(K_v M K_r)$. For MLRT, it needs to search the rotation angle and FT operation, while its computational complexity is $O(K_\theta M K_r)$. For SCIFT, it can use the symmetric autocorrelation function to estimate the parameters and avoid the 2-D searching. As a result, the computational complexity of this method is $O(M^2 K_r)$. The SCRFT needs to search target's velocity and range to continuously realize PC and CI, which requires the computational complexity of $O(K_v M K_r)$. For a clearer display, the computational complexity of all the methods is listed in Table 12.3.

Suppose $M = K_r = K_\theta = K_v$, the computational complexity of all the methods is $O(M^3)$. That is to say, all mentioned algorithms have the same order of magnitude in computational complexity, but SCRFT has better integration performance than the rest methods.

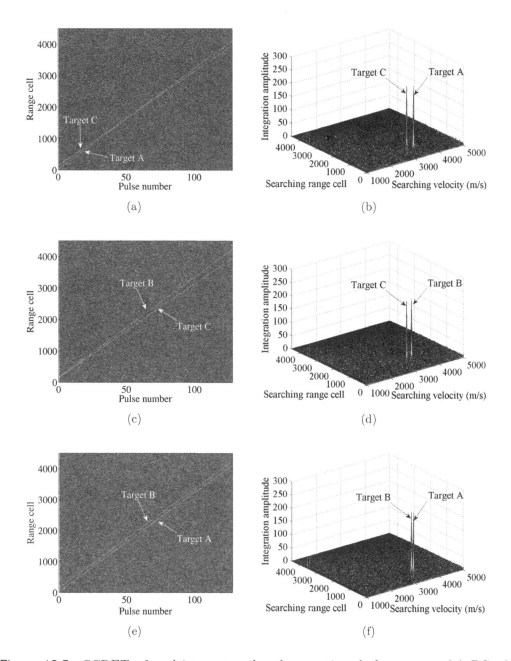

Figure 12.5 SCRFT of multi-target under the mentioned three cases. (a) PC of targets A and C. (b) CI of targets A and C. (c) PC of targets B and C. (d) CI of targets B and C. (e) PC of targets A and B. (f) CI of targets A and B.

TABLE 12.3   Computational Complexity of Different Algorithms

Algorithms	Computational complexity
RFT	$O\left(K_v M K_r\right)$
MLRT	$O\left(K_\theta M K_r\right)$
SCIFT	$O\left(M^2 K_r\right)$
SCRFT	$O\left(K_v M K_r\right)$

TABLE 12.4   Actual Processing Time of Different Algorithms

Algorithms	Time cost (s)
RFT	120.6562
MLRT	119.3235
SCIFT	110.4215
SCRFT	125.0364

Note: Main configuration of the computer: CPU: Intel(R) Xeon(R) E5-2650 v2 2.60 GHz; RAM: 64.00G; Operating System: Windows 7; Software: Matlab R2014a.

Besides, we contrast the actual processing time of the mentioned algorithms. The simulation parameters in Example 1 are used here, and Table 12.4 gives the results, where the mentioned four methods have similar time cost. Although SCRFT will spend longer processing time, it can achieve better detection and estimation performance than the rest methods.

12.4   SOME DISCUSSIONS ABOUT SCRFT

In this section, we discuss and compare the CI performance of SCRFT and RFT at first. Then, the influence of mismatched velocity on SCRFT is also discussed.

12.4.1   Discussion for CI Performance Improvement of SCRFT

The CI performance improvement of SCRFT compared with RFT (which has the best CI performance in the mentioned traditional methods) is given in this section.

In (12.11), the CI amplitude of RFT output can be written as

$$|S_{\mathrm{RFT}}\left(r', v'\right)| = \begin{cases} |A_{x1}| T_{sum} \left|\chi_1^*\right|, \left(\rho^2 < 1\right) \\ |A_{x1}| T_{sum} \left|\chi_2\right|, \left(\rho^2 > 1\right) \end{cases}. \tag{12.25}$$

Based on (12.25) and (12.10), the maximum peak value of RFT output is

$$|S_{\mathrm{RFT}}\left(r', v'\right)|_{\max} = \begin{cases} A_2 \sqrt{\dfrac{1}{2\mu|\triangle q|}} T_{sum} \left|\chi_1^*\right|, \left(\rho^2 < 1\right) \\ A_2 \sqrt{\dfrac{1}{2\mu|\triangle q|}} T_{sum} \left|\chi_2\right|, \left(\rho^2 > 1\right) \end{cases}. \tag{12.26}$$

The maximum peak value of SCRFT with matched velocity in (12.18) can be given as

$$|S_{\text{SCRFT}}(r', v')|_{\max} = A_3 T_{sum}. \tag{12.27}$$

Combining (12.26) and (12.27), the CI performance improvement of SCRFT in comparison with RFT can be expressed as

$$\triangle A_{\text{imp1}} = \frac{|S_{\text{SCRFT}}(r', v')|_{\max}}{|S_{\text{RFT}}(r', v')|_{\max}} = \begin{cases} \dfrac{A_3\sqrt{2\mu|\triangle q|}}{A_2\,|\chi_1^*|}, (\rho^2 < 1) \\ \dfrac{A_3\sqrt{2\mu|\triangle q|}}{A_2\,|\chi_2|}, (\rho^2 > 1) \end{cases}. \tag{12.28}$$

In the following, a simulation example (Example 4) is given to show the CI performance improvement of SCRFT in comparison with RFT when the scale effect appears.

Example 4: The radar and target parameters are identical to those of Example 1. Fig. 12.6 gives the CI results of SCRFT and RFT. To be specific, Fig. 12.6(a) shows PC of SCRFT, and the target energy of each pulse is focused well within the intra-pulse time. Fig. 12.6(b) shows PC of RFT, and the target energy within the intra-pulse time is dispersed due to the scale effect. Fig. 12.6(c) gives the CI output of SCRFT with a peak value of 506.6, while Fig. 12.6(d) shows that the CI peak value of RFT is reduced to 126.2. Therefore, the CI amplitude gain of SCRFT is more than four times that of RFT. Additionally, Fig. 12.6(e) shows the range cell slice of SCRFT's CI result (which is shown as a sinc envelope), while the range cell slice of RFT's CI output is shown in Fig. 12.6(f) (which is no longer a sinc envelope, mainly because of the scale effect).

12.4.2 Discussion of the Influence for Mismatched Velocity

The SCRFT requires to search for the target's velocity, and the searching velocity error (or mismatched velocity) may affect the PC and CI performance. To analyze the influence of the mismatched velocity on SCRFT, we provide some related derivations and discussions in the following.

As presented in (12.16), when the searching velocity is not equal to the actual velocity ($v' < v_0$ or $v' > v_0$), the PC result is mismatched (i.e., the PC result is a combination of Fresnel function rather than a sinc function). In this case, the CI of SCRFT with mismatched velocity can be written as

$$S_{\text{SCRFT}_{\text{mis}}}(r', v')$$
$$= \begin{cases} \displaystyle\int_0^{T_{sum}} A_{z1} \cdot \chi_3^* \exp\left(j4\pi f_c \dfrac{v' t_m}{c - 3v'}\right) dt_m, (v' < v_0) \\ \displaystyle\int_0^{T_{sum}} A_{z1} \cdot \chi_4 \exp\left(j4\pi f_c \dfrac{v' t_m}{c - 3v'}\right) dt_m, (v' > v_0) \end{cases}. \tag{12.29}$$

The CI amplitude of SCRFT with mismatched velocity in (12.29) is given as

$$|S_{\text{SCRFT}_{\text{mis}}}(r', v')| = \begin{cases} |A_{z1}| T_{sum} |\chi_3^*|, (v' < v_0) \\ |A_{z1}| T_{sum} |\chi_4|, (v' > v_0) \end{cases}. \tag{12.30}$$

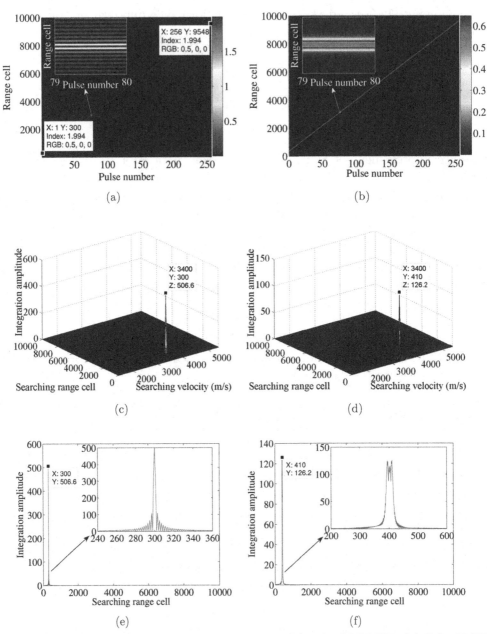

Figure 12.6 CI comparison of SCRFT and RFT. (a) PC of SCRFT. (b) PC of RFT. (c) CI of SCRFT. (d) CI of RFT. (e) CI slice result of SCRFT. (f) CI slice result of RFT.

On the basis of (12.30) and (12.17), the maximum peak value of SCRFT with mismatched velocity can be expressed as

$$|S_{\text{SCRFT}_{\text{mis}}}(r',v')|_{\max} = \begin{cases} A_2 T_{sum} \sqrt{\dfrac{1}{|2\mu\triangle q_0|}} \, |\chi_3^*| \,, (v' < v_0) \\[4mm] A_2 T_{sum} \sqrt{\dfrac{1}{|2\mu\triangle q_0|}} \, |\chi_4| \,, (v' > v_0) \end{cases} \qquad (12.31)$$

Combining (12.27) and (12.31), it could be seen that the CI performance loss of SCRFT with mismatched velocity is

$$\triangle A_{\text{imp2}} = \frac{|S_{\text{SCRFT}}(r',v')|_{\max}}{|S_{\text{SCRFT}_{\text{mis}}}(r',v')|_{\max}} = \begin{cases} \dfrac{A_3 \sqrt{2\mu|\triangle q_0|}}{A_2 |\chi_3^*|}, (v' < v_0) \\[4mm] \dfrac{A_3 \sqrt{2\mu|\triangle q_0|}}{A_2 |\chi_4|}, (v' > v_0) \end{cases} \qquad (12.32)$$

It can be seen from (12.32) that the larger the velocity mismatch error, the higher the SCRFT integration loss. Therefore, to obtain good CI performance, setting the searching intervals reasonably is necessary. Specifically, we can set the searching regions of initial slant range and velocity by radar observation time and relative prior information (e.g., the varieties and motion status of targets to be detected), which can be written as $r' \in [r'_{\min}, r'_{\max}]$ and $v' \in [v'_{\min}, v'_{\max}]$, respectively. In addition, the searching intervals of the slant range and target velocity can be determined according to radar system parameters [10], i.e.,

$$\rho_r = \frac{c}{2f_s}, \qquad (12.33)$$

$$\rho_v = \frac{\lambda}{2T_{sum}}, \qquad (12.34)$$

where $\lambda = \frac{c}{f_c}$ denotes the wave length and f_s represents the sampling frequency.

12.5 NUMERICAL RESULTS

In this section, numerical experiments are given to evaluate the SCRFT from several aspects, including CI of a single target, CI of multi-target, detection performance and velocity estimation performance.

12.5.1 CI of a Single Target

A single target is considered in this simulation, where its motion parameters are the same as those in Example 1 and the SNR before PC is −49 dB. The CI of SCIFT, MLRT, RFT and SCRFT is given in Fig. 12.7. In particular, Fig. 12.7(a) shows the CI output of the SCRFT algorithm. It can be seen the target energy is coherently integrated as a peak in the output, which is benefit to target detection and velocity estimation. The CI results of SCIFT, MLRT and RFT are given in Fig. 12.7(b)– 12.7(d), respectively. Comparing Fig. 12.7(a) with other figures, the SCRFT can

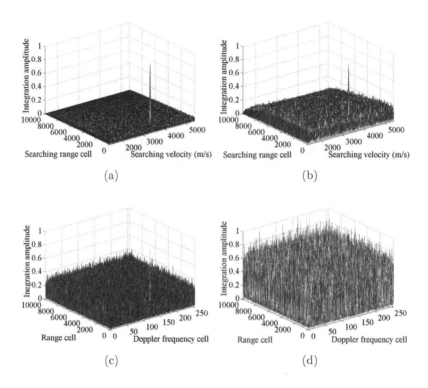

Figure 12.7 CI of a single target. (a) CI of SCRFT. (b) CI of RFT. (c) CI of MLRT. (d) CI of SCIFT.

obtain better performance than MLRT, RFT and SCIFT with respect to focusing target's energy and removing the noise. To compare the CI performance more clearly, the output SNRs after CI of the mentioned four methods are listed in Table 12.5. From this table, we can see that SCRFT achieves about 10.2, 11 and 21 dB SNR improvements compared with RFT, MLRT and SCIFT, respectively.

12.5.2 CI of Multi-Target

In this experiment, we analyze the CI performance of the SCRFT for multi-target. Three targets with different motion parameters, i.e., targets D, E and F, are used in the scene. In this simulation, the target parameters are shown in Table 12.6, and the radar parameters are the same with those of Example 3. Fig. 12.8(a), 12.8(c)

TABLE 12.5 The Output SNRs after CI for Different Algorithms

Algorithms	Output SNRs after CI (dB)
SCRFT	29.8114
RFT	19.6223
MLRT	18.7967
SCIFT	8.7887

TABLE 12.6   Motion Parameters of Targets D, E and F

Targets	Initial range cells	Radial velocities (m/s)	Input-SNRs (dB)
D	300	1000	−49.2
E	200	2000	−49.4
F	100	3000	−49

and 12.8(e) show the PC results when targets D, E and F are with matched velocity, respectively. Correspondingly, Fig. 12.8(b), 12.8(d) and 12.8(f) give the PC results without noise to present the match and mismatch between the searching velocities and the actual ones for the referred targets. To clearly show the difference between matched velocity and mismatched one, the 59th pulse of each target's echo is extracted and enlarged. Referring to the colorbars, we can see that the energy of the target with matched PC is stronger than that of the target with mismatched PC. Besides, Table 12.7 quantitatively gives the peak indexes of the 59th pulses of targets D, E and F, where the matched targets have the largest one (i.e., 1.994). Finally, Fig. 12.8(g) shows the CI results of the three targets, where the SCRFT-based algorithm can integrate the inter-pulse energy well for the multi-target.

12.5.3   Detection Performance

We evaluate the detection performance of traditional moving target detection (MTD), SCIFT, MLRT, RFT and SCRFT via 1000 times Monte Carlo simulations (the false alarm probability is set as $P_{fa} = 10^{-4}$) [13–15], where the target and radar parameters of Example 1 are used here. Note that the traditional MTD directly uses the range compression and slow time FT to achieve the accumulation of target signal [16]. The variation region of the input-SNRs is set as $[-75:1:-25]$ dB, and the detection curves of the aforementioned methods are shown in Fig. 12.9, where the result of ideal MTD is also provided. It could be noticed that the detection performance of SCRFT is superior to other four methods and very close to the ideal MTD. In particular, with detection probability $P_d = 0.8$, compared with SCRFT, the SNR loss of SCIFT, MLRT and RFT is, respectively, 16, 9 and 8 dB.

12.5.4   Velocity Estimation Performance

In order to evaluate the velocity estimation performance of SCRFT, RFT, MLRT and SCIFT, we perform 1000 times Monte-Carlo trials for each SNR to calculate the root-mean-square errors (RMSEs) of the velocity estimation of these methods. The variation range of input-SNRs is set between −75 and −40 dB. The target motion and radar parameters are set to be exactly the same as those of Section 12.5.3, and the RMSE results of the mentioned four methods are given in Fig. 12.10. The RMSE of the estimated velocity obtained by the SCRFT approaches to zero when SNR is greater than −62 dB. In comparison, the RMSEs obtained by the RFT and MLRT reach to zero when the SNR is greater than −54 and −53 dB, respectively. The SCIFT requires even higher SNR to obtain the accurate velocity estimation (i.e., the RMSE is close to zero). However, when the SNR is smaller than −65 dB, the estimation

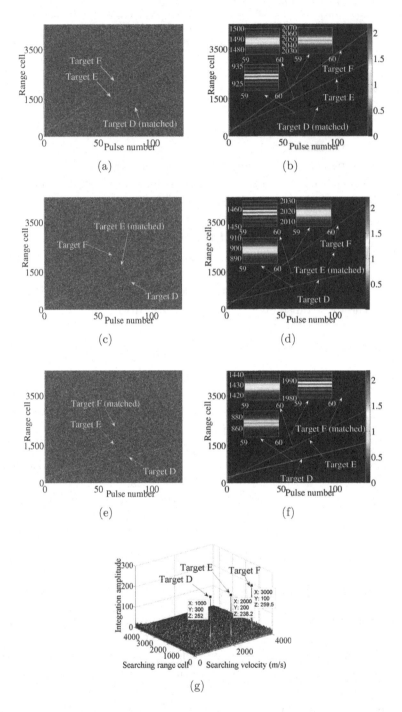

Figure 12.8 CI of multi-target. (a) PC when target D is with matched velocity. (b) PC without noise when target D is with matched velocity. (c) PC when target E is with matched velocity. (d) PC without noise when target E is with matched velocity. (e) PC when target F is with matched velocity. (f) PC without noise when target F is with matched velocity. (g) CI of targets D, E and F.

TABLE 12.7 Peak Indexes of the 59th Pulses of Target D, E and F

Radial velocities (m/s)	Target D	Target E	Target F
1000	1.994	1.255	0.593
2000	1.255	1.994	1.254
3000	0.592	1.255	1.994

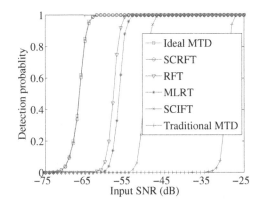

Figure 12.9 Detection performance of ideal MTD, SCRFT, RFT, MLRT, SCIFT and traditional MTD.

performance of all the methods becomes worse with the reducing SNR. For example, the RMSEs of all the methods are greater than 50 m/s when the SNR is −73 dB, i.e., the velocity could not be accurately estimated. This is because when the SNR is lower than −73 dB, the target signal after CI is still buried in noise and the target detection probabilities of the four methods are zero (as shown in Fig. 12.9), and thus the target velocity estimation could not be accurately obtained. Furthermore, it should be noted that because of the bounded velocity searching region, the RMSEs of the mentioned algorithms go to around 60 m/s as SNR goes to negative infinity in Fig. 12.10.

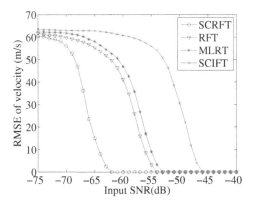

Figure 12.10 Velocity estimation performance of SCRFT, RFT, MLRT and SCIFT.

12.6 SUMMARY

In this chapter, we have addressed the coherent integration problem for the detection of hypersonic target with the scale effect. In summary:

- We established a motion model which considers the intra-pulse and inter-pulse motions simultaneously for the hypersonic target. Besides, we discussed the scale effect which may appear in the large time-bandwidth product condition.

- We discussed the SCRFT algorithm to achieve the CI for the hypersonic target with the scale effect and RW. In this algorithm, the PC and CI are successively realized via jointly searching in the range-velocity domain.

- We evaluated the performance of SCRFT via numerical experiments. The results have indicated that SCRFT has better CI performance than RFT, MLRT and SCIFT. Additionally, its detection and velocity estimation performance are also better than those of the traditional algorithms.

12.7 THE PROOF OF EQUATION (12.9)

Considering the intra-pulse motion and scale effect, the result of (12.9) has two cases: $\rho^2 < 1$ and $\rho^2 > 1$. Let $\triangle q = \frac{1}{\mu^2}\left(1 - \frac{1}{\rho^2}\right)$ and (12.9) can be recast as

$$
\begin{aligned}
s(t, t_m) = \int_{-\infty}^{+\infty} A_{x0}\text{rect}\left(\frac{f+\alpha}{\rho B}\right)\text{rect}\left(\frac{f}{B}\right) \\
\times \exp\left\{j\pi\mu\triangle q\left[f + \frac{1}{\mu\triangle q}\left(t - \frac{\alpha}{\mu\rho^2} - \tau_m\right)\right]^2\right\}df,
\end{aligned}
\tag{12.35}
$$

where

$$
\begin{aligned}
A_{x0} = A_1\exp\left(-j2\pi\frac{f\alpha}{\mu\rho^2}\right)\exp\left(-j2\pi f_c\tau_m\right) \\
\times \exp\left\{-j\pi\left[\frac{1}{\mu\triangle q}\left(t - \frac{\alpha}{\mu\rho^2} - \tau_m\right)\right]^2\right\}.
\end{aligned}
\tag{12.36}
$$

Let

$$
N(f) = \exp\left\{j\pi\mu\triangle q\left[f + \frac{1}{\mu\triangle q}\left(t - \frac{\alpha}{\mu\rho^2} - \tau_m\right)\right]^2\right\}.
\tag{12.37}
$$

Then, (12.35) can be rewritten as

$$
s(t, t_m) = \int_{-\infty}^{+\infty} A_{x0}\text{rect}\left(\frac{f+\alpha}{\rho B}\right)\text{rect}\left(\frac{f}{B}\right)N(f)df.
\tag{12.38}
$$

Case 1: When $\rho^2 < 1$, (12.38) can be recast as

$$
\begin{aligned}
s(t, t_m) = A_{x0}\text{rect}\left(\frac{t+\alpha}{\rho B}\right)\int_{-\alpha-\frac{\rho B}{2}}^{-\alpha+\frac{\rho B}{2}} N(f)df \\
+ \text{rect}\left[\frac{t + \frac{B}{4}(\rho - 1) - \frac{\alpha}{2}}{-\alpha + \frac{\rho B}{2} + \frac{B}{2}}\right]\int_{-\frac{B}{2}}^{-\alpha+\frac{\rho B}{2}} N(f)df.
\end{aligned}
\tag{12.39}
$$

Let

$$y = \sqrt{-2\mu\triangle q}\left[f + \frac{1}{\mu\triangle q}\left(t - \frac{\alpha}{\mu\rho^2} - \tau_m\right)\right]. \tag{12.40}$$

Then, we can obtain

$$\int_{-\alpha-\frac{\rho B}{2}}^{-\alpha+\frac{\rho B}{2}} N(f)df = \sqrt{-\frac{1}{2\mu\triangle q}}\int_{\omega_2}^{\omega_1}\exp\left(-j\frac{\pi y^2}{2}\right)dy$$

$$= \sqrt{-\frac{1}{2\mu\triangle q}}\chi^*(\omega_1,\omega_2), \tag{12.41}$$

where $\chi(\cdot)$ is the Fresnel function, which has the formula as

$$\chi(u,v) = \chi_{\exp}(u) - \chi_{\exp}(v), \tag{12.42}$$

$$\chi_{\exp}(z) = \int_0^z \exp\left(\frac{j\pi y^2}{2}\right)dy = \chi_{\cos}(z) + j\chi_{\sin}(z), \tag{12.43}$$

$$\omega_1 = \sqrt{-2\mu\triangle q}\left[-\alpha + \frac{\rho B}{2} + \frac{1}{\mu\triangle q}\left(t - \frac{\alpha}{\mu\rho^2} - \tau_m\right)\right]$$

$$= \sqrt{2\mu|\triangle q|}\left(-\alpha + \frac{\rho B}{2} + \frac{t - \frac{\alpha}{\mu\rho^2} - \tau_m}{\mu\triangle q}\right), \tag{12.44}$$

$$\omega_2 = \sqrt{2\mu|\triangle q|}\left(-\alpha - \frac{\rho B}{2} + \frac{t - \frac{\alpha}{\mu\rho^2} - \tau_m}{\mu\triangle q}\right). \tag{12.45}$$

Similarly

$$\int_{\frac{B}{2}}^{-\alpha+\frac{\rho B}{2}} N(f)df = \sqrt{-\frac{1}{2\mu\triangle q}}\chi^*(\omega_1,\omega_3), \tag{12.46}$$

where

$$\omega_3 = \sqrt{2\mu|\triangle q|}\left(-\frac{B}{2} + \frac{t - \frac{\alpha}{\mu\rho^2} - \tau_m}{\mu\triangle q}\right). \tag{12.47}$$

Therefore, (12.57) can be further written as

$$s(t,t_m) = A_{x1}\cdot\chi_1^*, \tag{12.48}$$

where

$$\chi_1^* = \left\{\text{rect}\left(\frac{t+\alpha}{\rho B}\right)\chi^*(\omega_1,\omega_2)\right.$$

$$\left. +\text{rect}\left[\frac{t + \frac{B}{4}(\rho-1) - \frac{\alpha}{2}}{-\alpha + \frac{\rho B}{2} + \frac{B}{2}}\right]\chi^*(\omega_1,\omega_3)\right\}, \tag{12.49}$$

$$A_{x1} = A_{x0}\sqrt{\frac{1}{2|\mu\triangle q|}}.$$

Case 2: When $\rho^2 > 1$, (12.38) can be recast as

$$s(t, t_m) = A_{x1} \cdot \chi_2, \tag{12.50}$$

where

$$\chi_2 = \left\{ \text{rect}\left(\frac{t}{B}\right) \chi(\omega_4, \omega_2) \right.$$

$$\left. + \text{rect}\left[\frac{t + \frac{B}{4}(\rho - 1) + \frac{\alpha}{2}}{\alpha + \frac{\rho B}{2} + \frac{B}{2}}\right] \chi(\omega_4, \omega_3) \right\}, \tag{12.51}$$

$$\omega_4 = \sqrt{2\mu|\triangle q|} \left(\frac{B}{2} + \frac{t - \frac{\alpha}{\mu\rho^2} - \tau_m}{\mu\triangle q}\right). \tag{12.52}$$

12.8   THE PROOF OF EQUATION (12.16)

The output of (12.16) can be divided into three situations: $v' < v_0$, $v' = v_0$ and $v' > v_0$. In this appendix, we will give the detailed derivation to obtain (12.16).

Let $\triangle q_0 = \frac{1}{\mu^2\rho'^2} - \frac{1}{\mu^2\rho^2}$, $\triangle q_1 = \frac{\alpha}{\mu^2\rho^2} - \frac{\alpha'}{\mu^2\rho'^2}$ and $\triangle q_2 = \frac{\alpha^2}{\mu^2\rho^2} - \frac{\alpha'^2}{\mu^2\rho'^2}$. Then, (12.13) can be further written as

$$s_c(t, t_m) = \int_{-\infty}^{+\infty} A_{z0}\text{rect}\left(\frac{f + \alpha}{\rho B}\right)\text{rect}\left(\frac{f + \alpha'}{\rho' B}\right)$$

$$\times \exp\left\{j\pi\mu\triangle q_0\left[f + \frac{1}{\mu\triangle q_0}(t - \mu\triangle q_1 - \tau_m)\right]^2\right\} df, \tag{12.53}$$

where

$$A_{z0} = A_1 \exp\left(-j\pi\mu\triangle q_2\right)\exp\left(-j2\pi f_c\tau_m\right)$$

$$\times \exp\left\{-j\pi\left[\frac{1}{\mu\triangle q_0}(t - \mu\triangle q_1 - \tau_m)\right]^2\right\}. \tag{12.54}$$

Let

$$L(f) = \exp\left\{j\pi\mu\triangle q_0\left[f + \frac{1}{\mu\triangle q_0}(t - \mu\triangle q_1 - \tau_m)\right]^2\right\}. \tag{12.55}$$

Then, (12.53) can be rewritten as

$$s_c(t, t_m) = \int_{-\infty}^{+\infty} A_{z0}\text{rect}\left(\frac{f + \alpha}{\rho B}\right)\text{rect}\left(\frac{f + \alpha'}{\rho' B}\right) L(f) df. \tag{12.56}$$

Situation 1: When $v' < v$, considering the actual parameters setting, we have $\alpha' < \alpha$, $\rho' > \rho$ and $\triangle q_0 < 0$. In this case, (12.56) can be recast as

$$s_c(t, t_m) = A_{z0}\text{rect}\left(\frac{t + \alpha}{\rho B}\right)\int_{-\alpha - \frac{\rho B}{2}}^{-\alpha + \frac{\rho B}{2}} L(f) df$$

$$+ \text{rect}\left[\frac{t + \alpha + \frac{1}{2}(\alpha' + \alpha) + \frac{B}{4}(\rho' - \rho)}{(\alpha' + \alpha) + \frac{B}{2}(\rho' + \rho)}\right]\int_{-\alpha' - \frac{\rho' B}{2}}^{-\alpha + \frac{\rho B}{2}} L(f) df. \tag{12.57}$$

Let

$$\eta = \sqrt{-2\mu\triangle q_0}\left[f + \frac{1}{\mu\triangle q_0}\left(t - \mu\triangle q_1 - \tau_m\right)\right]. \qquad (12.58)$$

Then, we can obtain

$$\int_{-\alpha-\frac{\rho B}{2}}^{-\alpha+\frac{\rho B}{2}} L(f)df = \sqrt{-\frac{1}{2\mu\triangle q_0}}\int_{\vartheta_2}^{\vartheta_1} \exp\left(-j\frac{\pi\eta^2}{2}\right)d\eta$$

$$= \sqrt{-\frac{1}{2\mu\triangle q_0}}\chi^*\left(\vartheta_1, \vartheta_2\right), \qquad (12.59)$$

where $\chi(\cdot)$ is the Fresnel function (defined in (12.42) and (12.43)).

$$\vartheta_1 = \sqrt{-2\mu\triangle q_0}\left[-\alpha + \frac{\rho B}{2} + \frac{1}{\mu\triangle q_0}\left(t - \mu\triangle q_1 - \tau_m\right)\right]$$

$$= \sqrt{2\mu|\triangle q_0|}\left(-\alpha + \frac{\rho B}{2} + \frac{t - \mu\triangle q_1 - \tau_m}{\mu\triangle q_0}\right), \qquad (12.60)$$

$$\vartheta_2 = \sqrt{2\mu|\triangle q_0|}\left(-\alpha - \frac{\rho B}{2} + \frac{t - \mu\triangle q_1 - \tau_m}{\mu\triangle q_0}\right). \qquad (12.61)$$

Similarly

$$\int_{-\alpha'-\frac{\rho' B}{2}}^{-\alpha+\frac{\rho B}{2}} L(f)df = \sqrt{-\frac{1}{2\mu\triangle q_0}}\chi^*\left(\vartheta_1, \vartheta_3\right), \qquad (12.62)$$

where

$$\vartheta_3 = \sqrt{2\mu|\triangle q_0|}\left(-\alpha' - \frac{\rho' B}{2} + \frac{t - \mu\triangle q_1 - \tau_m}{\mu\triangle q_0}\right). \qquad (12.63)$$

Therefore, (12.57) can be further written as

$$s_c\left(t, t_m\right) = A_{z1} \cdot \chi_3^*, \qquad (12.64)$$

where

$$\chi_3^* = A_{z1}\left\{\text{rect}\left(\frac{t+\alpha}{\rho B}\right)\chi^*\left(\vartheta_1, \vartheta_2\right) \right.$$

$$\left. + \text{rect}\left[\frac{t + \frac{1}{2}\left(\alpha + \alpha'\right) + \frac{B}{4}\left(\rho\prime - \rho\right)}{\left(\alpha' - \alpha\right) + \frac{B}{2}\left(\rho\prime + \rho\right)}\right]\chi^*\left(\vartheta_1, \vartheta_3\right)\right\}, \qquad (12.65)$$

$$A_{z1} = A_{z0}\sqrt{\frac{1}{|2\mu\triangle q_0|}}.$$

With $t = 2r'/(c - 3v_0)$, we can get another form of χ_1^*, i.e.,

$$
\begin{aligned}
\chi_3^* = A_{z1} \Bigg\{ &\text{rect} \left[\frac{2r'/(c - 3v_0) + \alpha}{\rho B} \right] \chi^* (\vartheta_1, \vartheta_2) \\
&+ \text{rect} \left[\frac{2r'/(c - 3v_0) + \frac{1}{2}(\alpha + \alpha') + \frac{B}{4}(\rho\prime - \rho)}{(\alpha' - \alpha) + \frac{B}{2}(\rho\prime + \rho)} \right] \\
&\times \chi^* (\vartheta_1, \vartheta_3) \Bigg\}.
\end{aligned}
\tag{12.66}
$$

Situation 2: When $v' = v_0$, we can get $\rho' = \rho$ and $\alpha' = \alpha$. Besides, it can be seen that $\triangle q_0 = 0$, $\triangle q_1 = 0$ and $\triangle q_2 = 0$. The matched PC will be achieved, which can be given as

$$
\begin{aligned}
s_c(t, t_m) = &A_2 \text{sinc} \left[2\pi B \left(t - \frac{2r_0}{c - v_0} - \frac{2v_0 t_m}{c - v_0} \right) \right] \\
&\times \exp \left(-j 2\pi f_c \tau_m \right) \\
= &A_2 \text{sinc} \left\{ \frac{2\pi B}{c - v_0} \left[(r' - r_0) - v_0 t_m \right] \right\} \\
&\times \exp \left[-j 4\pi f_c \frac{(r_0 + v_0) t_m}{c - 3v_0} \right],
\end{aligned}
\tag{12.67}
$$

where A_2 is the complex amplitude after matched PC.

Situation 3: When $v' > v_0$, similar to Situation 1, we can obtain

$$
s_c(t, t_m) = A_{z1} \cdot \chi_4,
\tag{12.68}
$$

where

$$
\begin{aligned}
\chi_4 = A_{z1} \Bigg\{ &\text{rect} \left(\frac{t + \alpha'}{\rho' B} \right) \chi(\vartheta_4, \vartheta_3) \\
&+ \text{rect} \left[\frac{t + \frac{1}{2}(\alpha + \alpha') + \frac{B}{4}(\rho\prime - \rho)}{(\alpha' - \alpha) + \frac{B}{2}(\rho\prime + \rho)} \right] \chi(\vartheta_4, \vartheta_2) \Bigg\},
\end{aligned}
\tag{12.69}
$$

$$
\vartheta_4 = \sqrt{2\mu |\triangle q_0|} \left(-\alpha' + \frac{\rho' B}{2} + \frac{t - \mu \triangle q_1 - \tau_m}{\mu \triangle q_0} \right).
\tag{12.70}
$$

Similarly, with $t = 2r'/(c - 3v_0)$, (12.69) can be recast as

$$
\begin{aligned}
\chi_4 = A_{z1} \Bigg\{ &\text{rect} \left[\frac{2r'/(c - 3v_0) + \alpha'}{\rho' B} \right] \chi(\vartheta_4, \vartheta_3) \\
&+ \text{rect} \left[\frac{2r'/(c - 3v_0) + \frac{1}{2}(\alpha + \alpha') + \frac{B}{4}(\rho\prime - \rho)}{(\alpha' - \alpha) + \frac{B}{2}(\rho\prime + \rho)} \right] \\
&\times \chi(\vartheta_4, \vartheta_2) \Bigg\}.
\end{aligned}
\tag{12.71}
$$

Bibliography

[1] J. B. Zheng, H. W. Liu, and Q. H. Liu, "Parameterized centroid frequency-chirp rate distribution for LFM signal analysis and mechanisms of constant delay introduction," *IEEE Transactions on Signal Processing*, vol. 65, no. 24, pp. 6435–6447, December 2017.

[2] P. H. Huang, X. G. Xia, G. S. Liao, Z. W. Yang, J. J. Zhou, and X. Z. Liu, "Ground moving target refocusing in SAR imagery using scaled GHAF," *IEEE Transactions on Geoscience and Remote Sensing*, vol. 56, no. 2, pp. 1030–1045, February 2018.

[3] X. L. Chen, J. Guan, Y. Huang, N. B. Liu, and Y. He, "Radon-linear canonical ambiguity function-based detection and estimation method for marine target with micromotion," *IEEE Transactions on Geoscience and Remote Sensing*, vol. 53, no. 4, pp. 2225–2240, April 2015.

[4] L. C. Qian, J. Xu, X. G. Xia, W. F. Sun, T. Long, and Y. N. Peng, "Wideband-scaled Radon-Fourier transform for high-speed radar target detection," *IET Radar Sonar and Navigation*, vol. 10, no. 9, pp. 1671–1682, December 2016.

[5] X. F. Xu, G. S. Liao, Z. W. Yang, and C. H. Wang, "Moving-in-pulse duration model-based target integration method for HSV-borne high-resolution radar," *Digital Signal Processing*, vol. 68, pp. 31–43, May 2017.

[6] I. G. Cumming, and F. H. Wong, "Digital processing of synthetic aperture radar data algorithms and implementation," Artech House, Boston, MA, USA, 2005.

[7] Z. Bao, M. D. Xing, and T. Wang, "Radar imaging technology," House of Electronics Industry, Beijing, China, 2005.

[8] Z. Sun, X. L. Li, W. Yi, G. L. Cui, and L. J. Kong, "A coherent detection and velocity estimation algorithm for the high-speed target based on the modified location rotation transform," *IEEE Journal of Selected Topics in Applied Earth Observation and Remote Sensing*, vol. 11, no. 7, pp. 2346–2361, July 2018.

[9] Z. Sun, X. L. Li, W. Yi, G. L. Cui, and L. J. Kong, "Detection of weak maneuvering target based on keystone transform and matched filtering process," *Signal Processing*, vol. 140, pp. 127–138, May 2017.

[10] X. L. Chen, J. Guan, N. B. Liu, and Y. He, "Maneuvering target detection via Radon-Fractional Fourier transform-based long-time coherent integration," *IEEE Transactions on Signal Processing*, vol. 62, no. 4, pp. 939–953, February 2014.

[11] J. Xu, J. Yu, Y. N. Peng, and X. G. Xia, "Radon-Fourier transform (RFT) for radar target detection (I): generalized Doppler filter bank processing," *IEEE Transactions on Aerospace and Electronic Systems*, vol. 47, no. 2, pp. 1186–1202, April 2011.

[12] J. B. Zheng, T. Su, W. T. Zhu, X. H. He, and Q. H. Liu, "Radar high-speed target detection based on the scaled inverse Fourier transform," *IEEE Journal of Selected Topics in Applied Earth Observation and Remote Sensing*, vol. 8, no. 3, pp. 1108–1119, March 2015.

[13] L. J. Kong, X. L. Li, G. L. Cui, W. Yi, and Y. C. Yang, "Coherent integration algorithm for a maneuvering target with high-order range migration," *IEEE Transactions on Signal Processing*, vol. 63, no. 17, pp. 4474–4486, September 2015.

[14] X. L. Li, Z. Sun, W. Yi, G. L. Cui, L. J. Kong, and X. B. Yang, "Computationally efficient coherent detection and parameter estimation algorithm for maneuvering target," *Signal Processing*, vol. 155, pp. 130–142, February 2019.

[15] X. L. Li, Z. Sun, and T. S. Yeo, "Computational efficient refocusing and estimation method for radar moving target with unknown time information," *IEEE Transactions on Computational Imaging*, vol. 6, pp. 544–557, January 2020.

[16] M. I. Skolnik, G. Linde, and K. Meads, "Senrad: an advanced wideband air-surveillance radar," *IEEE Transactions on Aerospace and Electronic Systems*, vol. 37, no. 4, pp. 1163–1175, October 2001.

CHAPTER 13

CLEAN-Based Coherent Integration Processing

For the high-speed multi-target scene, if the scattering intensities of different targets differ significantly, the weak one would be shadowed by the strong target and makes it difficult to achieve the coherent integration for the weak target. This chapter addresses the coherent integration problem for high-speed multi-target. A coherent integration method based on keystone transform and generalized dechirp processing (KTGDP) is designed to remove the migrations (range migration [RM] and Doppler frequency migration [DFM]) and realize the coherent accumulation. Two CLEAN techniques based on sinc-like point spread function and modified point spread function are presented to eliminate the strong target's effect and highlight the weak ones. By this way, the coherent integration of strong target and weak ones can be achieved iteratively.

13.1 SIGNAL MODEL

Suppose that the radar transmits a linear frequency modulated (LFM) signal, i.e.,

$$s_{trans}(t, t_m) = \text{rect}\left(\frac{t}{T_p}\right) \exp\left(j\pi\mu t^2\right) \exp(j2\pi f_c t_m), \qquad (13.1)$$

where

$$\text{rect}(x) = \begin{cases} 1 & |x| \leq \frac{1}{2}, \\ 0 & |x| > \frac{1}{2}, \end{cases}$$

T_p is the pulse duration; μ is the frequency modulated rate; f_c is the carrier frequency; $t_m = mT_r$ is the slow time, $m = 0, 1, \cdots, N - 1$; T_r is the pulse repetition time; N denotes the number of coherent integrated pulses and t is the fast time.

Assume that there are K targets in the scene. Neglecting the high-order components, the instantaneous slant range $r_k(t_m)$ of the kth target with complex motions satisfies [1, 2]

$$r_k(t_m) = r_{0,k} + v_{0,k}t_m + a_{0,k}t_m^2 + a_{1,k}t_m^3, \qquad (13.2)$$

DOI: 10.1201/9781003529101-13

where $r_{0,k}$ is the initial slant range from the radar to the kth target and $v_{0,k}$, $a_{0,k}$ and $a_{1,k}$ denote, respectively, the kth target's radial velocity, acceleration and jerk.

The received baseband signal of K targets can be stated as [1, 12]

$$s_r(t_m, t) = \sum_{k=1}^{K} A_{0,k}\text{rect}\left(\frac{t - 2r_k(t_m)/c}{T_p}\right) \exp\left(-j\frac{4\pi r_k(t_m)}{\lambda}\right)$$
$$\times \exp\left[j\pi\mu\left(t - \frac{2r_k(t_m)}{c}\right)^2\right], \tag{13.3}$$

where $A_{0,k}$ is the target reflectivity, c is the light speed and $\lambda = c/f_c$ denotes the wavelength.

After pulse compression, the compressed signal can be represented as

$$s_c(t_m, t) = \sum_{k=1}^{K} A_{1,k}\text{sinc}\left[B\left(t - \frac{2r_k(t_m)}{c}\right)\right]$$
$$\times \exp\left(-j\frac{4\pi r_k(t_m)}{\lambda}\right), \tag{13.4}$$

where B denotes the bandwidth of the transmitted signal.

Let $t = \frac{2r}{c}$, where r is the range corresponding to t. Then, (13.4) can be rewritten as

$$s_c(t_m, r) = \sum_{k=1}^{K} A_{1,k}\text{sinc}\left\{B\left[\frac{2r - 2r_k(t_m)}{c}\right]\right\}$$
$$\times \exp\left(-j\frac{4\pi r_k(t_m)}{\lambda}\right). \tag{13.5}$$

Equation (13.5) is the signal model in the slow time-range domain. On the one hand, we can see that the point spread function (PSF) of kth target is a sinc-like function which can be determined by five parameters, i.e., amplitude $A_{1,k}$, initial slant range $r_{0,k}$, radial velocity $v_{0,k}$, radial acceleration $a_{0,k}$ and radial jerk $a_{1,k}$. Thus, if we obtain the estimations of these five parameters, the PSF of kth target can be reconstructed, which is helpful to the CLEAN technique [3–5].

On the other hand, equation (13.5) shows that the envelope of kth target changes with the slow time after pulse compression. In order to remove the effect of target motion and realize the coherent integration, a Fourier transform (FT) to the variable r is employed to transform the compressed signal into slow time-range frequency $(t_m - f)$ domain, which can be written as

$$S_c(t_m, f) = \sum_{k=1}^{K} A_{2,k}\text{rect}\left(\frac{f}{B}\right)\exp\left(-j\frac{4\pi(f + f_c)r_k(t_m)}{\lambda f_c}\right), \tag{13.6}$$

where f is the range frequency corresponding to r.

Due to the high-speed characteristic of target and the low pulse repetition frequency f_p, undersampling would occur [6, 7]. Therefore, the velocity of the kth target can be expressed as

$$v_{0,k} = N_k v_{amb} + v_{unamb,k}, \tag{13.7}$$

where N_k is the fold factor of the kth target, $v_{amb} = \lambda f_p/2$ is the blind velocity and $v_{unamb,k} = \mathrm{mod}(v_{0,k}, v_{amb})$ is the unambiguous velocity, which satisfies $|v_{unamb,k}| < v_{amb}/2$.

Substituting (13.2) and (13.7) into (13.6) yields

$$S_r(t_m, f) = \sum_{k=1}^{K} A_{3,k} \exp\left[-j\frac{4\pi r_{0,k}}{\lambda}\left(1 + \frac{f}{f_c}\right)\right]$$
$$\times \exp\left[-j\frac{4\pi v_{unamb,k} t_m}{\lambda}\left(1 + \frac{f}{f_c}\right)\right]$$
$$\times \exp(-j4\pi N_k v_{amb} t_m f / f_c \lambda) \tag{13.8}$$
$$\times \exp\left[-j\frac{4\pi a_{0,k} t_m^2}{\lambda}\left(1 + \frac{f}{f_c}\right)\right]$$
$$\times \exp\left[-j\frac{4\pi a_{1,k} t_m^3}{\lambda}\left(1 + \frac{f}{f_c}\right)\right],$$

where $A_{3,k} = A_{2,k}\mathrm{rect}\left(\frac{f}{B}\right)$, and (13.8) is obtained by substituting $\exp(-j2\pi N_k f_p t_m) = 1$ [7] into (13.8).

Equation (13.8) shows the signal model in the slow time-range frequency domain. For the kth target, it can be seen that there are four exponential terms of slow time t_m, which are all coupled with the range frequency f. More specifically,

- $\exp\left[-j\frac{4\pi v_{unamb,k} t_m}{\lambda}\left(1 + \frac{f}{f_c}\right)\right]$ represents the first-order phase term due to the unambiguous velocity.

- $\exp(-j4\pi N_k v_{amb} t_m f / f_c \lambda)$ is the fold factor phase term because of undersampling.

- $\exp\left[-j\frac{4\pi a_{0,k} t_m^2}{\lambda}\left(1 + \frac{f}{f_c}\right)\right]$ indicates the quadratic phase term induced by the target's radial acceleration.

- $\exp\left[-j\frac{4\pi a_{1,k} t_m^3}{\lambda}\left(1 + \frac{f}{f_c}\right)\right]$ denotes the cubic phase term induced by the target's jerk motion.

The first-order terms of t_m will result in RM, while the quadratic phase term and cubic phase term will result in, respectively, linear Doppler frequency migration(LDFM) and quadratic Doppler frequency migration(QDFM), which would make the signal energy defocused. Both RM and DFM (i.e., LDFM and QDFM) will bring difficulties to the coherent integration of target energy.

13.2 MIGRATIONS CORRECTION AND PARAMETERS ESTIMATION

In this section, a coherent integration method based on KT, fold factor phase term compensation and generalized dechirp process (KTGDP), is introduced to remove the migrations (RM, LDFM and QDFM) and realize the coherent accumulation for a maneuvering target with jerk motion. Without loss of generality, the signal process procedure of the kth target is given and analyzed in the following.

13.2.1 RM Correction

13.2.1.1 *Keystone Transform*

First of all, the KT, which performs scaling $t_m = \frac{f_c}{f+f_c} t_n$ in the $t_m - f$ domain, is employed to correct the RM caused by the unambiguous velocity [8]. Substituting the scaling formula into (13.8) yields

$$
\begin{aligned}
S_{KT}(t_n, f) = \sum_{k=1}^{K} A_{3,k} \exp &\left[-j\frac{4\pi r_{0,k}}{\lambda} \left(1 + \frac{f}{f_c} \right) \right] \\
\times \exp &\left(-j\frac{4\pi v_{unamb,k} t_n}{\lambda} \right) \exp \left(-j\frac{4\pi a_{0,k} t_n^2}{\lambda} \frac{f_c}{f+f_c} \right) \\
\times \exp &\left[-j\frac{4\pi a_{1,k} t_n^3}{\lambda} \left(\frac{f_c}{f+f_c} \right)^2 \right] \\
\times \exp &\left(-j\frac{4\pi N_k v_{amb} t_n}{\lambda} \frac{f}{f+f_c} \right).
\end{aligned}
\tag{13.9}
$$

Under the narrow band environment $f \ll f_c$, we have $\frac{f_c}{f+f_c} \approx 1$ and $\frac{f}{f+f_c} \approx \frac{f}{f_c}$. Thus, (13.9) can be recast as

$$
\begin{aligned}
S_{KT}(t_n, f) \approx \sum_{k=1}^{K} A_{3,k} \exp &\left[-j\frac{4\pi r_{0,k}}{\lambda} \left(1 + \frac{f}{f_c} \right) \right] \\
\times \exp &\left(-j\frac{4\pi v_{unamb,k} t_n}{\lambda} \right) \exp \left(-j\frac{4\pi a_{0,k} t_n^2}{\lambda} \right) \\
\times \exp &\left(-j\frac{4\pi a_{1,k} t_n^3}{\lambda} \right) \exp \left(-j\frac{4\pi N_k v_{amb} t_n}{\lambda} \frac{f}{f_c} \right).
\end{aligned}
\tag{13.10}
$$

By (13.10), the coupling between $v_{unamb,k}$ and f is removed and the RM induced by the unambiguous velocity $v_{unamb,k}$ is corrected. However, the residual RM induced by integral multiples of blind velocity, i.e, the fold factor phase term, is still present.

13.2.1.2 *Fold Factor Phase Term Compensation*

Define the fold factor phase compensation function as follows:

$$
H_a(t_n, f; n') = \exp \left(j\frac{4\pi n' v_{amb} t_n}{\lambda} \frac{f}{f_c} \right),
\tag{13.11}
$$

where n' denotes the searching fold factor.

Multiplying (13.10) with (13.11) yields

$$
\begin{aligned}
S_{KT}(t_n, f; n') = \sum_{k=1}^{K} A_{3,k} \exp &\left[-j \frac{4\pi r_{0,k}}{\lambda} \left(1 + \frac{f}{f_c} \right) \right] \\
&\times \exp \left(-j \frac{4\pi v_{unamb,k} t_n}{\lambda} \right) \exp \left(-j \frac{4\pi a_{0,k} t_n^2}{\lambda} \right) \\
&\times \exp \left(-j \frac{4\pi (N_k - n') v_{amb} t_n}{\lambda} \frac{f}{f_c} \right) \\
&\times \exp \left(-j \frac{4\pi a_{1,k} t_n^3}{\lambda} \right).
\end{aligned}
\tag{13.12}
$$

Applying inverse Fourier transform (IFT) on (13.12) with respect to f, we have

$$
\begin{aligned}
s_{KT}(t_n, r; n') = \sum_{k=1}^{K} A_{1,k} \mathrm{sinc} &\left[\frac{2r - 2r_{0,k}}{c} - \frac{2(N_k - n') v_{amb} t_n}{c} \right] \\
&\times \exp \left(-j \frac{4\pi r_{0,k}}{\lambda} \right) \exp \left(-j \frac{4\pi v_{unamb,k} t_n}{\lambda} \right) \\
&\times \exp \left(-j \frac{4\pi a_{0,k} t_n^2}{\lambda} \right) \exp \left(-j \frac{4\pi a_{1,k} t_n^3}{\lambda} \right).
\end{aligned}
\tag{13.13}
$$

When $n' = N_k$, i.e, the searching fold factor equals to the fold factor of kth target, (13.13) can be written as

$$
\begin{aligned}
s_{KT}(t_n, r) = A_{1,k} \mathrm{sinc} &\left(\frac{2r - 2r_{0,k}}{c} \right) \exp \left(-j \frac{4\pi r_{0,k}}{\lambda} \right) \\
&\times \exp \left(-j \frac{4\pi v_{unamb,k} t_n}{\lambda} \right) \exp \left(-j \frac{4\pi a_{0,k} t_n^2}{\lambda} \right) \\
&\times \exp \left(-j \frac{4\pi a_{1,k} t_n^3}{\lambda} \right) + s_{other}(t_n, r),
\end{aligned}
\tag{13.14}
$$

where

$$
\begin{aligned}
s_{other}(t_n, r) = \sum_{l=1, l \neq k}^{K} A_{1,l} \mathrm{sinc} &\left(\frac{2r - 2r_{0,l}}{c} - \frac{2(N_l - N_k) v_{amb} t_n}{c} \right) \\
&\times \exp \left(-j \frac{4\pi r_{0,l}}{\lambda} \right) \exp \left(-j \frac{4\pi v_{unamb,l} t_n}{\lambda} \right) \\
&\times \exp \left(-j \frac{4\pi a_{0,l} t_n^2}{\lambda} \right) \exp \left(-j \frac{4\pi a_{1,l} t_n^3}{\lambda} \right).
\end{aligned}
\tag{13.15}
$$

From (13.14), we can see that the RM of kth target is removed, i.e., the kth target energy located in the same range cell (corresponding to the target's initial slant range $r_{0,k}$). Hence, the signal energy, which is accumulated along the slow time via the following equation, can be used to estimate fold factor

$$
E(n') = \sum_{n=0}^{N-1} |s_{KT}(t_n, r; n')|^2.
\tag{13.16}
$$

When $E(n')$ reaches its maximum value, the corresponding n' is the estimation of kth target's fold factor, i.e.,

$$\hat{N}_k = \arg\max_{n'} E(n').\tag{13.17}$$

With the estimated fold factor $\hat{N}_k$, we can compensate the fold factor phase term and obtain the signal (13.14). As a result, the RM of kth target is corrected, which is the foundation for coherent integration.

13.2.2 Acceleration and Jerk Estimation via GDP

From (13.14), it is observed that the phase of kth target signal is a cubic chirp signal after RM correction. Motivated by the dechirp process technique [9–11], which is an effective method to estimate the quadratic phase term of a chirp signal, we propose a method named GDP to estimate the acceleration and jerk.

Let denote $M_{d_2 d_3}(t_n)$ the phase compensation function, i.e.,

$$M_{d_2 d_3}(t_n) = \exp\left[j\frac{4\pi}{\lambda}\left(d_2 t_n^2 + d_3 t_n^3\right)\right],\tag{13.18}$$

where d_2 and d_3 represent the searching radial acceleration and jerk, respectively.

Multiplying (13.14) with (13.18), we have

$$
\begin{aligned}
s_{KT}(t_n, r) =& A_{1,k}\mathrm{sinc}\left[B\left(\frac{2r - 2r_{0,k}}{c}\right)\right]\\
&\times \exp\left(-j\frac{4\pi r_{0,k}}{\lambda}\right)\exp\left(-j\frac{4\pi v_{unamb,k}t_n}{\lambda}\right)\\
&\times \exp\left[-j\frac{4\pi(a_{0,k} - d_2)}{\lambda}t_n^2\right]\\
&\times \exp\left[-j\frac{4\pi(a_{1,k} - d_3)}{\lambda}t_n^3\right]\\
&+ \sum_{l=1,l\neq k}^{K} A_{1,l}\mathrm{sinc}\left(\frac{2r - 2r_{0,l}}{c} - \frac{2(N_l - N_k)v_{amb}t_n}{c}\right)\\
&\times \exp\left(-j\frac{4\pi r_{0,l}}{\lambda}\right)\exp\left(-j\frac{4\pi v_{unamb,l}t_n}{\lambda}\right)\\
&\times \exp\left[-j\frac{4\pi(a_{0,l} - d_2)}{\lambda}t_n^2\right]\\
&\times \exp\left[-j\frac{4\pi(a_{1,l} - d_3)}{\lambda}t_n^3\right].
\end{aligned}\tag{13.19}
$$

Equation (13.19) shows that when $d_2 = a_{0,k}$ and $d_3 = a_{1,k}$, the kth target signal becomes a complex sinusoid signal over t_n and reaches the best focusing in each range cell. Hence, $a_{0,k}$ and $a_{1,k}$ can be estimated as

$$(\hat{a}_{0,k}, \hat{a}_{1,k}) = \arg\max_{d_2,d_3}\left|\mathrm{FT}_{t_n}\left[s_{KT}(t_n, r)M_{d_2 d_3}(t_n)\right]\right|,\tag{13.20}$$

where $\mathrm{FT}_{t_n}(\cdot)$ denotes the FT over t_n dimension. Interestingly, the process in (13.18)−(13.20) is similar to the dechirp process [9–11], so we call it as GDP.

13.2.3 Integration via FT

With the estimated $\hat{a}_{0,k}$ and $\hat{a}_{1,k}$, a phase compensation function can be generated as follows:

$$G\left(t_n\right) = \exp\left[j\frac{4\pi}{\lambda}\left(\hat{a}_{0,k}t_n^2 + \hat{a}_{1,k}t_n^3\right)\right]. \tag{13.21}$$

Multiplying (13.14) with (13.21) yields

$$\begin{aligned} s_{KT}\left(t_n, r\right) =& A_{1,k}\mathrm{sinc}\left[B\left(\frac{2r - 2r_{0,k}}{c}\right)\right] \\ & \times \exp\left(-j\frac{4\pi r_{0,k}}{\lambda}\right)\exp\left(-j\frac{4\pi v_{unamb,k}t_n}{\lambda}\right) \\ & + s_{cross}\left(t_n, r\right), \end{aligned} \tag{13.22}$$

where

$$\begin{aligned} s_{cross}\left(t_n, r\right) =& \sum_{l=1, l\neq k}^{K} A_{1,l}\mathrm{sinc}\left(\frac{2r - 2r_{0,l}}{c} - \frac{2(N_l - N_k)v_{amb}t_n}{c}\right) \\ & \times \exp\left(-j\frac{4\pi r_{0,l}}{\lambda}\right)\exp\left(-j\frac{4\pi v_{unamb,l}t_n}{\lambda}\right) \\ & \times \exp\left[-j\frac{4\pi(a_{0,l} - a_{0,k})}{\lambda}t_n^2\right] \\ & \times \exp\left[-j\frac{4\pi(a_{1,l} - a_{1,k})}{\lambda}t_n^3\right]. \end{aligned} \tag{13.23}$$

Applying slow time FT to (13.22) with respect to t_n, we have

$$\begin{aligned} s_{KT}\left(f_{t_n}, r\right) =& P_k\mathrm{sinc}\left[B\left(\frac{2r - 2r_{0,k}}{c}\right)\right] \\ & \times \mathrm{sinc}\left[T\left(f_{t_n} + \frac{2v_{unamb,k}}{\lambda}\right)\right] + s_{cross}\left(f_{t_n}, r\right), \end{aligned} \tag{13.24}$$

where $T = NT_r$ denotes the integration time and P_k represents the amplitude after coherent integration.

From (13.24), it is observed that the coherent integration of kth target energy is achieved. The target is detected if the peak value of (13.24) is larger than a given threshold. Besides, we can obtain the estimations of kth target's initial slant range $(\hat{r}_{0,k})$ and unambiguous velocity $(\hat{v}_{unamb,k})$ based on the peak location of (13.24). Then, with the estimated fold factor $\hat{N}_k$ and unambiguous velocity $\hat{v}_{unamb,k}$, the kth target's radial velocity can be estimated as follows:

$$\hat{v}_{0,k} = \hat{N}_k v_{amb} + \hat{v}_{unamb,k}. \tag{13.25}$$

Furthermore, the amplitude of the kth target after pulse compression (i.e., $A_{1,k}$) can be achieved by

$$\hat{A}_{1,k} = \frac{|P_k|}{N}. \tag{13.26}$$

TABLE 13.1   Motion Parameters of Three Maneuvering Targets

Moving parameters	Target A	Target B	Target C
Initial slant range (km)	604.5	600.9	600
Radial velocity (m/s)	6000	3000	3600
Radial acceleration (m/s^2)	20	20	10
Radial jerk (m/s^3)	20	10	10

13.2.4 Discussions

From the analysis above, we can see that the KTGDP method can remove the migrations (RM and DFM) and achieve the coherent integration as well as the parameters estimation for the kth target, via KT, fold factor phase compensation and GDP. For multiple moving targets, if the scattering intensities are equivalent approximately, the KTGDP method is effective. However, if the scattering intensities of different targets differ significantly, the weak targets may be shadowed by the strong ones because of the maximizing operation shown in (13.17), and then we could not obtain the parameters estimation or coherent integration for the weak targets. In the following, two examples are given to illustrate this effect.

Example 1: Suppose that there are three targets in the scene. The parameters of the three targets and radar are listed in Tables 13.1 and 13.2, respectively. The signal-to-noise ratios (SNRs) of these three targets (i.e., targets A, B and C) are -7, -8 and -8 dB, respectively. Fig. 13.1(a) shows the result after pulse compression, which implies that serious RM occurs because of targets' high velocity. Fig. 13.1(b) shows the result of fold factor searching. It can be seen that there are three peaks in the output, which indicate the estimated fold factor of the three targets. With the estimated fold factor, the RM of targets can be corrected and the coherent integration can obtained via the KTGDP method.

Example 2: The SNRs of these three targets are -7, -19 and -19 dB. Other parameters of radar and targets are as the same as those of example 1. Fig. 13.2(a) shows the result after pulse compression, where the trajectory of target A is clearly visible, whereas those of targets B and C are buried. Fig. 13.2(b) shows the result of fold factor searching. Unfortunately, there is only one peak in the output, which corresponds to the estimated fold factor of target A. However, the fold factor estimations of the weak ones (i.e., targets B and C) cannot be obtained, which makes it difficult to obtain the coherent integration for weak targets.

TABLE 13.2   Simulation Parameters of Radar

Parameters	Value
Carrier frequency f_c	10 GHz
Bandwidth B	1 MHz
Sample frequency f_s	5 MHz
Pulse repetition frequency f_p	200 Hz
Pulse duration T_p	100 μs
Number of pulses N	201

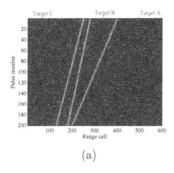 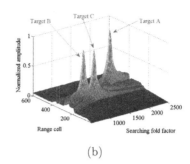

(a)   (b)

Figure 13.1   Maneuvering targets with similar scattering intensities. (a) Result after pulse compression. (b) Result of fold factor searching.

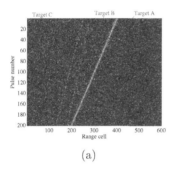 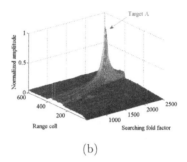

(a)   (b)

Figure 13.2   Maneuvering targets with different scattering intensities. (a) Result after pulse compression. (b) Result of fold factor searching.

In the next section, two CLEAN techniques are presented to eliminate the effect of the strong targets and highlight the weak targets. By this way, the coherent integration of strong moving targets and weak ones can be achieved iteratively. The following will discuss it in more detail.

13.3   TWO CLEAN TECHNIQUES

13.3.1   CLEAN Technique Using SPSF

With the five estimated parameters of kth target, i.e., radial velocity $\hat{v}_{0,k}$, radial acceleration $\hat{a}_{0,k}$, radial jerk $\hat{a}_{1,k}$, amplitude $\hat{A}_{1,k}$ and initial slant range $\hat{r}_{0,k}$, we reconstruct the sinc-like point spread function (SPSF) of the kth target in the slow time-range domain as follows:

$$\text{SPSF}_k\left(t_m, r\right) = \hat{A}_{1,k}\text{sinc}\left\{B\left[\frac{2r - 2\hat{r}_k(t_m)}{c}\right]\right\}$$
$$\times \exp\left(-j\frac{4\pi\hat{r}_k(t_m)}{\lambda}\right), \tag{13.27}$$

where

$$\hat{r}_k(t_m) = \hat{r}_{0,k} + \hat{v}_{0,k}t_m + \hat{a}_{0,k}t_m^2 + \hat{a}_{1,k}t_m^3. \tag{13.28}$$

Subtract the SPSF of kth target from multi-target's echo shown in (13.5), we have

$$s_{\text{CLEAN}}(t_m, r) \approx \sum_{l=1, l \neq k}^{K} A_{1,l} \text{sinc} \left[B \left(\frac{2r - 2r_l(t_m)}{c} \right) \right]$$
$$\times \exp \left[-j \frac{4\pi r_l(t_m)}{\lambda} \right]. \tag{13.29}$$

It can be seen that the cleaned signal in (13.29) is similar to (13.5). Repeating the processing procedure (13.6)−(13.27) on the next brightest target, we can obtain the parameters estimation and coherent integration result of the next brightest target. In this way, the coherent integration for strong targets and weak ones can be achieved iteratively.

13.3.2 CLEAN Technique Using MPSF

The CLEAN technique using SPSF can remove the strong target's effect without loss of performance. Nevertheless, the performance of the CLEAN technique with SPSF is sensitive to the motion parameter estimation, and the parameter estimation error may make it invalid. Hence, a suboptimal approach, named CLEAN technique using MPSF, is introduced in this section to eliminate the strong target's effect when parameter estimation error occurs.

With the estimated parameters of kth target, we define the modified point spread function (MPSF) in the slow time-range domain as follows:

$$\text{MPSF}_k(t_m, r) = \text{rect} \left[\frac{r - \hat{r}_k(t_m)}{W_L} \right] s_c(t_m, r), \tag{13.30}$$

where W_L denotes the length of the rectangular window and is a constant. Equation (13.30) shows that the MPSF contains two parts. The first part is the rectangular window function, which determines the non-zero range. The second part is the compressed signal .

Note that the compressed signal (13.5) can be rewritten as

$$s_c(t_m, r) = \sum_{k=1}^{K} A_{1,k} \text{sinc} \left\{ B \left[\frac{r - r_k(t_m)}{\rho_r} \right] \right\}$$
$$\times \exp \left(-j \frac{4\pi r_k(t_m)}{\lambda} \right), \tag{13.31}$$

where $\rho_r = c/2B$ denotes the range resolution. Neglecting the effect of sidelobe, it can be seen from (13.31) that the energy of kth target locates at the following range:

$$r_{\text{energy}} \in [r_k(t_m) - \rho_r, r_k(t_m) + \rho_r]. \tag{13.32}$$

Subtracting the MPSF in (13.30) from the multi-target's signal shown in (13.5), we have

$$s_{\text{CLEAN}}(t_m, r) = \left[1 - \text{rect}\left(\frac{r - \hat{r}_k(t_m)}{W_L} \right) \right] s(t_m, r) \qquad (13.33)$$

$$\approx \begin{cases} 0 & |r - \hat{r}_k(t_m)| \leq \frac{W_L}{2}. \\ s(t_m, r) & \text{else}. \end{cases}$$

From (13.33), it can be seen that the signal energy located in the rectangular window will be eliminated.

Note that

$$|r_{\text{energy}} - \hat{r}_k(t_m)| \leq \rho_r + \frac{c}{2f_s} + \lambda, \qquad (13.34)$$

where f_s denotes the sample frequency and the proof of (13.34) is given in Section 13.7. Hence, if the window length W_L satisfies

$$W_L \geq 2\left(\rho_r + \frac{c}{2f_s} + \lambda \right), \qquad (13.35)$$

then the energy of kth target will be removed via CLEAN technique using MPSF in spite of parameter estimation error occurs, which is helpful to the coherent integration for other weak targets.

It is certain that the CLEAN technique using MPSF would degrade the integration performance of the weak target when the trajectories of the strong target and weak one intersect. Suppose that the following condition is satisfied for $t_m \in [T_1, T_2]$

$$|r_l(t_m) - r_k(t_m)| \leq \frac{W_L}{2}, k \neq l. \qquad (13.36)$$

Then, during the elimination process for kth target's energy via the CLEAN technique using MPSF, about $\frac{T_2 - T_1}{T}$ energy of the lth target is also removed. Therefore, the coherent integration performance loss of lth target could be estimated by

$$\text{SNR}_{loss} \approx 20\log_{10}\left(\frac{T}{T - (T_2 - T_1)} \right). \qquad (13.37)$$

13.3.3 Algorithm Summary

To summarize, the coherent integration algorithm for multiple targets via CLEAN technique can be stated as follows:

Step 1) Perform pulse compression on the received signal of K targets.

Step 2) Applying FT on the compressed signal over range dimension and then perform keystone transform.

Step 3) Estimate the fold factor via fold factor searching.

Step 4) Compensate the fold factor phase term and estimate the radial acceleration and jerk via GDP.

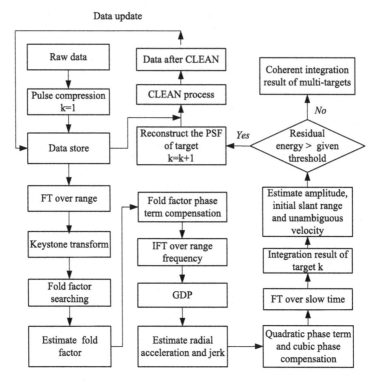

Figure 13.3 Flowchart of the integration method based on CLEAN technique.

Step 5) Compensate the quadratic phase term and cubic phase term with the estimated motion parameters and obtain the coherent integration via slow time FT.

Step 6) Achieve the estimation of initial slant range, unambiguous velocity and amplitude based on the integration result.

Step 7) Reconstruct the PSF (SPSF or MPSF) of the target with the estimated parameters and subtract the PSF from the multi-targets echo.

Step 8) Repeat Steps 2–7 until the residual signal energy is less than a given threshold. Finally, the coherent integration for multi-targets is achieved iteratively.

The detail flowchart of the algorithm is given in Fig. 13.3.

13.4 COMPUTATIONAL COMPLEXITY

In what follows, the computational complexity of major steps in the KTGDP method will be analyzed in terms of the number of operations, i.e., complex multiplications (Mc) and additions (Ac). Denote the number of range cells, echo pulses, searching velocity, searching acceleration, searching jerk and searching fold factor by M, N, N_{a_1}, N_{a_2}, N_{a_3} and N_F, respectively. For the RM correction, $NM\log_2 M/2 + N^2 M + N_F$

TABLE 13.3 Computational Complexity of the KTGDP Method and GRFT

Methods	Multiplications	Additions
KTGDP	$NM\log_2 M/2 + N^2 M$ $+N_{a_2}N_{a_3}MN\log_2 N/2$ $+N_F(M\log_2 M/2 + 2MN)$	$NM\log_2 M + N(N-1)M$ $+N_{a_2}N_{a_3}MN\log_2 N$ $+N_F[M\log_2 M + 2M(N-1)]$
GRFT	$NMN_{a_1}N_{a_2}N_{a_3}$	$(N-1)MN_{a_1}N_{a_2}N_{a_3}$

$(M\log_2 M/2 + 2MN)$ Mc and $NM\log_2 M + N(N-1)M + N_F(M\log_2 M + 2M(N-1))$ Ac are needed. As to radial acceleration and radial jerk estimation via GDP, $N_{a_2}N_{a_3}MN\log_2 N/2$ Mc and $N_{a_2}N_{a_3}MN\log_2 N$ Ac are needed.

On the other hand, $NMN_{a_1}N_{a_2}N_{a_3}$ Mc and $(N-1)MN_{a_1}N_{a_2}N_{a_3}$ Ac are needed for the GRFT algorithm [1]. The detailed computational complexities of the KTGDP algorithm and GRFT are given in Table 13.3. Suppose that $M = N_{a_1} = N_{a_2} = N_{a_3} = N_F = N_\Omega = N$, then the computational complexity of the method is $O(N^4\log_2 N)$, whereas the computational burden of GRFT is $O(N^5)$. As is evident in the above analysis, the computational cost of the KTGDP method is reduced significantly compared with the GRFT algorithm.

13.5 SIMULATED ANALYSIS

In this section, several numerical simulations are presented to demonstrate the effectiveness of the KTGDP coherent integration algorithm for maneuvering targets with jerk motion, where the parameters of the radar are listed in Table 13.2 and the window length $W_L = 2\left(\rho_r + \frac{c}{2f_s} + \lambda\right)$. Comparisons with other popular coherent integration methods, i.e., MTD, RFT [12], RFRFT [13] and GRFT [1], are also given.

13.5.1 Coherent Integration for a Single Target

We first analyze the coherent integration performance of the KTGDP method for a single target in Fig. 13.4, where the motion parameters of the target are initial slant range $r_0 = 300$ km, radial velocity $v_0 = 3600$ m/s, acceleration $a_0 = 20$ m/s^2 and jerk $a_1 = 10$ m/s^3. The SNR of the received echo is -12 dB. Fig. 13.4(a) shows the result after pulse compression. It can be seen that serious RM occurs because of target's high velocity. Fig. 13.4(b) shows the result after KT and fold factor phase term compensation, which indicates that the RM has been corrected. Fig. 13.4(c) shows the estimation result of a_2 and a_3 obtained via GDP, where the peak value represents the estimated parameters. With the estimations of a_2 and a_3, the motion compensation is made to eliminate the DFM, and the coherent integration result of the KTGDP method is shown in Fig. 13.4(d). We can see that the target energy is coherently accumulated as one obvious peak in the output, which is helpful to the target detection.

For the sake of comparison, the coherent integration results of MTD, RFT, RFRFT and GRFT are also given. More specifically, Fig. 13.4(e)−Fig. 13.4(g) show, respectively, the integration result of MTD, RFT and RFRFT. Because of the

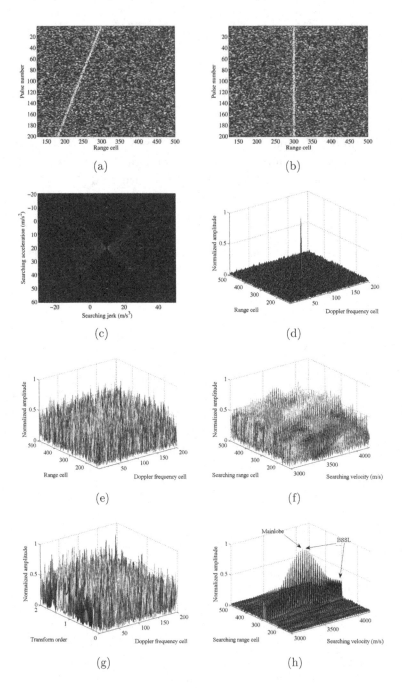

Figure 13.4 Coherent integration for a single target. (a) Result after pulse compression. (b) Result after KT and fold factor phase term compensation. (c) Result of GDP. (d) Integration result of the KTGDP method. (e) MTD. (f) RFT. (g) RFRFT. (h) GRFT.

TABLE 13.4 Motion Parameters of Targets in Situation 1

Moving parameters	Target A	Target B	Target C
Initial slant range (km)	600	602.1	597
Radial velocity (m/s)	6000	3000	3600
Radial acceleration (m/s^2)	20	20	20
Radial jerk (m/s^3)	20	10	10
SNR (before pulse compression) (dB)	-7	-10	-19

high-order DFM (i.e., QDFM) induced by target's jerk motion, MTD, RFT and RFRFT become invalid. In addition, the integration result of GRFT is shown in Fig. 13.4(h). Although GRFT is also able to achieve the coherent accumulation via four-dimensional searching, the BSSL with large peak value appears in the GRFT output, which may lead to serious false alarm and make it hard to obtain the number of target, especially for the scenario of multiple targets.

13.5.2 Coherent Integration for Multiple Targets with CLEAN Technique

In this subsection, we evaluate the coherent integration performance of the KTGDP method for multiple targets under two different situations, which are denoted as situation 1 and situation 2. 1) In situation 1, the parameters estimation is accurate and the trajectories of different targets are far enough apart. 2) In situation 2, the parameters estimation error occurs and the trajectories of some targets intersect.

13.5.2.1 Situation 1

The motion parameters of three maneuvering targets are listed in Table 13.4. Fig. 13.5(a) shows the result of pulse compression, where the trajectory of target C is weak and target A is the strongest. Fig. 13.5(b) shows the result of fold factor searching, where the peak value indicates the estimated fold factor of target A. In Fig. 13.5(c), the result after KT and fold factor phase term compensation shows that the RM of target A is corrected, while the envelopes of targets B and C still vary with slow time. The integration result of target A is given in Fig. 13.5(d) .

Fig. 13.5(e) is the signal after elimination of the strongest signal contribution for the echo via the CLEAN technique using SPSF. We can see that the CLEAN technique using SPSF is effective. Fig. 13.5(f) shows the coherent integration result of the three targets via the KTGDP method using SPSF. It is shown that the targets' energy is accumulated as three peaks in the output. Besides, Fig. 13.5(g) shows the result after elimination of target A's signal contribution for the echo via the CLEAN technique using MPSF, and the coherent integration result of the three targets is shown in Fig. 13.5(h).

13.5.2.2 Situation 2

The motion parameters of the three maneuvering targets are listed in Table 13.5. Fig. 13.6(a) shows the result after pulse compression, which indicates that the trajectories of targets A and C intersect. The coherent integration result of target A is given in

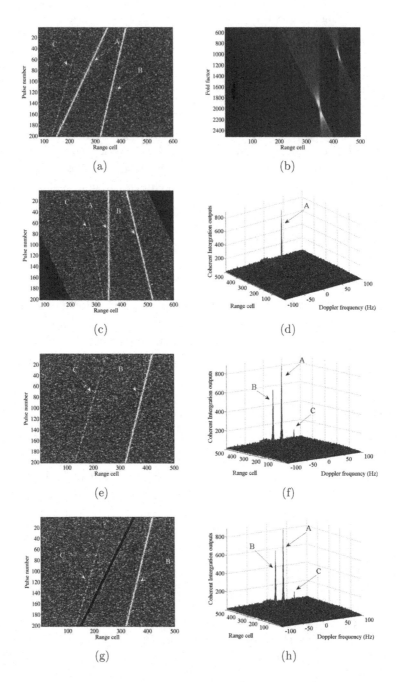

Figure 13.5   Coherent integration for multiple targets in situation 1. (a) Result after pulse compression. (b) Result of fold factor searching. (c) Result after RM correction. (d) Integration result of target A. (e) Result after CLEAN process using SPSF. (f) Integration result of the three targets using SPSF. (g) Result after CLEAN process using MPSF. (h) Integration result of the three targets using MPSF.

TABLE 13.5 Motion Parameters of Targets in Situation 2

Moving parameters	Target A	Target B	Target C
Initial slant range (km)	598.5	600.6	597
Radial velocity (m/s)	6001	3000	3600
Radial acceleration (m/s^2)	20	20	10
Radial jerk (m/s^3)	20	10	10
SNR (before pulse compression) (dB)	-5	-16	-16

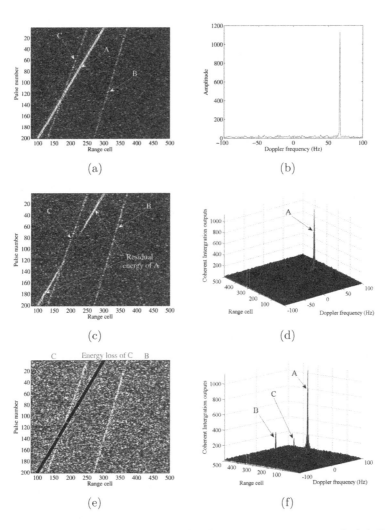

Figure 13.6 Coherent integration for multiple targets in situation 2. (a) Result after pulse compression. (b) Coherent integration result of target A (Doppler slice). (c) Result after CLEAN process using SPSF. (d) Integration result of the three targets using SPSF. (e) Result after CLEAN process using MPSF. (f) Integration result of the three targets using MPSF.

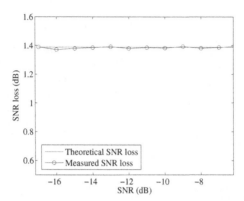

Figure 13.7 Coherent integration performance loss for target C induced by CLEAN technique using MPSF.

Fig. 13.6(b), and the radial velocity estimation is $\hat{v}_{0,1} = 6000.5$ m/s, i.e., velocity estimation error occurs. Fig. 13.6(c) shows the result after elimination of target A's signal contribution for the echo via CLEAN technique using SPSF. It can be seen that the CLEAN technique using SPSF is ineffective because of the radial velocity estimation error. The integration result of the three targets via the KTGDP method based on the CLEAN technique using SPSF is shown in Fig. 13.6(d). Unfortunately, only the energy of target A is effectively accumulated.

Fig. 13.6(e) shows the result after elimination of target A's signal contribution for the echo via the CLEAN technique using MPSF. It can be seen that the energy of target A has been eliminated in spite of radial velocity estimation error occurs, which is helpful to the coherent integration of targets B and C. Fig. 13.6(f) shows the coherent integration result of the three targets via the KTGDP method based on the CLEAN technique using MPSF, where the energy of the three targets is effectively integrated. Furthermore, the peak values of targets B and C after coherent integration are $P_2 = 328$ and $P_3 = 279.5$, respectively. Hence, the integration performance loss of target C due to the CLEAN process using MPSF is $20 * \log 10(P_2/P_3) = 1.3898$ dB, which matches the theoretical value 1.3888 dB (calculated via (13.37)) very well.

The SNR loss for the weak target induced by the CLEAN technique using MPSF in situation 2 is further investigated by Monte Carlo trials, where the SNR for the target C varied in $[-17, -6$ dB], and the other parameters are as the same as those in Fig. 13.6. For each SNR value, 500 simulations were performed. Both the measured SNR-loss curve and the theoretical SNR-loss curve of target C are illustrated in Fig. 13.7. It is shown that the experimental results agree with the theoretical curve very well.

13.5.3 Computational Cost and Detection Performance

1) The computational complexity and time cost of the KTGDP method and GRFT are also evaluated. Specifically, we first compare the computation complexities of the

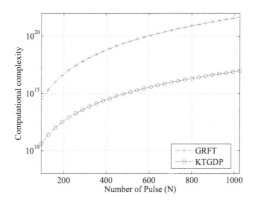

Figure 13.8 Computational complexity versus integration pulse number.

KTGDP method and GRFT in Fig. 13.8, where the computational complexity curves for different pulse numbers are plotted. It is evident that the computational cost of GRFT is much higher than the KTGDP method. For $N = 201$, the computational complexity of the KTGDP method is about 1.2×10^{-4} of that of GRFT. Furthermore, Table 13.6 illustrates the time cost of GRFT and the KTGDP algorithm. We can see that GRFT costs more time than the KTGDP method due to the four-dimensional searching.

2) The detection performances of MTD, RFT, RFRFT, GRFT and the KTGDP method are investigated by Monte Carlo trials. We combine the constant false alarm ratio (CFAR) detector [13] and the five methods as corresponding detectors. The Gaussian noises are added to the target echoes, and the false alarm ratio is set as $P_{fa} = 10^{-4}$. Fig. 13.9 shows the detection probability of the five detectors versus different SNR levels and in each case, and 1000 times of Monte Carlo simulations are done. The simulation results show that the probability of the detector based on the KTGDP method is superior to MTD, RFT and RFRFT thanks to its ability to deal with the high-order DFM. Moreover, it is worth pointing out that the KT-GDP method suffers some detection performance loss in comparison with the GRFT algorithm, which is the cost of the reduced computational complexity.

From the simulation results above (Fig. 13.4–Fig. 13.9 and Table 13.6), we can conclude that the coherent integration performance of the KTGDP method is superior to MTD, RFT and RFRFT for a maneuvering target with jerk motion. Besides,

TABLE 13.6 Time Cost of GRFT and KTGDP Methods

Methods	Time cost (s)
GRFT	482.329258
KTGDP	28.514873

Note: Main configuration of the computer: CPU: Intel Core i5-4430 3.0 GHz; RAM: 8.00G; Operating System: Windows 7; Software: Matlab 2012b.

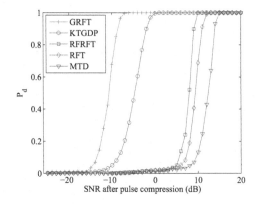

Figure 13.9 Detection probability of MTD, RFT, RFRFT, GRFT and the KTGDP method.

compared with the GRFT algorithm, the KTGDP method can avoid the BSSL effect and achieve a good balance between the computational cost and detection performance.

13.6 SUMMARY

In this chapter, we have addressed the coherent integration problem for high-speed multi-targets with different scattering intensities. We introduced a method based on KT, fold factor phase term compensation and GDP to realize the coherent integration for a maneuvering target. We considered the scene that target's scattering intensities differ significantly and presented two CLEAN techniques based on SPSF and MPSF to eliminate the strong target's effect and highlight the weak ones. In this way, the coherent accumulation of strong target and weak ones can be achieved iteratively.

13.7 PROOF OF (13.34)

In order to guarantee coherent integration performance, the searching interval of fold factor, radial acceleration and jerk is set as follows: $\Delta_{N_F} = 2f_c/Nf_s$ [7], $\Delta_{a_{0,k}} = \lambda/2T^2$ and $\Delta_{a_{1,k}} = \lambda/2T^3$ [13]. Hence, the estimation error of fold factor, radial acceleration and jerk for the kth target can be determined, i.e., $e_{N_F} \leq f_c/Nf_s$, $e_{a_{0,k}} \leq \lambda/4T^2$ and $e_{a_{1,k}} \leq \lambda/4T^3$.

Moreover, based on the radar system parameters (i.e., radar repetition frequency f_p and integration pulse number N), the Doppler frequency resolution can be determined, i.e., f_p/N. Thus, the unambiguous velocity estimation error $e_{v_{unamb,k}}$ satisfies

$$e_{v_{unamb,k}} \leq \frac{\lambda f_p}{2N}, \tag{13.38}$$

then the velocity estimation error $e_{v_{0,k}}$ satisfies

$$e_{v_{0,k}} = e_{N_F} v_a + e_{v_{0,k}} \leq \frac{\lambda f_p}{2N}\left(1 + \frac{f_c}{f_s}\right). \tag{13.39}$$

As a result, the slant range estimation error induced by the motion parameters estimation error satisfies

$$
\begin{aligned}
\Delta r &= |r_k(t_m) - \hat{r}_k(t_m)| \\
&= e_{v_{0,k}} t_m + e_{a_{0,k}} t_m^2 + e_{a_{1,k}} t_m^3 \\
&\ll e_{v_{0,k}} T + e_{a_{0,k}} T^2 + e_{a_{1,k}} T^3 \\
&\ll \frac{c}{2f_s} + \lambda.
\end{aligned}
\tag{13.40}
$$

Therefore,

$$
\begin{aligned}
|r_{\text{energy}} - \hat{r}_k(t_m)| &\leq |r_k(t_m) + \rho_r - \hat{r}_k(t_m)| \\
&\leq \rho_r + \Delta r \\
&\leq \rho_r + \frac{c}{2f_s} + \lambda.
\end{aligned}
\tag{13.41}
$$

Bibliography

[1] J. Xu, X. G. Xia, S. B. Peng, J. Yu, Y. N. Peng, and L. C. Qian, "Radar maneuvering target motion estimation based on Generalized Radon-Fourier transform," *IEEE Transactions on Signal Processing*, vol. 60, no. 12, pp. 6190–6201, December 2012.

[2] Y. Wang and Y. C. Jiang, "ISAR imaging of maneuvering target based on the L-Class of fourth-order complex-lag PWVD," *IEEE Transactions on Geoscience and Remote Sensing*, vol. 48, no. 3, pp. 1518–1527, March 2010.

[3] J. Tsao and B. D. Steinberg, "Reduction of sidelobe and speckle artifacts in microwave imaging: the CLEAN technique," *IEEE Transactions on Antennas and Propagation*, vol. 36, no. 4, pp. 543–556, April 1998.

[4] R. Bose, A. Freedman, and B. D. Steinberg, "Sequence CLEAN: a modified deconvolution technique for microwave images of contiguous targets," *IEEE Transactions on Aerospace and Electronic Systems*, vol. 38, no. 1, pp. 89–97, January 2002.

[5] H. Deng, "Effective CLEAN algorithms for performance-enhanced detection of binary coding radar signals," *IEEE Transactions on Signal Processing*, vol. 52, no. 1, pp. 72–78, January 2004.

[6] J. Tian, W. Cui, and S. Wu, "A novel method for parameter estimation of space moving targets," *IEEE Geoscience and Remote Sensing Letters*, vol. 11, no. 2, pp. 389–393, February 2014.

[7] M. D. Xing, J. H. Su, G. Y. Wang, and Z. Bao, "New parameter estimation and detection algorithm for high speed small target," *IEEE Transactions on Aerospace and Electronic Systems*, vol. 47, no. 1, pp. 214–224, January 2011.

[8] R. P. Perry, R. C. Dipietro, and R. L. Fante, "SAR imaging of moving targets," *IEEE Transactions on Aerospace and Electronic Systems*, vol. 35, no. 1, pp. 188–200, January 1999.

[9] W. P. Li, "Wigner distribution method equivalent to dechirp method for detecting a Chirp signal," *IEEE Transactions on Acoustics, Speech and Signal Processing*, vol. 35, no. 8, pp. 1210–1211, August 1987.

[10] A. Francos and M. Porat, "Analysis and synthesis of multicomponent signals using positive time-frequency distributions," *IEEE Transactions on Signal Processing*, vol. 47, no. 2, pp. 493–504, February 1999.

[11] J. Su, M. Xing, G. Wang, and Z. Bao, "High-speed multi-target detection with narrowband radar," *IET Radar Sonar and Navigation*, vol. 4, no. 4, pp. 595–603, August 2010.

[12] J. Xu, J. Yu, Y. N. Peng, and X. G. Xia, "Radon-Fourier transform (RFT) for radar target detection (I): generalized Doppler filter bank processing," *IEEE Transactions on Aerospace and Electronic Systems*, vol. 47, no. 2, pp. 1186–1202, April 2011.

[13] X. L. Chen, J. Guan, N. B. Liu, and Y. He, "Maneuvering target detection via Radon-Fractional Fourier transform-based long-time coherent integration," *IEEE Transactions on Signal Processing*, vol. 62, no. 4, pp. 939–953, February 2014.

Printed in the United States
by Baker & Taylor Publisher Services